IMPEDANCE SPECTROSCOPY

IMPEDANCE SPECTROSCOPY

EMPHASIZING SOLID MATERIALS AND SYSTEMS

Edited by

J. Ross Macdonald
Department of Physics and Astronomy
University of North Carolina
Chapel Hill, North Carolina

A WILEY-INTERSCIENCE PUBLICATION

JOHN WILEY & SONS
NEW YORK   CHICHESTER   BRISBANE   TORONTO   SINGAPORE

Library of Congress Cataloging in Publication Data:

Impedance spectroscopy.

   ''A Wiley-Interscience publication.''
   Bibliography: p.
   Includes index.
   1. Impedance spectroscopy.   2. Solids—Analysis.

I. Macdonald, J. Ross.
QD116.I57I47 1987   543'.0858   86-32582
ISBN 0-471-83122-0

Printed in the United States of America

10 9 8 7 6 5 4 3 2 1

CONTRIBUTORS

N. Bonanos
Department of Metallurgy and
  Materials Science
Imperial College of Science
  and Technology
London, England

E. P. Butler
ALCAN International Ltd.
Banbury Laboratories
Banbury, Oxon, England

Donald R. Franceschetti
Department of Physics
Memphis State University
Memphis, Tennessee

William B. Johnson
Department of Metallurgy
Ohio State University
Columbus, Ohio

Digby D. Macdonald
SRI International
Menlo Park, California

J. Ross Macdonald
Department of Physics and
  Astronomy
University of North Carolina
Chapel Hill, North Carolina

Michael C. H. McKubre
SRI International
Menlo Park, California

Ian D. Raistrick
Los Alamos National Laboratory
Los Alamos, New Mexico

B. C. H. Steele
Department of Metallurgy and
  Materials Science
Imperial College of Science
  and Technology
London, England

Wayne L. Worrell
Department of Materials
  Science and Engineering
University of Pennsylvania
Philadelphia, Pennsylvania

PREFACE

Impedance spectroscopy (IS) appears destined to play an important role in fundamental and applied electrochemistry and materials science in the coming years. In a number of respects it is the method of choice for characterizing the electrical behavior of systems in which the overall system behavior is determined by a number of strongly coupled processes, each proceeding at a different rate. With the current availability of commercially made, high-quality impedance bridges and automatic measuring equipment covering the millihertz to megahertz frequency range, it appears certain that impedance studies will become increasingly popular as more and more electrochemists, materials scientists, and engineers understand the theoretical basis for impedance spectroscopy and gain skill in the interpretation of impedance data.

This book is intended to serve as a reference and/or textbook on the topic of impedance spectroscopy, with special emphasis on its application to solid materials. The goal was to produce a text that would be useful to both the novice and the expert in IS. To this end, the book is organized so that each individual chapter stands on its own. It is intended to be useful to the materials scientist or electrochemist, student or professional, who is planning an IS study of a solid state system and who may have had little previous experience with impedance measurements. Such a reader will find an outline of basic theory, various applications of impedance spectroscopy, and a discussion of experimental methods and data analysis, with examples and appropriate references. It is hoped that the more advanced reader will also find this book valuable as a review and summary of the literature up to the time of writing, with a discussion of current theoretical and experimental issues. A considerable amount of the material in the book is applicable not only to solid ionic systems but also to the electrical response of liquid electrolytes as well as to solid ones, to electronic as well as to ionic conductors, and even to dielectric response.

The novice should begin by reading Chapter 1, which presents a broad overview of the subject and provides the background necessary to appreciate the power of the technique. He or she might then proceed to Chapter 4, where many different applications of the technique are presented. The emphasis in this chapter is on presenting specific applications of IS rather than extensive reviews; details of how and why the technique is useful in each area are presented. To gain a fuller appreciation of IS, the reader could then proceed to Chapters 2 and 3, which present the theory and measuring and analysis techniques.

For someone already familiar with IS, this text will also be useful. For those familiar with one application of the technique the book will provide both a convenient source for the theory of IS, as well as illustrations of applications in areas possibly unfamiliar to the reader. For the theorist who has studied IS, the applications discussed in Chapter 4 pose questions the experimentalist would like answered; for the experimentalist, Chapters 2 and 3 offer different (and better!) methods to analyze IS data. All readers should benefit from the presentation of theory, experimental data, and analysis methods in one source. It is our hope that this widened perspective of the field will lead to a more enlightened and thereby broadened use of IS.

In format and approach, the present book is intended to fall somewhere between the single-author (or few-author) text and the "monograph" of many authors and as many chapters. Although the final version is the product of 10 authors' labors, considerable effort has been made to divide the writing tasks so as to produce a unified presentation with consistent notation and terminology and a minimum of repetition. To help reduce repetition, all authors had available to them copies of Sections 1.1–1.3, 2.2, and 3.2 at the beginning of their writing of the other sections. We believe that whatever repetition remains is evidence of the current importance to IS of some subjects, and we feel that the discussion of these subjects herein from several different viewpoints is worthwhile and will be helpful to the readers of the volume.

J. ROSS MACDONALD

Chapel Hill, North Carolina
March 1987

ACKNOWLEDGMENTS

The idea for this book originated with Dr. Henk de Bruin, then of Flinders University, Adelaide, Australia. At his initiative a number of the present contributors, as well as others, were invited to attend a three-month workshop to be held at Flinders in the spring of 1984. This workshop was supported by the U.S. National Science Foundation, the Australian Department of Science and Energy, and Flinders University. Further, Dr. de Bruin expected to be the editor of the book that would grow out of the workshop, with myself as coeditor. Although I was the only external participant able to attend for the full three months, Wayne Worrell participated at Flinders for two weeks and made important contributions to the planning of the book.

As it turned out, Dr. de Bruin was unable to serve as editor of the work. Nevertheless, he made valuable contributions to its initial planning and organization and, as well, contributed useful material to Section 1.1. All the present contributors owe him a strong debt of gratitude for perceiving a need and taking the initial steps toward filling it. I want to thank him and his wife Hilary, particularly for their warm hospitality to me and my wife during our stay in Australia.

Finally, we the contributors acknowledge with thanks the support of the U.S. National Science Foundation, the U.S. Army Research Office, Flinders University, the Australian Department of Science and Energy, Research Corporation, Memphis State University, and Sandia National Laboratory. Without their invaluable aid, this book would not exist.

On a more personal note, I want to thank Ms. Jody Gollan for her unflagging, invaluable, and cheerful aid to me in the various demanding tasks involved in editing the present work. Finally, I want to thank all the other contributors for their long and hard work, and I especially thank Ms. Bea Shube, Senior Editor of Wiley–Interscience, and Ms. Diana Cisek, Senior Production Supervisor, for their invaluable help and dedication to helping us make this as good a book as possible.

J. ROSS MACDONALD

Chapel Hill, North Carolina
March 1987

CONTENTS

3. MEASURING TECHNIQUES AND DATA ANALYSIS   133

3.1. Impedance Measurement Techniques   133

Michael C. H. McKubre and Digby D. Macdonald

IMPEDANCE
SPECTROSCOPY

CHAPTER ONE

FUNDAMENTALS OF IMPEDANCE SPECTROSCOPY

J. Ross Macdonald
William B. Johnson

1.1 BACKGROUND, BASIC DEFINITIONS, AND HISTORY

1.1.1 The Importance of Interfaces

Since the end of World War II we have witnessed the development of solid state batteries as rechargeable high-power-density energy storage devices, a revolution in high-temperature electrochemical sensors in environmental, industrial, and energy efficiency control, and the introduction of fuel cells to avoid the Carnot inefficiency inherent in noncatalytic energy conversion. The trend away from corrosive aqueous solutions and toward solid state technology was inevitable in electrochemical energy engineering, if only for convenience and safety in bulk handling. As a consequence, the characterization of systems with solid–solid or solid–liquid interfaces, often involving solid ionic conductors and frequently operating well above room temperature, has become a major concern of electrochemists and materials scientists.

At an interface, physical properties—crystallographic, mechanical, compositional, and, particularly, electrical—change precipitously and heterogeneous charge distributions (polarizations) reduce the overall electrical conductivity of a system. Proliferation of interfaces is a distinguishing feature of solid state electrolytic cells, where not only is the junction between electrode and electrolyte considerably more complex than in aqueous cells, but the solid electrolyte is commonly polycrystalline. Each interface will polarize in its unique way when the system is subjected to an applied potential difference. The rate at which a polarized region will change when the applied voltage is reversed is characteristic of the type of interface: slow

1

for chemical reactions at the triple phase contacts between atmosphere, electrode, and electrolyte, appreciably faster across grain boundaries in the polycrystalline electrolyte. The emphasis in electrochemistry has consequently shifted from a time/concentration dependency to frequency-related phenomena, a trend toward small-signal ac studies. Electrical double layers and their inherent capacitive reactances are characterized by their relaxation times, or more realistically by the distribution of their relaxation times. The electrical response of a heterogeneous cell can vary substantially depending on the species of charge present, the microstructure of the electrolyte, and the texture and nature of the electrodes.

Impedance spectroscopy (IS) is a relatively new and powerful method of characterizing many of the electrical properties of materials and their interfaces with electronically conducting electrodes. It may be used to investigate the dynamics of bound or mobile charge in the bulk or interfacial regions of any kind of solid or liquid material: ionic, semiconducting, mixed electronic–ionic, and even insulators (dielectrics). Although we shall primarily concentrate in this monograph on solid electrolyte materials—amorphous, polycrystalline, and single crystal in form—and on solid metallic electrodes, reference will be made, where appropriate, to fused salts and aqueous electrolytes and to liquid–metal and high-molarity aqueous electrodes as well. We shall refer to the experimental cell as an electrode-material system. Similarly, although much of the present work will deal with measurements at room temperature and above, a few references to the use of IS well below room temperature will also be included.

In this chapter we aim to provide a working background for the practical materials scientist or engineer who wishes to apply IS as a method of analysis without needing to become a knowledgeable electrochemist. In contrast to the subsequent chapters, the emphasis here will be on practical, empirical interpretations of materials problems, based on somewhat oversimplified electrochemical models. We shall thus describe approximate methods of data analysis of IS results for simple solid state electrolyte situations in this chapter and discuss more detailed methods and analyses later. Although we shall concentrate on intrinsically conductive systems, most of the IS measurement techniques, data presentation methods, and analysis functions and methods discussed herein apply directly to lossy dielectric materials as well.

1.1.2   The Basic Impedance Spectroscopy Experiment

Electrical measurements to evaluate the electrochemical behavior of electrode and/or electrolyte materials are usually made with cells having two identical electrodes applied to the faces of a sample in the form of a circular cylinder or rectangular parallelepiped. However, if devices such as chemical sensors or living cells are investigated, this simple symmetrical geometry is often not feasible. Vacuum, a neutral atmosphere such as argon, or an oxidizing atmosphere are variously used. The general approach is to apply an electrical stimulus (a known voltage or current) to the electrodes and observe the response (the resulting current or voltage). It is virtually always assumed that the properties of the electrode–material system are time-invariant, and it is one of the basic purposes of IS to determine these properties, their interrelations, and their dependences on such controllable variables as temperature, oxygen partial pressure, applied hydrostatic pressure, and applied static voltage or current bias.

A multitude of fundamental microscopic processes take place throughout the cell when it is electrically stimulated and, in concert, lead to the overall electrical response. They include the transport of electrons through the electronic conductors, the transfer of electrons at the electrode–electrolyte interfaces to or from charged or uncharged atomic species which originate from the cell materials and its atmospheric environment (oxidation or reduction reactions), and the flow of charged atoms or atom agglomerates via defects in the electrolyte. The flowrate of charged particles (current) depends on the ohmic resistance of the electrodes and the electrolyte and on the reaction rates at the electrode–electrolyte interfaces. The flow may be further impeded by band structure anomalies at any grain boundaries present (particularly if second phases are present in these regions) and by point defects in the bulk of all materials. We shall usually assume that the electrode-electrolyte interfaces are perfectly smooth, with a simple crystallographic orientation. In reality of course, they are jagged, full of structural defects, electrical short and open circuits, and they often contain a host of adsorbed and included foreign chemical species that influence the local electric field.

There are three different types of electrical stimuli which are used in IS. First, in transient measurements a step function of voltage [$V(t) = V_0$ for $t > 0$, $V(t) = 0$ for $t < 0$] may be applied at $t = 0$ to the system and the resulting time-varying current $i(t)$ measured. The ratio $V_0/i(t)$, often called the indicial impedance or the time-varying resistance, measures the impedance resulting from the step function voltage perturbation at the electrochemical interface. This quantity, although easily defined, is not the usual impedance referred to in IS. Rather, such time-varying results are generally Fourier-transformed into the frequency domain, yielding frequency-dependent impedance. Such transformation is only valid when $|V_0|$ is sufficiently small that system response is linear. The advantages of this approach are that it is experimentally easily accomplished and that the independent variable, voltage, controls the rate of the electrochemical reaction at the interface. Disadvantages include the need to Fourier-analyze the results and the fact that the frequency spectrum is not directly controlled, so the impedance may not be well determined over all desired frequencies.

A second technique in IS is to apply a signal $v(t)$ composed of random (white) noise to the interface and measure the resulting current. Again, one generally Fourier-transforms the results to pass into the frequency domain and obtain an impedance. This approach offers the advantage of fast data collection because only one signal is applied to the interface for a short time. The technique has the disadvantages of requiring true white noise and then the need to carry out a Fourier analysis, which can be computationally difficult and time-consuming. Often a microcomputer is used for both the generation of white noise and the subsequent analysis. Limitations in the digital-to-analog electronics currently available generally restrict measurements to frequency ranges below 50–100 kHz. In addition, algorithms for

Fourier analysis on microcomputers at frequencies below 10 Hz are difficult to implement.

The third approach, the most common and standard one, is to measure impedance directly in the frequency domain by applying a single-frequency voltage to the interface and measuring the phase shift and amplitude, or real and imaginary parts, of the resulting current at that frequency. Commercial instruments are available which measure the impedance as a function of frequency automatically in the frequency ranges of about 1 mHz to 1 MHz and which are easily interfaced to laboratory microcomputers. The advantages of this approach are the availability of these instruments and the ease of their use, as well as the fact that the experimentalist can control the frequency range to examine the domain of most interest. In addition to these three approaches, one can combine them to generate other types of stimuli. The most important of these, ac polarography, combines the first and third techniques by simultaneously applying a linearly varying unipolar transient signal and a much smaller single-frequency sinusoidal signal (Smith [1966]).

Any intrinsic property that influences the conductivity of an electrode–materials system, or an external stimulus, can be studied by IS. The parameters derived from an IS spectrum fall generally into two categories: (a) those pertinent only to the material itself, such as conductivity, dielectric constant, mobilities of charges, equilibrium concentrations of the charged species, and bulk generation–recombination rates; and (b) those pertinent to an electrode–material interface, such as adsorption–reaction rate constants, capacitance of the interface region, and diffusion coefficient of neutral species in the electrode itself.

It is useful and not surprising that modern advances in electronic automation have included IS. Sophisticated automatic experimental equipment has been developed to measure and analyze the frequency response to a small-amplitude ac signal between about 10^{-4} and $> 10^6$ Hz, interfacing its results to computers and their peripherals (see Section 3.1). A revolution in the automation of an otherwise difficult measuring technique has moved IS out of the academic laboratory and has begun to make it a technique of significant importance in the areas of industrial quality control of paints, emulsions, electroplating, thin-film technology, materials fabrication, mechanical performance of engines, corrosion, and so on.

Although this book has a strong physicochemical bias, the use of IS to investigate polarization across biological cell membranes has been pursued by many investigators since 1925. Details and discussion of the historical background of this important branch of IS are given in the books of Cole [1972] and Schanne and Ruiz-Ceretti [1978].

1.1.3 Response to a Small-Signal Stimulus in the Frequency Domain

A monochromatic signal $v(t) = V_m \sin(\omega t)$, involving the single frequency $\nu \equiv \omega/2\pi$, is applied to a cell and the resulting steady state current $i(t) = I_m \sin(\omega t + \theta)$ measured. Here θ is the phase difference between the voltage and the current; it is zero for purely resistive behavior. One can now define the conventional impedance $Z(\omega) \equiv v(t)/i(t)$. Its magnitude or modulus is $|Z(\omega)| = V_m/I_m(\omega)$ and its phase angle is $\theta(\omega)$.

The concept of electrical impedance was first introduced by Oliver Heaviside in the 1880s and was soon after developed in terms of vector diagrams and complex representation by A. E. Kennelly and especially C. P. Steinmetz. Impedance is a more general concept than resistance because it takes phase differences into account, and it has become a fundamental and essential concept in electrical engineering. Impedance spectroscopy is thus just a specific branch of the tree of electrical measurements. The magnitude and direction of a planar vector in a right-hand orthogonal system of axes can be expressed by the vector sum of the components a and b along the axes, that is, by the complex number $Z = a + jb$. The imaginary number $j \equiv \sqrt{-1} \equiv \exp(j\pi/2)$ indicates an anticlockwise rotation by $\pi/2$ relative to the x axis. Thus, the real part of Z, a, is in the direction of the real axis x, and the imaginary part b is along the y axis. An impedance $Z(\omega) = Z' + jZ''$ is such a vector quantity and may be plotted in the plane with either rectangular or polar coordinates, as shown in Figure 1.1.1. Here the two rectangular coordinate values are clearly

$$\text{Re}(Z) \equiv Z' = |Z|\cos(\theta) \quad \text{and} \quad \text{Im}(Z) \equiv Z'' = |Z|\sin(\theta) \qquad (1)$$

with the phase angle

$$\theta = \tan^{-1}(Z''/Z') \qquad (2)$$

and the modulus

$$|Z| = [(Z')^2 + (Z'')^2]^{1/2} \qquad (3)$$

This defines the Argand diagram or complex plane, widely used in both mathematics and electrical engineering. In polar form, Z may now be written as $Z(\omega) = |Z|\exp(j\theta)$, which may be converted to rectangular form through the use of

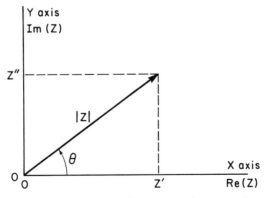

FIGURE 1.1.1. The impedance Z plotted as a planar vector using rectangular and polar coordinates.

the Euler relation $\exp(j\theta) = \cos(\theta) + j\sin(\theta)$. It will be noticed that the original time variations of the applied voltage and the resulting current have disappeared, and the impedance is time-invariant (provided the system itself is time-invariant).

In general, Z is frequency-dependent, as defined above. Conventional IS consists of the (nowadays often automated) measurement of Z as a function of ν or ω over a wide frequency range. It is from the resulting structure of the $Z(\omega)$ vs. ω response that one derives information about the electrical properties of the full electrode–material system.

For nonlinear systems, i.e., most real electrode–material systems, IS measurements in either the time or the frequency domain are useful and meaningful in general only for signals of magnitude such that the overall electrode–material system response is electrically linear. This requires that the response to the sum of two separate input-measuring signals applied simultaneously be the sum of the responses of the signals applied separately. A corollary is that the application of a monochromatic signal, one involving $\sin(\omega t)$, results in no, or at least negligible, generation of harmonics in the output, that is, components with frequencies $n\nu$ for $n = 2, 3, \ldots$. Both solid and liquid electrochemical systems tend to show strong nonlinear behavior, especially in their interfacial response, when applied voltages or currents are large. But so long as the applied potential difference (p.d.) amplitude V_m is less than the thermal voltage, $V_T \equiv RT/F \equiv kT/e$, about 25 mV at 25°C, it can be shown that the basic differential equations which govern the response of the system become linear to an excellent approximation. Here k is Boltzmann's constant, T the absolute temperature, e the proton charge, R the gas constant, and F the faraday. Thus if the applied amplitude V_m is appreciably less than V_T, the system will respond linearly. Note that in the linear regime it is immaterial as far as the determination of $Z(\omega)$ is concerned whether a known $v(\omega t)$ is applied and the current measured or a known $i(\omega t)$ applied and the resulting voltage across the cell measured. When the system is nonlinear, this reciprocity no longer holds.

1.1.4 Impedance-Related Functions

The impedance has frequently been designated as the ac impedance or the complex impedance. Both these modifiers are redundant and should be omitted. Impedance without a modifier always means impedance applying in the frequency domain and usually measured with a monochromatic signal. Even when impedance values are derived by Fourier transformation from the time domain, the impedance is still defined for a set of individual frequencies and is thus an alternating-current impedance in character.

Impedance is by definition a complex quantity and is only real when $\theta = 0$ and thus $Z(\omega) = Z'(\omega)$, that is, for purely resistive behavior. In this case the impedance is completely frequency-independent. When Z' is found to be a variable function of frequency, the Kronig–Kramers (Hilbert integral transform) relations (Macdonald and Brachman [1956]), which holistically connect real and imaginary parts

TABLE 1.1.1. Relations Between the Four Basic Immittance Functions[a]

	M	Z	Y	ε
M	M	μZ	μY^{-1}	ϵ^{-1}
Z	$\mu^{-1} M$	Z	Y^{-1}	$\mu^{-1}\epsilon^{-1}$
Y	μM^{-1}	Z^{-1}	Y	$\mu\epsilon$
ϵ	M^{-1}	$\mu^{-1} Z^{-1}$	$\mu^{-1} Y$	ϵ

[a] $\mu \equiv j\omega C_c$, where C_c is the capacitance of the empty cell.

with each other, ensure that Z'' (and θ) cannot be zero over all frequencies but must vary with frequency as well. Thus it is only when $Z(\omega) = Z'$, independent of frequency, so $Z' = R$, an ordinary linear resistance, that $Z(\omega)$ is purely real.

There are several other measured or derived quantities related to impedance which often play important roles in IS. All of them may be generically called immittances. First is the admittance, $Y \equiv Z^{-1} \equiv Y' + jY''$. In the complex domain where v, i, and Z are all taken complex, we can write $v = Zi$ or alternatively $i = Yv$. It is also customary in IS to express Z and Y in terms of resistive and capacitance components as $Z = R_s(\omega) - jX_s(\omega)$ and $Y = G_p(\omega) + jB_p(\omega)$, where the reactance $X_s \equiv [\omega C_s(\omega)]^{-1}$ and the susceptance $B_p \equiv \omega C_p(\omega)$. Here the subscripts s and p stand for "series" and "parallel."

The other two quantities are usually defined as the modulus function $M = j\omega C_c Z = M' + jM''$ and the complex dielectric constant or dielectric permittivity $\epsilon = M^{-1} \equiv Y/(j\omega C_c) \equiv \epsilon' - j\epsilon''$. In these expressions $C_c \equiv \epsilon_0 A_c/l$ is the capacitance of the empty measuring cell of electrode area A_c and electrode separation length l. The quantity ϵ_0 is the dielectric permittivity of free space, 8.854×10^{-12} F/m. The dielectric constant ϵ is often written elsewhere as ϵ^* or $\hat{\epsilon}$ to denote its complex character. Here we shall reserve the superscript asterisk to denote complex conjugation; thus $Z^* = Z' - jZ''$. The interrelations between the four immittance functions are summarized in Table 1.

The modulus function $M = \epsilon^{-1}$ was apparently first introduced by Schrama [1957] and has been used appreciably by McCrum et al. [1967], Macedo et al. [1972], and Hodge et al. [1975, 1976]. The use of the complex dielectric constant goes back much further but was particularly popularized by the work of Cole and Cole [1941], who were the first to plot ϵ in the complex plane.

Some authors have used the designation *modulus spectroscopy* to denote small-signal measurement of M vs. ν or ω. Clearly, one could also define admittance and dielectric permittivity spectroscopy. The latter is just another way of referring to ordinary dielectric constant and loss measurements. Here we shall take the general term *impedance spectroscopy* to include all these other very closely related approaches. Thus IS also stands for *immittance spectroscopy*. The measurement and use of the complex $\epsilon(\omega)$ function is particularly appropriate for dielectric materials, those with very low or vanishing conductivity, but all four functions are valuable in IS, particularly because of their different dependence on and weighting with frequency.

1.1.5 Early History

Impedance spectroscopy is particularly characterized by the measurement and analysis of some or all of the four impedance-related functions Z, Y, M, and ϵ and the plotting of these functions in the complex plane. Such plotting can, as we shall see, be very helpful in interpreting the small-signal ac response of the electrode-material system being investigated. Historically, the use of Z and Y in analyzing the response of electrical circuits made up of lumped (ideal) elements (R, L, and C) goes back to the beginning of the discipline of electrical engineering. An important milestone for the analysis of real systems, that is, ones distributed in space, was the plotting by Cole and Cole [1941] of ϵ' and ϵ'' for dielectric systems in the complex plane, now known as a Cole–Cole plot, an adaption at the dielectric constant level of the circle diagram of electrical engineering (Carter [1925]), exemplified by the Smith–Chart impedance diagram (Smith [1939, 1944]). Further, Z and/or Y have been widely used in theoretical treatments of semiconductor and ionic systems and devices from at least 1947 (see, e.g., Randles [1947], Jaffé [1952], Chang and Jaffé [1952], Macdonald [1953], and Friauf [1954]). Complex plane plots have sometimes been called Nyquist diagrams. This is a misnomer, however, since Nyquist diagrams refer to transfer function (three- or four-terminal) response, while conventional complex plane plots involve only two-terminal input immittances.

On the experimental side, one should mention the early work of Randles and Somerton [1952] on fast reactions in supported electrolytes; no complex plane plotting appeared here. But complex plane plotting of G_p/ω vs. C_p was used by Macdonald [1955] for experimental results on photoconducting alkali halide single crystals. Apparently the first plotting of impedance in the impedance plane for aqueous electrolytes was that of Sluyters [1960] (theory) and Sluyters and Oomen [1960] (experiment). The use of admittance plane plotting for accurate conductivity determination of solid electrolytes was introduced by Bauerle [1969], the first important paper to deal with IS for ionic solids directly. Since then, there have been many pertinent theoretical and experimental papers dealing with IS and complex plane plots. Many of them will be cited later, and we conclude this short survey of early history pertinent to IS with the mention of three valuable reviews: Sluyters–Rehbach and Sluyters [1970], Armstrong et al. [1978], and Archer and Armstrong [1980]. The first and second of these deal almost entirely with liquid electrolytes but are nevertheless somewhat pertinent to IS for solids.

1.2 ADVANTAGES AND LIMITATIONS

Although we believe that the importance of IS is demonstrated throughout this monograph by its usefulness in the various applications discussed, it is of some value to summarize the matter briefly here. IS is becoming a popular analytical tool in materials research and development because it involves a relatively simple electrical measurement that can readily be automated and whose results may often be correlated with many complex materials variables: from mass transport, rates of chemical reactions, corrosion, and dielectric properties, to defects, microstructure, and compositional influences on the conductance of solids. IS can predict aspects of the performance of chemical sensors and fuel cells, and it has been used extensively to investigate membrane behavior in living cells. It is useful as an empirical quality control procedure, yet it can contribute to the interpretation of fundamental electrochemical and electronic processes.

A flow diagram of a general characterization procedure using IS is presented in Fig. 1.2.1. Here CNLS stands for complex nonlinear least squares fitting (see Section 3.2.2). Experimentally obtained impedance data for a given electrode-materials system may be analyzed by using an exact mathematical model based on a plausible physical theory that predicts theoretical impedance $Z_t(\omega)$ or by a relatively empirical equivalent circuit whose impedance predictions may be denoted by $Z_{ec}(\omega)$. In either the case of the relatively empirical equivalent circuit or of the exact mathematical model, the parameters can be estimated and the experimental $Z_e(\omega)$ data compared to either the predicted equivalent circuit impedance $Z_{ec}(\omega)$

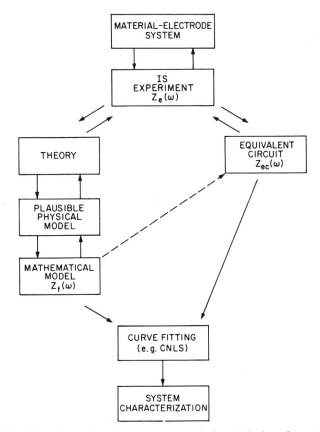

FIGURE 1.2.1. Flow diagram for the measurement and characterization of a material–electrode system.

or to the theoretical impedance $Z_t(\omega)$. Such fitting is most accurately accomplished by the CNLS method described and illustrated in Section 3.2.2.

An analysis of the charge transport processes likely to be present in an experimental cell (the physical model) will often suggest an equivalent circuit of ideal resistors and capacitors (even inductors or negative capacitors in some instances) and may account adequately for the observed IS response. For example, Schouler et al. [1983] found that the effects of densification by sintering a polycrystalline electrolyte will reduce the magnitude of the resistance across the grain boundaries and simultaneously decrease the surface area associated with the interface capacitance. These components will clearly be electrically in parallel in this situation. Their combination will be in series with other similar subcircuits representing such processes as the ionization of oxygen at the electrodes.

In another example, the oxidation–reduction reaction for the Zn^{2+} couple in an aqueous solution with a dropping mercury electrode (Sluyters and Oomen [1960]) can be represented by a reaction resistance R_R, arising from the transfer of electrons between the electrode and the solution, in parallel with a capacitor C_R associated with the space charge diffuse double layer near the electrode surface. It is not difficult to calculate the theoretical impedance for such a circuit in terms of the parameters R_R and C_R. From an analysis of the parameter values in a plausible equivalent circuit as the experimental conditions are changed, the materials system can be characterized by analysis of its observed impedance response, leading to estimates of its microscopic parameters such as charge mobilities, concentrations, and electron transfer reaction rates.

The disadvantages of IS are primarily associated with possible ambiguities in interpretation. An important complication of analyses based on an equivalent circuit (e.g., Bauerle [1969]) is that ordinary ideal circuit elements represent ideal lumped-constant properties. Inevitably, all electrolytic cells are distributed in space, and their microscopic properties may be also independently distributed. Under these conditions, ideal circuit elements may be inadequate to describe the electrical response. Thus, it is often found that $Z_e(\omega)$ cannot be well approximated by the impedance of an equivalent circuit involving only a finite number of ordinary lumped-constant elements. It has been observed by many in the field that the use of distributed impedance elements [e.g., constant-phase elements (CPEs) (see Section 2.2.2.2)] in the equivalent circuit greatly aids the process of fitting observed impedance data for a cell with distributed properties.

There is a further serious potential problem with equivalent circuit analysis, not shared by the direct comparison with $Z_t(\omega)$ of a theoretical model: What specific equivalent circuit out of an infinity of possibilities should be used if one is necessary? An equivalent circuit involving three or more circuit elements can often be rearranged in various ways and still yield exactly the same $Z_{ec}(\omega)$. For the different interconnections the values of the elements will have to be different to yield the same $Z_{ec}(\omega)$ for all ω, but an essential ambiguity is present. An example is presented in Fig. 1.2.2. In these circuits the impedance Z_i is arbitrary and may be made up of either lumped elements, distributed elements, or a combination of these types. Examples of other circuits which demonstrate this type of ambiguity will be presented in Section 2.2.2.3. Which one of two or more circuits which all

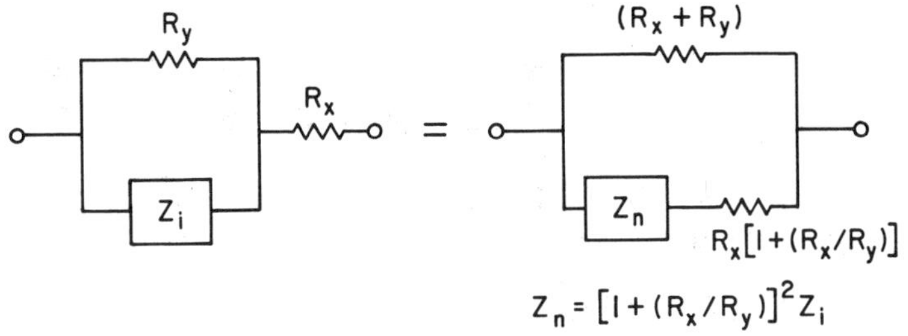

FIGURE 1.2.2. An example of different circuits with the same overall impedance at all frequencies.

yield exactly the same $Z_{ec}(\omega)$ for all ω should be used for physicochemical analysis and interpretation? This question cannot be answered for a single set of $Z_e(\omega)$ data alone. An approach to its solution can only be made by employing physical intuition and by carrying out several $Z_e(\omega)$ sets of measurements with different conditions, as discussed in Section 2.2.2.3.

1.2.1   Differences Between Solid State and Aqueous Electrochemistry

The electrochemist who works with aqueous electrolytes has available to him at least one major strategem not accessible to those who work with solid electrolytes. If he is interested in the interfacial behavior of a particular charged species, he is usually free to add to the solution an excess of a second electrolyte, the ions of which are neither adsorbed nor react at the interface, but which by sheer numbers are able to screen the interior of the electrolyte from any electric field and cause nearly all the potential drop to occur within a few angstroms of the interface. The investigator is thus (at least by assumption) freed from having to take into account the effect of a nonuniform electric field on the transport of the electroactive species through the bulk electrolyte and need not (again by assumption) puzzle over the fraction of the applied signal which directly governs the exchange of ions or electrons between the electrode surface and the adjacent layer of electrolyte. The added electrolyte species which thus simplifies the interpretation of the experimental results is termed the *indifferent* or *supporting electrolyte*, and systems thus prepared are termed *supported systems*. Solid electrolytes must necessarily be treated as unsupported systems, even though they may display some electrical characteristics usually associated with supported ones. The distinction between unsupported and supported situations is a crucial one for the interpretation of IS results.

It is thus unfortunate that there has been a tendency among some workers in the solid electrolyte field to take over many of the relatively simple theoretical results derived for supported conditions and use them uncritically in unsupported situations, situations where the supported models and formulas rarely apply adequately. For example, the expression for the Warburg impedance for a redox reaction in a supported situation is often employed in the analysis of data on unsupported situations where the parameters involved are quite different (see, e.g., Sections 2.2.3.2 and 2.2.3.3).

There are a few other important distinctions between solid and liquid electrolytes. While liquid electrolytes and many solid electrolytes have negligible electronic conductivity, quite a number of solid electrolytes can exhibit substantial electronic conductivity, especially for small deviations from strict stoichiometric composition. Solid electrolytes may be amorphous, polycrystalline, or single-crystal, and charges of one sign may be essentially immobile (except possibly for high temperatures and over long time spans). On the other hand, all dissociated charges in a liquid electrolyte or fused salt are mobile, although the ratio between the mobilities of positive and negative charges may differ appreciably from unity. Further, in solid electrolytes mobile ions are considered to be able to move as close to an electrode as permitted by ion-size steric considerations. But in liquid electrolytes there is usually present a compact inner or Stern layer composed of solvent molecules, for example, H_2O, immediately next to the electrode. This layer may often be entirely devoid of ions and only has some in it when the ions are specifically adsorbed at the electrode or react there. Thus capacitative effects in electrode interface regions can be considerably different between solid and liquid electrolyte systems.

1.3   ELEMENTARY ANALYSIS OF IMPEDANCE SPECTRA

1.3.1   Physical Models for Equivalent Circuit Elements

A detailed physicoelectrical model of all the processes which might occur in investigations on an electrode–material system may be unavailable, premature, or perhaps too complicated to warrant its initial use. One then tries to show that the experimental impedance data $Z_e(\omega)$ may be well approximated by the impedance $Z_{ec}(\omega)$ of an equivalent circuit made up of ideal resistors, capacitors, perhaps inductances, and possibly various distributed circuit elements. In such a circuit a resistance represents a conductive path, and a given resistor in the circuit might account for the bulk conductivity of the material or even the chemical step associated with an electrode reaction (see, e.g., Randles [1947] or Armstrong et al. [1978]). Similarly, capacitances and inductances will be generally associated with space charge polarization regions and with specific adsorption and electrocrystallization processes at an electrode. It should be pointed out that ordinary circuit elements, such as resistors and capacitors, are always considered as lumped-constant quantities which involve ideal properties. But all real resistors are of finite size and are thus disturbed in space; they therefore always involve some inductance, capacitance, and time delay of response as well as resistance. These residual properties are unimportant over wide frequency ranges and therefore usually allow a physical resistor to be well approximated in an equivalent circuit by an ideal resistance, one which exhibits only resistance over all frequencies and yields an immediate rather than a delayed response to an electrical stimulus.

The physical interpretation of the distributed elements in an equivalent circuit is somewhat more elusive. They are, however, essential in understanding and interpreting most impedance spectra. There are two types of distributions with which we need to be concerned. Both are related, but in different ways, to the finite spatial extension of any real system. The first is associated directly with nonlocal processes, such as diffusion, which can occur even in a completely homogeneous material, one whose physical properties, such as charge mobilities, are the same everywhere. The other type, exemplified by the constant-phase element (CPE), arises because microscopic material properties are themselves often distributed. For example, the solid electrode–solid electrolyte interface on the microscopic level is not the often presumed smooth and uniform surface. It contains a large number of surface defects such as kinks, jags, and ledges, local charge inhomogeneities, two- and three-phase regions, adsorbed species, and variations in composition and stoichiometry. Reaction resistance and capacitance contributions differ with electrode position and vary over a certain range around a mean, but only their average effects over the entire electrode surface can be observed. The macroscopic impedance which depends, for example, on the reaction rate distribution across such an interface is measured as an average over the entire electrode. We account for such averaging in our usual one-dimensional treatments (with the dimension of interest perpendicular to the electrodes) by assuming that pertinent material properties are continuously distributed over a given range from minimum to maximum values. For example, when a given time constant, associated with an interface or bulk processes, is thermally activated with a distribution of activation energies, one passes from a simple ideal resistor and capacitor in parallel or series to a distributed impedance element, for example, the CPE, which exhibits more complicated frequency response than a simple undistributed RC time constant process (Macdonald [1984, 1985a, c, d], McCann and Badwal [1982]).

Similar property distributions occur throughout the frequency spectrum. The classical example for dielectric liquids at high frequencies is the bulk relaxation of dipoles present in a pseudoviscous liquid. Such behavior was represented by Cole and Cole [1941] by a modification of the Debye expression for the complex dielectric constant and was the first distribution involving the CPE. In normalized form the complex dielectric constant for the Cole–Cole distribution may be written

$$(\epsilon - \epsilon_\infty)/(\epsilon_s - \epsilon_\infty) = \left[1 + (j\omega\tau_0)^{1-\alpha} \right]^{-1} \tag{1}$$

where ϵ is the dielectric constant, ϵ_s and ϵ_∞ are the static and high-frequency limiting dielectric constants, τ_0 the mean relaxation time, and α a parameter describing the width of the material property distribution (in this case a distribution of dielectric relaxation times in frequency space).

1.3.2 Simple RC Circuits

Figure 1.3.1 shows two RC circuits common in IS and typical Z and Y complex plane responses for them. The response of Fig. 1.3.1a is often present (if not always measured) in IS results for solids and liquids. Any electrode–material system in a measuring cell has a geometrical capacitance $C_g \equiv C_\infty = C_1$ and a bulk

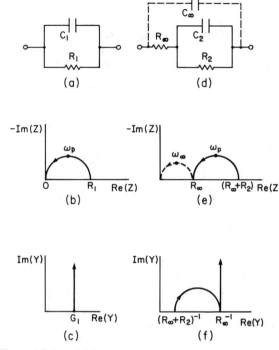

FIGURE 1.3.1. Figures 1.3.1a and d show two common RC circuits. Parts b and e show their imped-ance plane plots and c and f their admittance plane plots. Arrows indicate the direction of increasing frequency.

resistance $R_b \equiv R_\infty = R_1$ in parallel with it. These elements lead to the time constant $\tau_D = R_\infty C_\infty$, the dielectric relaxation time of the basic material. Usually, τ_D is the smallest time constant of interest in IS experiments. It is often so small ($< 10^{-7}$ s) that for the highest angular frequency applied, ω_{max}, the condition $\omega_{max} \tau_D \ll 1$ is satisfied and little or nothing of the impedance plane curve of Figure 1.3.1b is seen. It should be noted, however, that lowering the temperature will often increase τ_D and bring the bulk arc within the range of measurement. Since the peak frequency of the complete semicircle of Figure 1.3.1b, ω_p, satisfies $\omega_p \tau_D = 1$, it is only when $\omega_{max} \tau_D \gg 1$ that nearly the full curve of Figure 1.3.1b is obtained. Although the bulk resistance is often not appreciably distributed, par-ticularly for single crystals, when it is actually distributed the response of the circuit often leads to a partial semicircle in the Z plane, one whose center lies below the real axis instead of to a full semicircle with its center on the real axis. Since this distributed element situation is frequently found for processes in the $\omega \ll \tau_D^{-1}$ frequency range, however, we shall examine in detail one simple repre-sentation of it shortly.

Besides $R_1 = R_\infty$ and $C_1 = C_\infty$, one often finds parallel R_1, C_1 response asso-ciated with a heterogeneous electrode reaction. For such a case we would set $R_1 = R_R$ and $C_1 = C_R$, where R_R is a reaction resistance and C_R is the diffuse double-layer capacitance of the polarization region near the electrode in simplest cases. The circuit of Figure 1.3.1d combines the above possibilities when $R_2 = R_R$ and $C_2 = C_R$. The results shown in Figure 1.3.1e and f are appropriate for the well-separated time constants, $R_\infty C_\infty \ll R_2 C_2$. It is also possible that a parallel RC combination can rise from specific adsorption at an electrode, possibly associated with delayed reaction processes. The response arising from R_∞ and C_∞ in Figure 1.3.1e is shown dotted to remind one that it often occurs in too high a frequency region to be easily observed. Incidentally, we shall always assume that the capacitance and resistance of leads to the measuring cell have been subtracted out (e.g., by using the results of a preliminary calibration of the system with the cell empty or shorted) so that we always deal only with the response of the material–electrode system alone.

In the complex plane plots the arrows show the direction of increasing frequency. Further, $G_1 \equiv R_1^{-1}$, $G_\infty \equiv R_\infty^{-1}$, $G_2 \equiv R_2^{-1}$. Because IS results usually involve capacitance and rarely involve inductance, it has become customary to plot impedance in the $-\mathrm{Im}(Z)$, $\mathrm{Re}(Z)$ plane rather than the $\mathrm{Im}(Z)$, $\mathrm{Re}(Z)$ plane, thereby ensuring that the vast majority of all curves fall in the first quadrant, as in Figure 1.3.1b. This procedure is also equivalent to plotting $Z^* = Z' - iZ''$ rather than Z, so we can alternatively label the ordinate $\mathrm{Im}(Z^*)$ instead of $-\mathrm{Im}(Z)$. Both choices will be used in the rest of this work.

The admittance of the parallel RC circuit of Figure 1.3.1a is just the sum of the admittances of the two elements, that is,

$$Y_a = G_1 + j\omega C_1 \qquad (2)$$

It immediately follows that

$$Z_a = Y_a^{-1} = R_1/(R_1 Y_a) = R_1/[1 + j\omega R_1 C_1] \qquad (3)$$

This result can be rationalized by multiplying by $[1 - j\omega R_1 C_1]$, the complex conjugate of $[1 + j\omega R_1 C_1]$, in both numerator and denominator. The response of the Figure 1.3.1a circuit is particularly simple when it is plotted in the Y plane, as in Figure 1.3.1c. To obtain the overall admittance of the Figure 1.3.1d circuit, it is simplest to add R_∞ to the expression for Z_a above with $R_1 \rightarrow R_2$ and $C_1 \rightarrow C_2$, convert the result to an admittance by inversion, and then add the $j\omega C_\infty$ admittance. The result is

$$Y_d = j\omega C_\infty + [1 + j\omega R_2 C_2]/[(R_2 + R_\infty) + j\omega C_2 R_2 R_\infty]. \qquad (4)$$

Although complex plane data plots, such as those in Figures 1.3.1b, c, e, and f in which frequency is an implicit variable, can show response patterns which are often very useful in identifying the physicochemical processes involved in the electrical response of the electrode–material system, the absence of explicit frequency dependence information is frequently a considerable drawback. Even when frequency values are shown explicitly in such two-dimensional (2-D) plots, it is usually found that with either equal intervals in frequency or equal frequency ratios the frequency points fall very nonlinearly along the curves. The availability of computerized plotting procedures makes the plotting of all relevant information in a single graph relatively simple. For example, three-dimensional (3-D) perspective plotting, as introduced by Macdonald, Schoonman, and Lehnen [1981], displays the frequency dependence along a new log (ν) axis perpendicular to the complex plane (see Section 3.2). For multi-time-constant response in particular, this method is particularly appropriate. The full response information can alternately be plotted with orthographic rather than perspective viewing.

1.3.3 Analysis of Single Impedance Arcs

Analysis of experimental data that yield a full semicircular arc in the complex plane, such as that in Figure 1.3.1b, can provide estimates of the parameters R_1 and C_1 and hence lead to quantitative estimates of conductivity, faradic reaction rates, relaxation times, and interfacial capacitance (see detailed discussion in Section 2.2.3.3). In practice, however, experimental data are only rarely found to yield a full semicircle with its center on the real axis of the complex plane. There are three common perturbations which may still lead to at least part of a semicircular arc in the complex plane:

1. The arc does not pass through the origin, either because there are other arcs appearing at higher frequencies and/or because $R_\infty > 0$.

2. The center of an experimental arc is frequently displaced below the real axis because of the presence of distributed elements in the material–electrode system. Similar displacements may also be observed in any of the other complex planes plots (Y, M, or ϵ). The relaxation time τ is then not single-valued but is distributed continuously or discretely around a mean, $\tau_m = \omega_m^{-1}$. The angle θ by which such a semicircular arc is depressed below the real axis is related to the width of the relaxation time distribution and as such is an important parameter.

3. Arcs can be substantially distorted by other relaxations whose mean time constants are within two orders of magnitude or less of that for the arc under consideration. Many of the spectra shown in following chapters involve overlapping arcs.

We shall begin by considering simple approximate analysis methods of data yielding a single, possibly depressed, arc. Suppose that IS data plotted in the impedance plane (actually the Z^* plane) show typical depressed circular arc behavior, such as that depicted in Figure 1.3.2. Here we have included R_∞ but shall initially ignore any effect of C_∞. We have defined some new quantities in this figure which will be used in the analysis to yield estimates of the parameters R_∞, $R_R \equiv R_0 - R_\infty$, τ_R and the fractional exponent ψ_{ZC}, parameters which fully characterize the data when they are well represented by the distributed-element ZARC impedance expression (see Section 2.2.2.2),

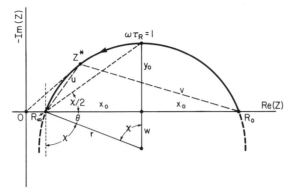

FIGURE 1.3.2. Impedance plane plot for a depressed circular arc showing definitions of quantities used in its analysis.

$$Z - R_\infty \equiv Z_{ZARC} \equiv (R_0 - R_\infty)\, I_Z \tag{5}$$

where

$$I_Z \equiv \left[1 + (j\omega\tau_R)^{\psi_{ZC}}\right]^{-1} \equiv \left[1 + (js)^{\psi_{ZC}}\right]^{-1} \tag{6}$$

Here $s \equiv \omega\tau_R$ is a normalized frequency variable, and I_Z is the normalized, dimensionless form of Z_{ZARC}. Notice that it is exactly the same as the similarly normalized Cole–Cole dielectric response function of Eq. (1) when we set $\psi_{ZC} = 1 - \alpha$. We can also alternatively write the ZARC impedance as the combination of the resistance R_R in parallel with the CPE impedance Z_{CPE} (see Section 2.2.2.2). The CPE admittance is (Macdonald [1984])

$$Y_{CPE} = Z_{CPE}^{-1} \equiv A_0(j\omega)^{\psi_{ZC}} \equiv (j\omega\tau_R)^{\psi_{ZC}} \tag{7}$$

Then Eq. (5) may be expressed as

$$Z_{ZARC} = R_R / \left[1 + B_0(j\omega)^{\psi_{ZC}}\right], \tag{8}$$

where $B_0 \equiv \tau_R^{\psi_{ZC}} \equiv R_R A_0$. The fractional exponent ψ_{ZC} satisfies $0 \leqslant \psi_{ZC} \leqslant 1$.

Let us start by considering two easy-to-use approximate methods of estimating the parameters, methods often adequate for initial approximate characterization of the response. The estimates obtained by these approaches may also be used as initial values for the more complicated and much more accurate CNLS method described and illustrated in Section 3.2.2. Note that the single $R_R C_R$ situation, that where $\theta = 0$ and $\psi_{ZC} = 1$, is included in the analysis described below.

From the figure, $-Z''$ reaches its maximum value, y_0, when $\omega = \omega_m = \tau_R^{-1}$ and thus $s = 1$. At this point the half-width of the arc on the real axis is $Z' - R_\infty = x_0 \equiv R_R/2$. Now from the data, the complex plane plot, and estimated values of x_0, y_0, and ω_m, one can immediately obtain estimates of R_∞, R_0, R_R, and τ_R. In order to obtain θ, one must, of course, find the direction of the circle center. The easiest graphical method is to draw on the Z^* plane plot several lines perpendicular to the semicircle; the center will be defined by their intersection. Two other more accurate approaches will be described below. Incidentally, when there is more than one arc present and there is some overlap which distorts the right, lower-frequency side of the arc, the present methods can still be used without appreciable loss of accuracy provided overlap distortion is only significant for $\omega < \omega_m$, that is, on the right side of the center of the left arc. Then all parameters should be estimated from the left side of the arc, that is, for $\omega \geqslant \omega_m$. A similar approach may be used when data are available only for $\omega \leqslant \omega_m$. From Figure 1.3.2 and Eq. (5) we readily find that $\theta = \pi/2 - \chi \equiv (\pi/2)(1 - \psi_{ZC})$; thus when $\psi_{ZC} = 1$ there is no depression and one has simple single-time-constant ($\tau_R \equiv R_R C_R$) Debye response with $A_0 \equiv C_R$. When $\psi_{ZC} < 1$, $\tau_R = (R_R A_0)^{1/\psi_{ZC}}$, but an ideal C_R capacitor cannot be directly defined, reflecting the distributed nature of the response.

The rest of the analysis proceeds as follows. First, one may obtain an estimate of ψ_{ZC} from the θ value using $\psi_{ZC} = 1 - 2\theta/\pi$. But a superior alternative to first obtaining θ by finding the circle center approximately is to use the values of x_0 and y_0 defined on the figure. For simplicity, it will be convenient to define

$$q \equiv (\omega \tau_R)^{\psi_{ZC}} \equiv (s)^{\psi_{ZC}} \tag{9}$$

$$\chi \equiv (\pi/2) - \theta \equiv (\pi/2)\psi_{ZC} \tag{10}$$

and note that

$$x_0 \equiv (R_0 - R_\infty)/2 \equiv R_R/2 \tag{11}$$

We may now rewrite Eq. (6) for I_Z as

$$I_Z(q, \chi) = \frac{\left[1 + q \cos(\chi)\right] - jq \sin(\chi)}{1 + 2q \cos(\chi) + q^2} \tag{12}$$

For $q = 1$, the peak point, one finds

$$I_Z(1, \chi) = 0.5\left[1 - j \tan(\chi/2)\right] \tag{13}$$

Let us further define for later use the quantity

$$\psi_J \equiv \tan(\chi/2) = \tan(\pi\psi_{ZC}/4) \tag{14}$$

Now in general from Eq. (12) we may write

$$-I_Z''/I_Z' = q \sin(\chi)/\left[1 + q \cos(\chi)\right] \tag{15}$$

which becomes, for $q = 1$,

$$-I_Z''/I_Z'\big|_{q=1} = y_0/x_0 = \tan(\chi/2) \equiv \psi_J \tag{16}$$

Thus from knowledge of y_0 and x_0 one can immediately calculate χ, ψ_J, ψ_{ZC}, and θ. For completeness, it is worth giving expressions for w and r which follow from the figure. One finds

$$w = x_0 \operatorname{ctn}(\chi) = x_0 \tan(\theta) = x_0\left[(1 - \psi_J^2)/2\psi_J\right] \tag{17}$$

and

$$r = y_0 + w = x_0 \csc(\chi) = x_0 \sec(\theta) = x_0\left[(1 + \psi_J^2)/2\psi_J\right] \tag{18}$$

A further method of obtaining ψ_{ZC} and θ is to first estimate R_∞ and plot $(Z - R_\infty)^{-1}$ in the Y plane. Then a spur inclined at the angle $[(\pi/2) - \theta] = \chi$ will appear whose $\omega \to 0$ intercept is $(R_0 - R_\infty)^{-1}$. A good estimate of ψ_{ZC} may be obtained from the χ value when the spur is indeed a straight line. Now at $\omega = \omega_m$, it turns out that $B_0\omega_m^{\psi_{ZC}} = 1$. Thus one may obtain an estimate of B_0 from $\omega_m^{-\psi_{ZC}}$. Then $\tau_R = B_0^{1/\psi_{ZC}} = \omega_m^{-1}$ and $A_0 = R_R^{-1}B_0$. Thus all the parameters of interest have then been estimated.

The above simple methods of estimating ψ_{ZC} depend only on the determination of x_0 and y_0 from the impedance complex plane arc or on the use of a few points in the admittance plane. Although they are often adequate for initial investigation, it is worth mentioning a relatively simple alternative procedure which can be used to test the appropriateness of Eqs. (5) and (6) and obtain the parameter estimates of interest. Consider the point Z^* on the arc of Figure 1.3.2, a point marking a specific value of Z. It follows from the figure and Eq. (5) that $Z^* - R_\infty = (R_0 - R_\infty)I_Z^* \equiv u$ and $R_0 - Z^* = (R_0 - R_\infty)(1 - I_Z^*) \equiv v$. Therefore,

$$\ln|v/u| = \ln\left|(I_Z^*)^{-1} - 1\right| = \ln(q) = \psi_{ZC}\left[\ln(\omega) + \ln(\tau_R)\right] \tag{19}$$

If one assumes that R_0 and R_∞ may be determined adequately from the complex plane plot, not always a valid assumption, then v and u may be calculated from experimental Z data for a variety of frequences. A plot of $\ln|v/u|$ vs. $\ln(\omega)$ will yield a straight line with a slope of ψ_{ZC} and an intercept of $\psi_{ZC}\ln(\tau_R)$ provided Eq. (19) holds. Ordinary linear least squares fitting may then be used to obtain estimates of ψ_{ZC} and $\ln(\tau_R)$.

Although a more complicated nonlinear least squares procedure has been described by Tsai and Whitmore [1982] which allows analysis of two arcs with some overlap, approximate analysis of two or more arcs without much overlap does not require this approach and CNLS fitting is more appropriate for one or more arcs with or without appreciable overlap when accurate results are needed. In this section we have discussed some simple methods of obtaining approximate estimates of some equivalent circuit parameters, particularly those related to the common symmetrical depressed arc, the ZARC. An important aspect of material–electrode characterization is the identification of derived parameters with specific physico-chemical processes in the system. This matter is discussed in detail in Sections 2.2 and 3.2 and will not be repeated here. Until such identification has been made, however, one cannot relate the parameter estimates, such as R_R, C_R, and ψ_{ZC}, to specific microscopic quantities of interest such as mobilities, reaction rates, and activation energies. It is this final step, however, yielding estimates of parameters immediately involved in the elemental processes occurring in the electrode–ma-terial system, which is the heart of characterization and an important part of IS.

1.4 SELECTED APPLICATIONS OF IS

In this section two applications will be presented which illustrate the power of the IS technique when it is applied to two very diverse areas, aqueous electrochemistry and fast ion transport in solids. These particular examples were chosen because of their historical importance and because the analysis in each case is particularly simple. Additional techniques and applications of IS to more complicated systems will be presented in Chapter 4 as well as throughout the text.

The first experimental use of complex plane analysis in aqueous electrochem-istry was performed in 1960 (Sluyters and Oomen [1960]). This study is a classic illustration of the ability of impedance spectroscopy to establish kinetic parameters in an aqueous electrochemical system. Using a standard hanging mercury drop cell, the impedance response of the $Zn(Hg)/Zn^{2+}$ couple in a 1M $NaClO_4$ + 10^{-3}M $HClO_4$ electrolyte was examined at 298K. For this couple, the reaction rate is such that in the frequency range of 20 Hz to 20 kHz the kinetics of charge transfer is slower than ion diffusion in the electrolyte. The results (Fig. 1.4.1) show a single

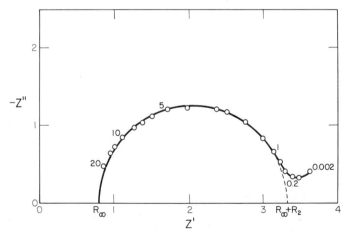

FIGURE 1.4.1. The impedance results of a $Zn(Hg)/Zn^{2+}$ couple in 1M $NaClO_4$ + 10^{-3}N $HClO_4$ with C_{Zn} = 8 × 10^{-6} moles/cm^3 and $C_{Zn^{2+}}$ = 8 × 10^{-6}. The numbers represent the frequency in kilohertz; the axes are in arbitrary scale units. (Sluyters and Oomen [1960])

semicircle characteristic of kinetic control by an electrochemical charge transfer step at the electrode–electrolyte interface. The physical model appropriate to this system is the same as that presented in Figure 1.3.1d. The semicircle beginning at the origin in Figure 1.3.1e is not observed in Figure 1.4.1 because the frequency range was limited to below 20 kHz. Thus, in Figure 1.4.1, R_∞ is the solution resistance, R_2 is the charge transfer resistance, and C_2 is the double-layer capacitance.

By solving the standard current–potential equation for an electrochemical reaction (see, for example, Bard and Faulkner [1980]) under the conditions of kinetic control (i.e., the rate of charge transfer is much slower than diffusive processes in the system), the value of R_2 can be evaluated. For a known concentration of Zn at the amalgam–electrolyte interface, $C_{Zn(Hg)}$, and a known concentration of Zn^{2+} at the electrolyte–electrode interface, $C_{Zn^{2+}}$, the value of R_2 is given by Eq. (1):

$$R_2 = \frac{RT}{n^2 F^2 k \left(C_{Zn^{2+}}\right)^\alpha \left(C_{Zn(Hg)}\right)^{1-\alpha}} \qquad (1)$$

where n is the number of electrons transferred, F is Faraday's constant, k is the rate constant for the electrochemical charge transfer reaction, α is the electrochemical transfer coefficient, R is the ideal gas constant, and T is the absolute temperature. When the concentration of Zn in the amalgam is equal to the concentration of Zn ions in the solution, then the rate constant k can be determined. Results at several different equal concentrations of Zn and Zn^{2+} (Table 1.4.1) gave a mean value of $k = 3.26 \times 10^3$ cm/s. By using different concentrations of Zn and Zn^{2+} the transfer coefficient α (Tables 1.4.2 and 1.4.3) was found to be 0.70. In addition, the value of the double-layer capacitance could be easily determined in each of the experiments.

In a similar experiment, the Hg/Hg^{2+} reaction in 1M $HClO_4$ has also been investigated (Sluyters and Oomen [1960]) using IS in the frequency range of 20 Hz to 20 kHz and for concentrations between 2×10^{-6} and 10×10^{-6} moles/cm^3

TABLE 1.4.1. Calculation of Rate Constant of Zn(Hg)/Zn^{2+} Couple

$C_{Zn} = C_{Zn}{}^{2+}$ (moles/cm^3)	R_2 (Ω-cm^2)	$R_2 \times C_{Zn}$ (moles-Ω/cm)	k (cm/sec)^a
2×10^{-6}	10.17	20.3×10^{-6}	
4	4.95	19.8	
5	4.26	21.3	
8	2.41	19.3	$3.26 \times 10^{-3} \pm 3.6\%$
10	2.13	21.3	
16	1.27	20.3	
16	1.28	20.5	

^{a}Calculated from the average value of $R_2 \times C_{Zn} = 20.4 \times 10^{-6}$ by $k = (R_2 C_{Zn} n^2 F^2)^{-1} RT$ according to Eq. (1).
Source: Sluyters and Oomen [1960].

TABLE 1.4.2. Calculation of Transfer Coefficient α of Zn(Hg)/Zn^{2+} Couple

C_{Zn} (moles/cm^3)	$C_{Zn^{2+}}$ (moles/cm^3)	R_2 (Ω-cm^2)	$\log R_2$	$-\log C_{Zn^{2+}}$	α^a
16×10^{-6}	16×10^{-6}	1.28	0.107	4.796	
16	8	2.00	0.301	5.097	0.70
16	4	3.29	0.517	5.398	
16	2	5.37	0.730	5.699	

^{a}From slope of $-\log C_{Zn^{2+}}$ vs. $\log R_2$ plot.
Source: Sluyters and Oomen [1960].

TABLE 1.4.3. Calculation of Transfer Coefficient 1 − α of Zn(Hg)/Zn^{2+} Couple

$C_{Zn^{2+}}$ (moles/cm^3)	C_{Zn} (moles/cm^3)	R_2 (Ω-cm^2)	$\log R_2$	$-\log C_{Zn}$	$1 - \alpha^a$
16×10^{-6}	16×10^{-6}	1.28	0.107	4.796	
16×10^{-6}	8	1.56	0.193	5.097	0.29
16×10^{-6}	4	1.93	0.286	5.398	

^{a}From slope of $-\log C_{Zn}$ vs. $\log R_2$ plot.
Source: Sluyters and Oomen [1960].

Hg^{2+}. The results (Fig. 1.4.2) show linear behavior in the complex plane with an angle of 45° to the real axis. Such a response is indicative of a distributed element as discussed in the previous section. In this case, the system is under diffusion control as the kinetics of the charge transfer at the electrode–electrolyte interface

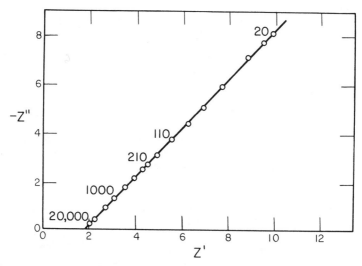

FIGURE 1.4.2. The impedance results of a Hg_2^{2+}/Hg couple in 1M $HClO_4$ electrolyte with $C_{Hg_2^{2+}} = 2 \times 10^{-6}$ moles/cm^3. The numbers represent the frequency in kilohertz; the axes are in arbitrary scale units. (Sluyters and Oomen [1960])

is much faster than the diffusion of the Hg^{2+} ions in the solution. Solution of the diffusion equation with the appropriate boundary conditions under a small ac perturbation gives the diffusional contribution to the impedance in the complex plane as (see Chapter 2 for a detailed discussion)

$$W = \sigma\omega^{-1/2} - j\sigma\omega^{-1/2} \qquad (2)$$

where the impedance W, is generally called the Warburg impedance, ω is the angular frequency, j is equal to $(-1)^{1/2}$ and σ is a constant given by

$$\sigma = \frac{RT}{n^2 F^2 \sqrt{2}} \left[\frac{1}{C_{Hg_2^{2+}}(D_{Hg_2^{2+}})^{1/2}} + \frac{1}{[C_{Hg}(D_{Hg})^{1/2}]} \right] \qquad (3)$$

where $D_{Hg_2^{2+}}$ and D_{Hg} are the diffusivity of mercurous ions in solution and mercury in amalgam, respectively, and the other terms are defined as above. This impedance is to be added (see Sluyters [1960]) and the discussion in Chapter 2) in series with R_2 of Figure 1.3.1d. When the impedance of this circuit is plotted in the complex plane, one obtains a semicircle combined with a straight line at an angle of 45° to the real axis. The line, when extended to the real axis, has an intercept of $R_\infty + R_2 - 2\sigma C_{dl}$. If $2\sigma C_{dl}$ is small, as in the present case, the semicircle is suppressed and the product of the imaginary part of W, Im (W) and $\omega^{1/2}$ will be equal to σ at all frequencies.

The experimental results in Figure 1.4.2 are thus consistent with a system under diffusion control. The diffusivity of Hg_2^{2+} ions in solution can be easily calculated (Table 1.4.4) at several different concentrations of Hg_2^{2+} in the solution from the value of σ. No further information can be obtained from this data because the time constant associated with the kinetics is too fast to be measured at frequencies below 20 kHz.

TABLE 1.4.4. Calculation of Diffusion Coefficient of Hg^{2+} in 1N $HClO_4$

$C_{Hg_2^{2+}}$ (moles/cm^3)	σ^a (Ω-sec$^{-1/2}$ cm^2)	$D_{Hg_2^{2+}}^b$ (cm^2/sec)	$R_\infty + R_2$ (Ω-cm^2)
10×10^{-6}	2.09	0.241×10^{-5}	0.190
5	4.10	0.251	0.188
4	4.99	0.264	0.188
3	6.60	0.268	0.195
2	9.73	0.277	0.193

[a] $\sigma =$ Im (W) $\omega^{1/2}$ was found to be independent of frequency within 2%.

[b] $D_{Hg_2^{2+}} = [RT(\sigma n^2 F^2 \sqrt{2} C_{Hg_2^{2+}})^{-1}]^2$ according to Eq. (3) with $1/[C_{Hg}(D_{Hg})^{-1/2}] \ll 1/[C_{Hg^{2+}}(D_{Hg_2^{2+}})^{1/2}]$, as is the case here with a pure Hg electrode.

Source: Sluyters and Oomen [1960].

The frequency range chosen in the above experiments was dictated by the limited electronics available in 1960 and the cumbersome experimental approach associated with it, which required that the impedance be measured independently at each frequency. The introduction of automated impedance analysis instruments removes this restriction and allows the experimenter to choose the most appropriate frequency range for a given experiment. This choice should be determined by the nature of the interfaces in the experiment and the time constants that are associated with them. For example, corrosion studies, which often involve a slow aqueous diffusion process, generally have relatively large time constants (on the order of 0.1–10 s), and thus most impedance studies of corroding systems use frequencies between a few millihertz and 100 kHz. On the other hand, studies of solid ionic conductors require higher frequencies to measure the time constant associated with ionic motion (milli- to microseconds), which is generally smaller than those found in aqueous diffusion processes. Thus, frequencies between a few hertz and 15 MHz are most appropriate here.

That is not to say that the frequency range should always be restricted based upon predetermined expectations. In the above studies, a wider frequency range would probably have allowed a determination of additional information. For the Zn/Zn^{2+} couple, lower frequencies would have allowed the measurement of the diffusivity of zinc ions in the solution. For the study of the Hg/Hg^{2+} couple, the kinetics of the electrochemical reaction at the interface could have been explored by using higher frequencies. Nevertheless, an understanding of the relationship between the time constant in an experiment and the frequencies with which to measure it provides an intelligent starting point in the choice of the most appropriate frequency range.

A second example which illustrates the utility of IS to solid state chemists is the application of impedance analysis to zirconia–yttria solid electrolytes (Bauerle [1969]). At elevated temperatures solid solution zirconia–yttria compounds are known to be oxygen-ion conductors which function by transport of oxygen ions through vacancies introduced by the dopant yttria. By examining cells of the form

$$Pt,O_2 \,|\, (ZrO_2)_{0.9} \, (Y_2O_3)_{0.1} \,|\, O_2,Pt \tag{4}$$

using IS, admittance plots were obtained (Fig. 1.4.3a). The equivalent circuit proposed to fit this data is shown in Figure 1.4.3b. By a careful examination of the effect of the electrode-area-to-sample-length ratio, and by measuring the dc conductivity of the samples, the high-frequency semicircle (the one on the right in Fig. 1.4.3a) was ascribed to bulk electrolyte behavior, while the low-frequency semicircle (on the left in Fig. 1.4.3a) corresponded to the electrode polarization. In the terminology of Figure 1.4.3b, R_1 and C_1 correspond to electrode polarization phenomena, while R_2, R_3, and C_2 describe processes which occur in the bulk of the electrolyte specimen. Furthermore, by varying temperature, oxygen partial pressure, and electrode preparation, the role of each component in the overall conduction mechanism was determined. In particular, R_1 represents an effective resistance for the electrode reaction:

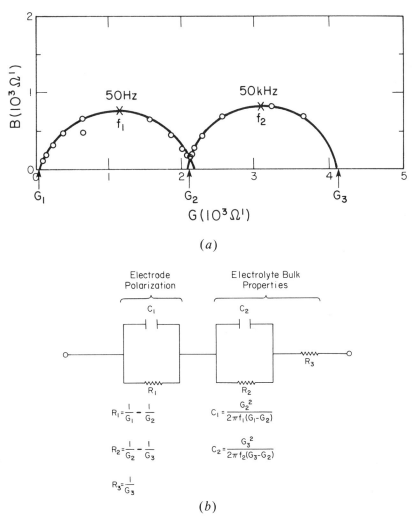

FIGURE 1.4.3. (*a*) Admittance behavior of the electrochemical cell given in 1.4.4 at 873K for a specimen with naturally porous electrodes (sputtered Pt). (*b*) The equivalent circuit for the behavior in part *a* showing the two impedance elements associated with each semicircicle. (Bauerle [1969])

$$\tfrac{1}{2} O_{2(g)} + 2e^- = O^{2-} \; (\text{electrolyte}) \tag{5}$$

where C_1 is the double-layer capacitance of the electrode; R_2 is a "constriction" or intergranular resistance corresponding to resistance of conduction across two different grains, primarily due to impurities located there; C_2 is the capacity across the intergranular region; and R_3 is the resistance to conduction within the grains. Electron microprobe studies supported the theory of impurities at the grain boundary. Thus, in a system as electrochemically complex at this, with many different effects interacting, one can still obtain fundamental information about processes occurring at each interface and in the bulk specimen.

This second study illustrates a very important point about IS. Although it is an extremely powerful technique in its own right, the analysis of complicated systems must be correlated with other experimental information to verify that the chosen circuit is physically reasonable. Furthermore, agreement between independently determined experimental values and those determined in a fitting procedure of the complex plane results can only strengthen the IS results and thus should never be overlooked.

CHAPTER TWO

THEORY

Ian D. Raistrick
J. Ross Macdonald
Donald R. Franceschetti

2.1 THE ELECTRICAL ANALOGS OF PHYSICAL AND CHEMICAL PROCESSES

2.1.1 Introduction

One of the most attractive aspects of impedance spectroscopy as a tool for investigating the electrical and electrochemical properties of materials and systems is the direct connection that often exists between the behavior of a real system and that of an idealized model circuit consisting of discrete electrical components. The investigator typically compares or fits the impedance data to an equivalent circuit, which is representative of the physical processes taking place in the system under investigation. The main objective of the present section is to define and discuss the analogies between circuit elements and electrochemical processes, so that the results of data fitting can be more easily converted into physical understanding. That such a close connection exists between electrochemistry and the behavior of idealized circuit elements is not surprising, since the fundamental laws which connect charge and potential and which define the properties of linear systems are unchanged in passing from electronic to ionic materials.

There are, however, dangers in the indiscriminate use of analogies to describe electrochemical systems. The first point to be made is that equivalent circuits are seldom unique. Only the simplest circuits can be said to be unambiguous in their description of experimental data; in complex situations, choices based upon other physical data are often necessary. It should also be remembered that electrolytes, interfaces, and so on are only *approximately* modeled by idealized circuit elements over a limited range of experimental conditions. One general condition, which will be assumed through much of this volume, is that we are dealing with small signals; that is, linear behavior is implied. The impedance is supposed to be independent of the amplitude of the applied signal, or at least approach a constant finite limit as the amplitude of the signal is decreased. Electrochemical systems, of course, can be highly nonlinear, and response to large signals includes rectification and higher harmonic generation. In Section 2.1.4 we discuss the linearization of interfacial kinetics to produce a charge transfer resistance. Interfacial capacitances are also voltage-dependent and mass transport will also be nonlinear if diffusion coefficients or thermodynamic terms, present in the diffusion expression, are a function of concentration (see Section 2.1.3.2). The use of small signals, however, is in general a distinct advantage of the impedance approach as compared to cyclic voltammetry, for example, where the wealth of information contained in a single experiment may prove too difficult to deconvolute. Usually, the voltage dependence of the electrochemical parameters is rather slow, and a linear expansion of the ac current, in terms of the variation of the perturbed concentrations and so on is well justified. Higher-order effects are not discussed in this chapter, but a recent discussion of electrochemical applications may be found in the recent review by Sluyters-Rehbach and Sluyters [1984].

Two other limitations on the exact correspondence between equivalent circuits and electrochemical systems are also addressed. The first of these is the effect of geometry on the current distribution. The effect of this on the frequency dispersion of the impedance has been little explored by those interested in impedance methods, but will prove to be important as the technique is extended to more complex geometries and small structures. A number of problems where current distribution is undoubtedly important, for example, in the behavior of polycrystalline solid electrolytes and the effect of roughness on interfacial impedance, have not yet been correctly treated in this respect.

A further limitation is the often observed anomalous frequency dependence of both bulk and interface parameters. Several electrochemical properties, for example, conductivity and interface capacitance, are predicted to be independent of frequency, whereas, in fact, they often show significant deviations from this behavior. This type of phenomenon has achieved recognition only since the application of ac techniques to a wide variety of problems, since a small degree of frequency dispersion is difficult to recognize in transient (time domain) experiments. Although good parameterization of this frequency dispersion has been achieved, and certain general or "universal" forms suggested, a validated microscopic description has not yet emerged. Some of the aspects of this phenomenology are discussed in Sections 2.1.2.3 and 2.1.2.7, and we have brought together some of the various attempts which have appeared in a wide variety of fields to deal with this problem.

The general approach adopted in this section is to treat bulk and interfacial phenomena separately. First the electrical properties of homogeneous phases are discussed. There are two aspects to this treatment, relating respectively to dielectric relaxation and long-range dc conductivity.

Although the well-established measurement of dielectric loss is not, in its narrowest sense, strictly impedance spectroscopy, a discussion of relaxation behavior is central to the family of techniques that use the interaction of a time-varying electromagnetic signal with a material to deduce microscopic detail. The generalization of the treatment of systems with a single relaxation time (Debye behavior) to those with multiple relaxations or distributions of relaxation times is discussed in Section 2.1.2.3. Recently, the application of impedance methods to disordered, condensed phases, such as organic polymers and glasses, warrants a general appreciation of the concepts involved. Dielectric loss measurements are also important and are used extensively to study the energetics of relaxations of complex ionic defects, such as those found in the fluorite family of materials.

The determination of dc ionic conductivity is perhaps the most widespread and also the simplest application of impedance spectroscopy. By using ac methods, electrode polarization can be *correctly* eliminated from an electrochemical system, and other sources of spurious frequency dispersion, such as grain boundary effects, may also be removed under certain circumstances. Electrodes may be inert foreign metals, thus eliminating the need for demonstrating the reversibility of parent metal electrodes.

Conductivity is of course closely related to diffusion in a concentration gradient, and impedance spectroscopy has been used to determine diffusion coefficients in a variety of electrochemical systems, including membranes, thin oxide films, and alloys. In materials exhibiting a degree of disorder, perhaps in the hopping distance or in the depths of the potential wells, *simple* random walk treatments of the statistics are no longer adequate; some modern approaches to such problems are introduced in Section 2.1.2.7.

The above mentioned sections deal with bulk phenomena. The other important area about which impedance spectroscopy gives important information is that of the electrochemical interface. This is usually a junction between an electronic and an ionic conductor; electrochemical devices utilize the charge transfer that occurs at this interface. The kinetics of this process as well as the electrical nature of the interface region are discussed in Section 2.1.4.

The emphasis of this section is on solid systems; therefore, several important aspects of electrical response appropriate to liquid electrochemistry are either neglected or given little emphasis. Examples are the omission of convection in the treatment of mass transport and the related neglect of ac impedance at a rotating disk electrode. Similarly, porous electrodes are not discussed, although related "rough" electrodes are briefly considered. Complex electrochemical mechanisms at solid–solid interfaces have been hardly mentioned in the literature; the treatment of the topic here reflects that. However, some attempt has been made to give a sufficiently general approach to interface kinetics and the development of expressions for the faradic impedance so that solid state scientists may be aware of the advanced state of development of the theory used by aqueous electrochemists.

2.1.2 The Electrical Properties of Bulk Homogeneous Phases.

2.1.2.1 Introduction

In this section we are concerned with the electrical response of solids with a uniform composition. Mass and charge transport in the presence of concentration gradients are discussed in Section 2.1.3.

On the time scale of interest to electrochemists (i.e., greater than ~ 1 μs) an electric field can interact with a solid in two principal ways. These are, respectively, the reorientation of defects having electric dipole moments (usually complex defects) and the translative motion of charge carriers (usually simple defects such as vacancies, ionic interstitials, and defect electronic species).

The first interaction leads to a displacement current,

$$i = d\mathbf{D}/dt \tag{1}$$

where $\mathbf{D}$ is the electric displacement (defined as the total charge density on the electrodes),

$$\mathbf{D} = \epsilon_0 \mathbf{E} + \mathbf{P} \tag{2}$$

where $\mathbf{E}$ is the electric field, ϵ_0 is the permittivity of free space, and $\mathbf{P}$ is the polarization of the dielectric material.

The second type of interaction leads to a purely real (dc) conductivity σ,

$$i = \sigma\mathbf{E} \tag{3}$$

We therefore deal first with the phenomenon of dielectric relaxation in materials with a single time constant and an absence of conductivity and later with materials that show long-range conductivity. As we are primarily concerned with developing the electrical analogs of these processes, little consideration is given to the characteristics of individual defects or materials. Similarly, the thermodynamics of the formation of defects, which determines their concentration, is also ignored.

2.1.2.2   Dielectric Relaxation in Materials with a Single Time Constant

When an electric field $\mathbf{E}$, is applied to an insulating material, the resulting polarization $\mathbf{P}$ may be divided into two parts according to the time constant of the response:

1. An almost instantaneous polarization due to the displacement of the electrons with respect to the nuclei. This defines the high-frequency dielectric constant ϵ_∞ related to the refractive index.

$$\epsilon_\infty - 1 = \mathbf{P}_\infty/\mathbf{E}\epsilon_0 \tag{4}$$

The time constant of this process is about 10^{-16} s and therefore occurs in the UV region of the electromagnetic spectrum. Ionic vibrations have a time constant which usually occurs in the infrared and are also therefore instantaneous as far as electrochemical experiments are concerned.

2. A time-dependent polarization $\mathbf{P}'(t)$ due to the orientation of dipoles in the electric field. If the field remains in place for an infinitely long time, the resulting total polarization $\mathbf{P}_s$ defines the static dielectric constant ϵ_s:

$$\epsilon_s - 1 = \mathbf{P}_s / \mathbf{E}\epsilon_0 \tag{5}$$

$$\mathbf{P}_s = \mathbf{P}_\infty + \mathbf{P}'(t = \infty) \tag{6}$$

The simplest assumption allowing calculation of the properties of such a system is that $\mathbf{P}'(t)$ is governed by first-order kinetics, that is, a single-relaxation time τ, such that

$$\tau d\mathbf{P}'(t)/dt = \mathbf{P}_s - \mathbf{P} \tag{7}$$

In other words, the rate at which $\mathbf{P}$ approaches $\mathbf{P}_s$ is proportional to the difference between them. Referring to Figure 2.1.1, on application of a unit step voltage $u_0(t)$

$$\mathbf{P} = \mathbf{P}_\infty u_0(t) + \mathbf{P}'(t) \tag{8}$$

If we take the Laplace transforms of the last two equations and solve for $\{\mathbf{P}\}$, we obtain

$$\{\mathbf{P}\} = \frac{\mathbf{P}_\infty}{(p + \omega_0)} + \frac{\omega_0 \mathbf{P}_s}{p(p + \omega_0)} \tag{9}$$

where $\{\mathbf{P}\}$ is the Laplace transform of the polarization and $\omega_0 = \tau^{-1}$; p is the complex frequency variable.

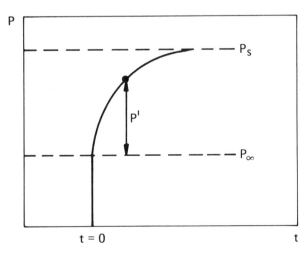

FIGURE 2.1.1. Time dependence of the polarization P after the application of an electric field to an insulator at $t = 0$.

The current density is obtained using the relation

$$\{i\} = p\{\mathbf{P}\} - \mathbf{P}(t = 0) \tag{10}$$

or by differentiating in the time domain, to give

$$\{i\} = \mathbf{P}_\infty + (\mathbf{P}_s - \mathbf{P}_\infty)\frac{\omega_0}{(p + \omega_0)} \tag{11}$$

and therefore

$$i(t) = \mathbf{P}_\infty\delta(t) + (\mathbf{P}_s - \mathbf{P}_\infty)\tau^{-1}\exp(-t/\tau) \tag{12}$$

This is the same result as that obtained for the circuit of Figure 2.1.2, with the identities

$$\tau = RC_2$$
$$C_2 = (\epsilon_s - \epsilon_\infty)\epsilon_0 \tag{13}$$
$$C_1 = \epsilon_\infty\epsilon_0$$

The admittance due the relaxation process is, since $\{\mathbf{E}\} = 1/p$:

$$Y* = \frac{\{i\}}{\{\mathbf{E}\}} = \epsilon_\infty\epsilon_0 p + (\epsilon_s - \epsilon_\infty)\epsilon_0\frac{\omega_0 p}{(p + \omega_0)} \tag{14}$$

or, separating the real and imaginary parts ($p = j\omega$),

$$Y* = \frac{\omega^2 RC_2}{1 + \omega^2 R^2 C_2^2} + j\frac{\omega C_2}{1 + \omega^2 R^2 C_2^2} + j\omega C_1 \tag{15}$$

The expression may be rewritten in terms of the complex dielectric constant $\epsilon*$ $= Y*/j\omega\epsilon_0$:

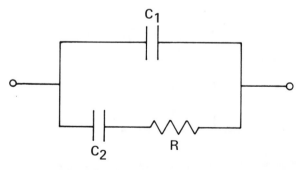

FIGURE 2.1.2. The Debye equivalent circuit.

$$\epsilon^* - \epsilon_\infty = \frac{\epsilon_s - \epsilon_\infty}{1 + \omega^2\tau^2} - j\,\frac{\omega\tau(\epsilon_s - \epsilon_\infty)}{1 + \omega^2\tau^2} \qquad (16)$$

The real and imaginary parts of this expression are the Debye dispersion relations, which have remained the basic model of dielectric relaxation since their inception (Debye [1929]). They are plotted against one another in Figure 2.1.3 and in Figure 2.1.4 are separately plotted against the normalized frequency ω/ω_0. The dielectric loss peak ϵ'', which corresponds to the real part of the admittance, has been widely used in solid state measurement for the characterization of relaxation processes. As will be seen later, the equivalent circuit of Figure 2.1.2 is also used in the interpretation of ac impedance data for solid electrolyte systems even though the physical phenomena describing the relaxation processes, that is, conductivity and space charge accumulation and depletion, are quite different.

The principal difference between a dielectric loss experiment and an impedance spectrum is that the former usually utilizes temperature as the independent variable and measurements are made at several fixed frequencies. A typical example of the use of dielectric loss measurements to obtain data about the relaxations of defects in crystalline solids is the recent paper by Wapenaar et al. [1982], who studied LaF_3-doped BaF_2. Two main types of relaxation are found in this material, corresponding to dipole moments caused by association of substitutional lanthanum with an interstitial fluorine along the $\langle 100 \rangle$ and $\langle 111 \rangle$ crystal axes, respectively. Loss peaks are seen at low levels of doping corresponding to both defects. Calculation of the respective dipole moments allows calculation of the concentration of defects from the strengths of the losses (i.e., from the associated values of C_2).

In practice, very few systems obey the Debye equations with accuracy; an extensive literature exists on the real properties of dielectric materials. It is often found, especially in disordered materials (e.g., glasses and amorphous thin films),

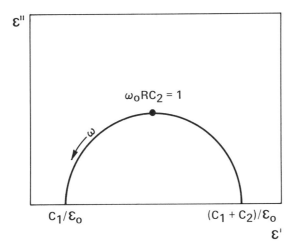

FIGURE 2.1.3. Nyquist plot of the frequency dependence of the complex permittivity modeled by the circuit of Figure 2.1.2.

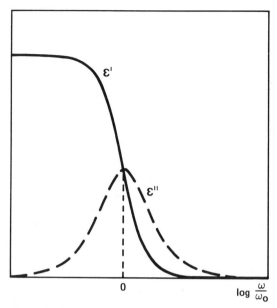

FIGURE 2.1.4. Real and imaginary parts of the complex permittivity as a function of normalized radial frequency.

that the Debye peak is considerably broadened over its theoretical half-width of $\log [(2 + \sqrt{3})/(2 - \sqrt{3})]$ decades. A number of empirical relaxation functions have been proposed to parameterize the observations, usually interpreted in terms of a distribution of relaxation times. This theme is discussed in the next section.

2.1.2.3 Distributions of Relaxation Times

The Debye dispersion relations were derived above for a process with a single relaxation time. Figure 2.1.4 showed that the dielectric loss function ϵ'' corresponding to this situation is symmetric about a central frequency, with a characteristic shape and width. The current flow in such a material, after the imposition of a voltage step function, decays exponentially with time. In view of the observations mentioned at the end of the previous section, attempts have been made to extend the Debye model by including processes with more than one relaxation time. By choosing a distribution of relaxation times with appropriate strengths and frequencies, it should prove possible to parameterize the broad response of many dielectric materials. Recently, the distribution of relaxation times approach has also been suggested as the origin of the "constant-phase elements" that are often seen in impedance studies of solid electrolytes and the solid–solid interface. In this section, some of the main features of this line of reasoning are presented. The dielectric literature in this area is extensive; no attempt is made here to be comprehensive.

Assuming linear relaxation processes of the type modeled by the series -RC-branch, the principle of superposition allows the dielectric function $\epsilon^* - \epsilon_\infty$ to be generalized:

$$\epsilon^* - \epsilon_\infty = \int_0^\infty \frac{(\epsilon_s - \epsilon_\infty)G(\tau)\, d\tau}{1 + p\tau} \tag{17}$$

where p is the complex frequency variable and $G(\tau)$ represents a distribution of relaxation times. The distribution should be normalizable,

$$\int_0^\infty G(\tau)\, d\tau = 1 \tag{18}$$

and should have upper and lower limits. Here $G(\tau)$ represents the fraction of the total dispersion $(\epsilon_s - \epsilon_\infty)$, which is contributed by processes having relaxation times between τ and $\tau + d\tau$.

Now ϵ^* may be divided into real and imaginary parts corresponding to the frequency $j\omega$,

$$\epsilon' - \epsilon_\infty = \int_0^\infty \frac{G(\tau)(\epsilon_s - \epsilon_\infty)\, d\tau}{1 + (\omega\tau)^2} \tag{19}$$

$$\epsilon'' = \int_0^\infty \frac{\omega\tau G(\tau)(\epsilon_s - \epsilon_\infty)\, d\tau}{1 + (\omega\tau)^2} \tag{20}$$

Kirkwood and Fuoss [1941] first showed that $G(\tau)$ could be recovered by integration from a set of ϵ'' values. A general treatment has been given by Macdonald and Brachman [1956], who provided a useful set of relations between the various functions used to describe networks and systems as well as between responses to various types of input.

Using the notation of those authors, the network function is defined as

$$Q(p) = \int_0^\infty \frac{G(\tau)\, d\tau}{1 + p\tau} \tag{21}$$

where $Q(p)$ corresponds to $(\epsilon^* - \epsilon_\infty)/(\epsilon_s - \epsilon_\infty)$.

The admittance is related to $Q(p)$:

$$Y(p) = pQ(p) \tag{22}$$

The step function response $A(t)$ and the impulse response $B(t)$ are

$$A(t) = L^{-1}[Q(p)] \tag{23}$$

$$B(t) = L^{-1}[Y(p)] \tag{24}$$

where L^{-1} is the inverse Laplace transform operator.

It was shown that $G(\tau)$ was derivable from these quantities through the relations

$$\tau G(\tau) = D(\lambda) = L^{-1}L^{-1}Q(p) \tag{25}$$

or

$$D(\lambda) = L^{-1}A(t), \tag{26}$$

where $D(\lambda)$ is a distribution function of the new variable $\lambda = \tau^{-1}$. Other relationships may also be derived. The authors give useful examples of various types of network functions and derived distributions. The simplest is

$$G(\tau) = \delta(\tau - \tau_0) \tag{27}$$

which corresponds to a single relaxation time and leads to the simple Debye dispersion equations. Rewriting $G(\tau)$ in terms of λ,

$$D(\lambda) = \tau G(\tau) = \tau\delta(\tau - \tau_0) \tag{28}$$

and therefore

$$\begin{aligned} D(\lambda) &= \lambda^{-1}\delta(\lambda^{-1} - \lambda_0^{-1}) \\ &= \lambda_0\delta(\lambda - \lambda_0) \end{aligned} \tag{29}$$

Hence,

$$A(t) = L\{D(\lambda)\} \tag{30}$$

$$= \lambda_0 \exp(-\lambda_0 t) \tag{31}$$

$$= (1/\tau_0) \exp(-t/\tau_0) \tag{32}$$

and

$$Q(p) = 1/(1 + \tau_0 p) \tag{33}$$

Observed relaxation times may occur over many orders of magnitude, and it seems reasonable that such a range of variation would, for a thermally activated process of the type

$$\tau = \tau^* \exp(E^*/kT) \tag{34}$$

correspond to a distribution of activation energies rather than to a distribution of τ^*. We may therefore define a distribution $K(E^*)$ such that

$$K(E^*)\, dE^* = G(\tau)\, d\tau \tag{35}$$

Evidently, from Eq. (34)

$$K(E^*) = (\tau/kT)G(\tau) \tag{36}$$

Macdonald [1962] has pointed out that if $K(E^*)$ is independent of T, then $G(\tau)$ cannot be so independent. Both the midpoint τ_0 and the width of the distribution will change with temperature. Not all of the $G(\tau)$ proposed in the literature are consistent with this postulate. See Section 2.2.3.4 for further discussion of activation energy distributions.

Van Weperen et al. [1977] noted that in fluorite-structure materials the dielectric and ionic thermocurrent peaks broadened with increasing concentration of dipoles (see, for example, Johnson et al. [1969]) and developed a theory of dipole–dipole interactions which predicted an almost Gaussian distribution of activation energies:

$$K(E^*) = \frac{1}{\sigma\sqrt{(2\pi)}} \exp\left[-\frac{(E^* - E_0^*)^2}{2\sigma^2}\right] \tag{37}$$

The corresponding distribution of τ is lognormal, the Wagner distribution

$$G(\tau) = \frac{b}{\tau\sqrt{\pi}} \exp\left[-b^2\left(\ln\frac{\tau}{\tau_0}\right)^2\right] \tag{38}$$

where $\sigma = kT/b\sqrt{2}$; σ^2 is the variance of the $K(E^*)$. If $K(E^*)$ is to be invariant with T, then b should be proportional to T.

The importance of this distribution, apart from being well defined in a physical sense, is its behavior for large σ, that is, wide distributions. As b becomes small, $G(\tau)$ becomes proportional to $1/\tau$, and $A(t)$, the current response to a unit step function, becomes proportional to $1/t$. The power spectrum may mimic $1/f$ behavior over several decades, and the dielectric function will show a very gradual frequency dispersion.

One of the most widely used distributions is that proposed by Cole and Cole [1941] to describe the occurrence of depressed semicircular arcs in the ϵ''–ϵ' plots obtained for a wide variety of polar liquids and solids. The dielectric constant behavior was described by the equation

$$\epsilon^* - \epsilon_\infty = \frac{(\epsilon_s - \epsilon_\infty)}{\left[1 + (j\omega\tau_0)^{1-\alpha}\right]} \tag{39}$$

where $\alpha\pi/2$ is the angle between the real axis and the line to the center of the circle from the high-frequency intercept. Now ϵ^* may be separated into real and imaginary parts

$$\frac{\epsilon' - \epsilon_\infty}{\epsilon_s - \epsilon_\infty} = \frac{1}{2}\left[1 - \frac{\sinh(1-\alpha)x}{\cosh(1-\alpha)x + \cos\alpha\pi/2}\right] \tag{40}$$

$$\frac{\epsilon''}{\epsilon_s - \epsilon_\infty} = \frac{1}{2} \frac{\cos \alpha\pi/2}{\cosh (1 - \alpha)x + \sin \alpha\pi/2} \tag{41}$$

where $x = \log \omega\tau_0$. These expressions reduce to the Debye relationships for $\alpha \to 0$. The derived distribution function of time constants is

$$G(\tau) = \frac{1}{2\pi\tau} \frac{\sin \alpha\pi}{\cosh (1 - \alpha) \log (\tau/\tau_0) - \cos \alpha\pi} \tag{42}$$

from which the distribution of activation energies would be, using Eq. (36),

$$K(E^*) = \frac{1}{2\pi kT} \frac{\sin \alpha\pi}{\cosh (1 - \alpha)(E^* - E_0^*)/kT - \cos \alpha\pi} \tag{43}$$

Unlike the Wagner distribution, $K(E^*)$ cannot be rendered temperature-independent except in limiting cases where $\alpha \to 0$ or 1. The Cole–Cole distribution, like the lognormal distribution, is symmetrical with respect to a central frequency or relaxation time. The distribution of time constants is plotted as a function of the variable $s \equiv \log (\tau/\tau_0)$ in Figure 2.1.5, and the complex plane plots for various values of α are given in Figure 2.1.6.

As was pointed out by Cole and Cole, dielectric response corresponding to the function of Eq. (39) may be decomposed into the circuit shown in Figure 2.1.7,

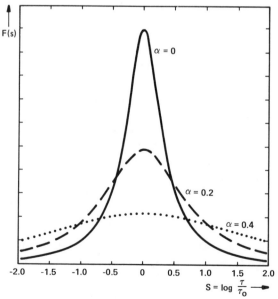

FIGURE 2.1.5. The distribution function $F(s)$ associated with the Cole–Cole distribution of relaxation times [Eq. (42)] for different values of α.

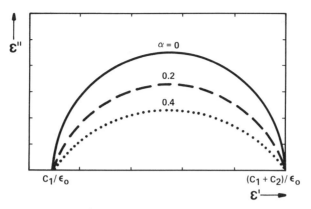

FIGURE 2.1.6. Complex permittivity associated with the Cole–Cole expression [Eq. (39)].

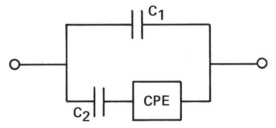

FIGURE 2.1.7. The analog of Figure 2.1.2, which models the electrical response associated with the Cole–Cole expression.

which contains a constant-phase element (CPE). The CPE is an empirical impedance function of the type

$$Z^*_{\text{CPE}} = A(j\omega)^{-\alpha} \tag{44}$$

which has proved of considerable value in data fitting. The admittance of this circuit may be expressed as

$$Y^* = j\omega C_1 + \frac{j\omega C_2}{\left[1 + C_2 A(j\omega)^{1-\alpha}\right]} \tag{45}$$

Dividing by $j\omega\epsilon_0$ and comparing with Eq. (39), we find that

$$A = \frac{\tau_0^{1-\alpha}}{\left(\epsilon_s - \epsilon_\infty\right)\epsilon_0} \tag{46}$$

It is interesting to enquire about the distribution of relaxation times implied by the presence of Z^*_{CPE} alone. If

$$Z_{CPE}^* = Ap^{-\alpha} \tag{47}$$

then the admittance is

$$Y(p) = A^{-1}p^\alpha \tag{48}$$

Therefore,

$$A(t) = L^{-1}\{A^{-1}p^{\alpha-1}\} \tag{49}$$

$$= \frac{A^{-1}t^{-\alpha}}{\Gamma(1-\alpha)} \tag{50}$$

where Γ is the gamma function. And

$$D(\lambda) = \frac{A^{-1}\lambda^{\alpha-1}}{\Gamma(1-\alpha)\Gamma\alpha} \tag{51}$$

or

$$G(\tau) = \frac{\sin \alpha\pi}{\pi} \cdot A^{-1}\tau^{-\alpha} \tag{52}$$

Thus, the distribution of relaxation times is proportional to $1/\tau^\alpha$. It has often been pointed out that this distribution is nonnormalizable. Physical acceptability may be restored by truncating the distribution at upper and lower limits of τ. The resulting distribution has been discussed by Matsumoto and Higasi [1962].

Assuming an expression for τ of the form given by Eq. (34), we find a distribution of activation energies:

$$K(E^*) \propto \exp\left[(1-\alpha)E^*/kT\right] \tag{53}$$

The important point to be made here is that the assumption of (1) an exponential distribution of activation energies and (2) an exponential form for τ lead directly to CPE behavior. The exponential distribution of activation energies has been further discussed by Macdonald [1963]. See also Section 2.2.3.4.

Two other distribution functions are due to Kirkwood and Fuoss [1941] and Davidson and Cole [1951] (see also Davidson [1961]).

The first of these is symmetric, and is again based on an extension of the Debye theory functions. In the Debye theory,

$$\epsilon''/\epsilon''_{max} = \text{sech } x \tag{54}$$

where $x = \log \omega/\omega_0$. Instead of this, Kirkwood and Fuoss wrote

$$\epsilon''/\epsilon''_{max} = \text{sech } \alpha x \tag{55}$$

which leads to a distribution of the form

$$G(s) = \frac{2}{\pi} \cdot \frac{\cos(\alpha\pi/2)\cosh \alpha s}{\cos^2(\alpha\pi/2) + \sinh^2 \alpha s} \tag{56}$$

where s is again equivalent to $\log(\tau/\tau_0)$.

The Davidson–Cole equation

$$\frac{\epsilon^* - \epsilon_\infty}{\epsilon_s - \epsilon_\infty} = \frac{1}{(1 + j\omega\tau_0)^\beta} \tag{57}$$

leads to a skewed arc in the ϵ''–ϵ' plane. It is a semicircle at low frequency, but asymptotic to $\beta\pi/2$ at high frequencies (Fig. 2.1.8).

The real and imaginary parts are

$$\epsilon' - \epsilon_\infty = (\epsilon_s - \epsilon_\infty)\cos \beta y (\cos y)^\beta \tag{58}$$

$$\epsilon'' = (\epsilon_s - \epsilon_\infty)\sin \beta y (\cos y)^\beta \tag{59}$$

where $y = \tan^{-1}\omega\tau_0$.

The current response to the application of a step function potential difference is

$$i(t) = \mathbf{P}_\infty \delta(t) + \frac{\mathbf{P}_s - \mathbf{P}_\infty}{\tau_0\Gamma(\beta)}\left(\frac{t}{\tau_0}\right)^{\beta-1}\exp-\left(\frac{t}{\tau_0}\right) \tag{60}$$

which may be compared with the equivalent Cole–Cole expression

$$i(t) = \mathbf{P}_\infty \delta(t) + \frac{\mathbf{P}_s - \mathbf{P}_\infty}{\tau_0\Gamma(1+\alpha)}\left(\frac{t}{\tau_0}\right)^{(1\pm\alpha)} \tag{61}$$

and the Debye expression, [Eq. (12)].

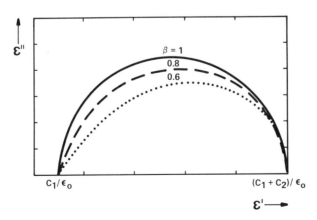

FIGURE 2.1.8. Complex permittivity associated with the Cole–Davidson expression [Eq. (57)].

The distribution of relaxation times is highly asymmetric:

$$G(\tau) = \frac{\sin \beta \pi}{\pi} \left(\frac{\tau}{\tau_0 - \tau} \right)^{\beta}, \qquad \tau < \tau_0$$

$$= 0 \qquad\qquad\qquad \tau > \tau_0 \tag{62}$$

Thus, the distribution ends abruptly at its most probable value.

A further generalization of the Debye approach was made by Williams and Watts [1970] and Williams et al. [1970] who introduced the use of the following fractional exponential form to describe the decay of polarization after the removal of a constant field,

$$\phi(t) = \mathbf{P}(t)/(\mathbf{P}_s - \mathbf{P}_\infty) = \exp - (t/\tau)^{\alpha}, \qquad 0 < \alpha < 1 \tag{63}$$

neglecting the instantaneous drop in polarization. The equivalent Debye expression has $\alpha = 1$.

Thus,

$$A(t) = -d\phi/dt = (\alpha t^{\alpha - 1}/t^{\alpha}) \exp - (t/\tau)^{\alpha} \tag{64}$$

The Laplace transform of $A(t)$, that is $Q(p)$, may be evaluated by series expansions, although care is needed because of slow convergence for certain ranges of α (Williams et al. [1970]). For the particular case of $\alpha = \frac{1}{2}$, an analytical expression is easily derived:

$$Q(p) = \frac{\sqrt{\pi}}{2\sqrt{\tau_0}} \exp \frac{1}{\tau_0} \operatorname{erfc} \frac{1}{\sqrt{(\tau_0 p)}} \tag{65}$$

The inverse Laplace transform of functions of this type, which gives the distribution of relaxation times, is given by Montroll and Bendler [1984]. A simple expression for $\alpha = \frac{1}{2}$ may be derived:

$$G(\tau) = \frac{1}{2\sqrt{(\pi t \tau_0)}} \exp \left[- (t/4\tau_0) \right] \tag{66}$$

Like the Cole–Davidson function, the Williams–Watts approach gives asymmetric plots in the complex ϵ^* plane. A detailed comparison of these two forms has been made by Lindsey and Patterson [1980].

Recent interest in the Williams–Watts approach has arisen, not only because of its empirical success in fitting dielectric data, but also because of its relation to certain types of diffusion and random walk problems. The mechanistic relation between diffusion and relaxation was introduced by Glarum [1960], who suggested a process in which a mobile defect enabled a "frozen in" dipole to relax. Further aspects of random walk processes and their relation to CPEs and other empirical functions are discussed in a later section.

2.1.2.4 *Conductivity and Diffusion in Electrolytes (Ibl [1983a], Newman [1973])*

In the previous sections the expressions for the admittance of materials were developed on the assumption that they had no dc conductivity. The real part of the admittance arose from the dissipative process of dipole reorientation. Energy was absorbed by the system when the orientation of dipoles was changed with respect to the electric field vector.

Dissipation may also occur by mass transport of particles in the bulk of the phase. Work must be done against the frictional forces of the medium through which the particle moves. In solids, migration and diffusion are usually important; in liquids and membranes, hydrodynamic mass transport must also be considered.

A convenient starting point for discussion of transport properties in electrolytes is a consideration of the physical laws which connect charge and electric potential. In a medium of uniform dielectric constant we may write Poisson's equation, which connects the gradient of the electric field with the charge density

$$\nabla^2 \Phi = -\rho / \epsilon_s \epsilon_0 \tag{67}$$

Here Φ is the electric potential. The charge density ρ is equal to the sum over the local concentrations of species multiplied by their charges.

$$\rho = F \sum z_i c_i \tag{68}$$

Because of the magnitude of the constant F / ϵ_0, very large electric fields result from very small deviations from electroneutrality. It is, therefore, a very good approximation to write for the interior of an electrolyte

$$\sum z_i c_i = 0 \tag{69}$$

In other words, the bulk of the electrolyte is electrically neutral. This condition is not true, of course, where there exists the possibility of large electric fields, for example, in the neighborhood of interfaces. It also follows that, in general, Laplace's equation

$$\nabla^2 \Phi = 0 \tag{70}$$

is a good approximation.

A second basic equation expresses the conservation of mass in the system

$$\partial c_i / \partial t = -\nabla \cdot j_i + R_i \tag{71}$$

This equation states that the rate of accumulation of a species i in a given volume element is equal to the negative of the divergence of the flux plus any terms that lead to the production or deletion of i, such as chemical reactions or recombination in the bulk of the material.

Third, we can write an equation for the electric current density in terms of the fluxes of charged species

$$i = F \sum z_i j_i \tag{72}$$

We now need an expression for the flux of species i in terms of the forces acting on the particles. The assumption that is usually made is that a particle has a characteristic mobility u_i which is the proportionality constant between its velocity and the force causing it to move. The driving force is supposed to be the gradient in electrochemical potential η_i of the species, so that when the mobility is multiplied by the driving force and the concentration, we obtain the flux of i:

$$j_i = -c_i u_i \nabla \eta_i \tag{73}$$

The problem with this equation lies in the formulation of the force term. In general, a particle may move in response to gradients in the electrochemical potentials of other species, leading to cross-terms in the flux equation. In principle, the presence of cross-terms will occur whenever a component is present whose chemical potential may vary independently of that of species i. Thus, the motion of i may depend not only on $\nabla \eta_i$ but also on $\nabla \eta_j$ if η_j is independent of η_i (i.e., it is not coupled through a Gibbs–Duhem relation). The flux equation may therefore be generalized as

$$j_i = -c_i u_i \left[\nabla \eta_i + \sum_j \alpha_{ij} \nabla \eta_j \right] \tag{74}$$

where the α_{ij} are the coefficients expressing the influence of $\nabla \eta_j$ on i.

It may be shown that this equation is equivalent to the phenomenological equations derived from irreversible thermodynamics, as well as the multicomponent diffusion equations derived from the Stefan–Maxwell equations, which were first used to describe diffusion in multicomponent gases.

Further development of transport theory involves solution of Eqs (71) and (73), subject to the appropriate initial and boundary conditions to give currents, concentration profiles, and so on.

The simplest approach, often adopted in practice, particularly in solution electrochemistry, may be termed the dilute solution approximation. We can write, for the gradient in electrochemical potential for a dilute solution,

$$\nabla \eta_i = RT \nabla c_i / c_i + z_i F \nabla \Phi \tag{75}$$

Thus,

$$j_i = -RT u_i \nabla c_i - z_i F c_i u_i \nabla \Phi \tag{76}$$

The quantity $RT u_i$ is called the diffusion coefficient (Nernst–Einstein relation)

$$j_i = -D_i \nabla c_i - z_i F c_i u_i \nabla \Phi \tag{77}$$

and the current density is given by the expression

$$i = -F \sum z_i \nabla c_i D_i - F^2 \sum z_i^2 c_i u_i \nabla \Phi \tag{78}$$

Substituting Eq. (77) into Eq. (71), we obtain

$$\partial c_i / \partial t = z_i F \nabla \cdot (u_i c_i \nabla \Phi) + \nabla \cdot (D_i \nabla c_i) + R_i \tag{79}$$

or, in one dimension,

$$\partial c_i / \partial t = z_i F u_i \frac{\partial c_i}{\partial x} \mathbf{E} + D_i \frac{\partial^2 c_i}{\partial x^2} + R_i \tag{80}$$

if we assume $\partial \mathbf{E} / \partial x = 0$, that is, electroneutrality. This is the classical Nernst–Planck equation.

Instead of assuming dilute or ideal behavior, it is possible to write

$$\nabla \eta_i = \nabla \mu_i + z_i F \nabla \Phi \tag{81}$$

$$j_i = -c_i u_i \nabla \mu_i - z_i F c_i u_i \nabla \Phi \tag{82}$$

$$= -D_k c_i \nabla \mu_i / RT - z_i F c_i u_i \nabla \Phi \tag{83}$$

Here D_k is known as the component diffusion coefficient. The importance of this definition lies in the fact that Nernst–Einstein proportionality between a diffusion coefficient and a mobility has been retained, even though the condition of ideality has been relaxed. This is important since the apparent violation of the Nernst–Einstein equation in nonideal solutions is not a failure of the proportionality between mobility and mean displacement; it is a weakness in the method of formulating the driving force for diffusion in terms of a concentration gradient (Fick's law) rather than in terms of an activity or chemical potential gradient.

2.1.2.5 Conductivity and Diffusion—A Statistical Description

In the previous section, the Nernst–Planck equation was developed from the macroscopic flux density and mass conservation equations [(71) and (73), respectively]. The same equation can also be derived by statistical methods, which describe the probability of finding a particle within a volume region at a time t, given an initial distribution and a set of jump probabilities. For the simplest case, in one dimension, with equal probabilities of the particle making a jump to the right or to the left, the time evolution of an initial delta function in concentration at $x = 0$ is Gaussian:

$$n(x, t) \, \Delta x = \frac{1}{2\sqrt{(\pi D t)}} \exp\left(-x^2 / 4 D t\right) \Delta x \tag{84}$$

where D, the diffusion coefficient, is equal to $\nu l^2/2$, where ν is the number of steps of length l the particle makes per unit time. A further assumption involved in the use of the statistical arguments which lead to this equation is that the jumps are statistically independent. $n(x, t)$ is plotted for several different times in Figure 2.1.9. The total area under the curve is constant.

The mean displacement of the particles is zero:

$$\langle x \rangle = \int x\, n(x, t)\, dx = 0 \tag{85}$$

as long as the jump probabilities are symmetrical. The second moment is

$$\langle x^2 \rangle = \int x^2\, n(x, t)\, dx = 2Dt \tag{86}$$

The probability density $n(x, t)$, given by Eq. (84), is a solution of the diffusion equation

$$\partial n(x, t)/\partial t = D\, \partial^2 n(x, t)/\partial x^2 \tag{87}$$

This is identical to the Nernst–Planck Eq. (80) in the absence of an electric field term and a generation–recombination term.

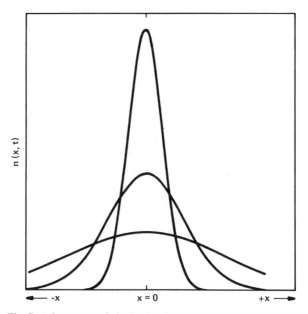

FIGURE 2.1.9. The Gaussian or normal distribution function $n(x, t)$ plotted as a function of distance from the origin for different times.

If the jump probabilities are not symmetrical—for example, in the presence of an electric field—then $\langle x \rangle$ is no longer equal to zero, and the probability distribution is

$$n(x, t) \, \Delta t = \frac{1}{2\sqrt{(\pi Dt)}} \exp\left[\left(-x - \langle v \rangle t\right)^2 / 4Dt\right] \Delta x \qquad (88)$$

Here $\langle v \rangle$ is the mean drift velocity $\langle x \rangle / t$. The mean drift velocity per unit field is the drift mobility b_i, and the conductivity σ_i is defined by

$$\sigma_i = b_i c_i z_i F \qquad (89)$$

$$= \langle v \rangle c_i z_i F / \mathbf{E} \qquad (90)$$

Note that b_i is equal to $z_i F u_i$.

Now $n(x, t)$ from Eq. (88) is the solution to Eq. (80) with R_i set to zero and describes a propagating Gaussian packet. The ratio of the dispersion $\sqrt{(\langle x^2 \rangle - \langle x \rangle^2)}$ to distance traveled is inversely proportional to the square root of time.

A more general approach via the master equation, leading to Fokker–Planck equations, may also be followed and may be found in texts on statistical physics (see, for example, Reichl [1980]).

The type of diffusion discussed here may be termed "*normal*" or *Gaussian diffusion*. It arises simply from the statistics of a process with two possible outcomes, which is attempted a very large number of times. In Section 2.1.2.7, the statistical basis of diffusion is enlarged to include random walks in continuous rather than discrete time, and also situations where different distributions of jump distances occur.

2.1.2.6 *Migration in the Absence of Concentration Gradients*

Under certain circumstances, the passage of electric current through an electrolyte does not lead to a concentration gradient, and Eq. (78) becomes

$$i = -F^2 \sum c_i z_i^2 u_i \nabla \Phi \qquad (91)$$

The term $F^2 \sum c_i z_i^2 u_i$ is called the conductivity σ, and under these conditions Ohm's law is obeyed by the electrolyte. Examples of this behavior are found where only one electrolyte species is mobile, for example, in a solid electrolyte, with reversible electrodes (see Section 2.1.3) or at high frequencies where several carriers may move, but where neither electrode nor concentration polarizations have time to build up.

For a simple hopping conductivity process, in the absence of long-range interactions, the conductivity is expected to be independent of frequency. Here, a single particle is presumed to move along an infinite lattice of identical potential wells (Fig. 2.1.10a). This might be contrasted with the case of a single particle hopping

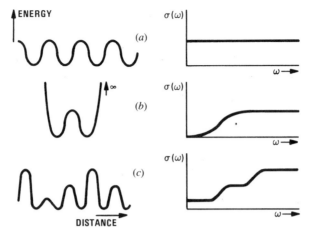

FIGURE 2.1.10. Frequency dependence of the hopping conductivity for different potential energy profiles: (*a*) Periodic constant activation energy, (*b*) a single bi-well, and (*c*) a potential profile with multiple activation energies.

backward and forward in double well, where the low-frequency conductivity is zero, and a Debye-like transition region is followed by a constant high-frequency conductivity (Fig. 2.1.10*b*).

Apart from the most dilute solutions, however, we do expect interactions between unassociated defect species, and in general this will lead to a frequency-dependent conductivity. This has been recognized for many years by electrochemists (Onsager [1926, 1927], Debye and Falkenhagen [1928]). The form of the frequency dependence, however, is of considerable interest.

Following the work of Jonscher (see, for example, Jonscher [1977, 1980]), who showed that a large number of different types of conductors exhibited a frequency dispersion of the CPE type, the presence of interactions has been invoked to explain the observed frequency dispersion in the conductivity of a number of solid electrolytes by Almond, West, and coworkers (Almond and West [1983a, b], Almond et al. [1982, 1983, 1984]). These authors expressed the real part of the ionic conductivity of a number of materials by an expression of the type

$$\sigma(\omega) = \sigma_0 + A\omega^n \qquad (92)$$

where σ_0 is a "dc" or frequency-independent part, and the second term is of the CPE type. Making use of Jonscher's empirical expressions, Eq. (92) was rewritten:

$$\sigma(\omega) = K\omega_p + K\omega_p^{1-n}\omega^n \qquad (93)$$

where ω_p is the hopping frequency and K depends on the concentration of the mobile charge carriers. The high-temperature limiting value for ω_p should be equal to the attempt frequency, which is independently accessible using IR spectroscopy.

For the case of sodium beta alumina, good agreement was found. Once ω_p is known, then the carrier concentration and activation entropy can also be deduced. For the case of beta alumina the hopping rate calculation has been confirmed by mechanical relaxation measurements. In a number of materials, however, including β''-alumina, the low-frequency region is not independent of frequency and a second CPE term must be included. The work of Almond and West is also discussed beginning at Eq. 78 of Section 2.2.3.4.

According to Jonscher, the origin of the frequency dependence of the conductivity was due to relaxation of the ionic atmosphere after the movement of the particle. This idea, and the earlier concepts of Debye, Onsager, and Falkenhagen, have been developed into a quantitative model suitable for solids by Funke [1986]. It is assumed that immediately after an ion hops to a new site (a new minimum in lattice potential energy) it is still displaced from the true minimum in potential energy, which includes a contribution from other mobile defects. At long times the defect cloud relaxes, until the true minimum coincides with the lattice site. The model predicts upper and lower frequency-limiting conductivities and a region in between of power law (CPE) behavior.

In general, both conductivity and dipolar relaxation processes may be present in the same material, and the total conductivity is given by

$$\sigma_\Sigma = \sigma + j\epsilon_0\epsilon^*\omega \tag{94}$$

where ϵ^* is given by Eq. (16)

The equivalent circuit for such a combination of processes is shown in Figure 2.1.11. An —RC— series combination will be present for each relaxation process present in the material. The dielectric loss peaks will be superimposed on a background loss due to the long-range conductivity process. Due to interactions, for example, dipole–dipole, or lattice relaxations, as discussed above, a distribution of relaxation times is to be expected, and Wapenaar and Schoonman [1981] have included Cole–Cole branches (series —CZ_{CPE}— combinations) rather than Debye

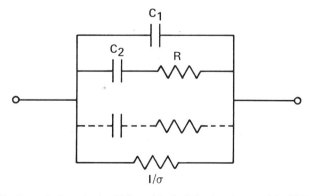

FIGURE 2.1.11. An equivalent circuit which models the behavior of a material which has both long-range conductivity (σ) and a number of discrete relaxation processes.

branches to fit data on $Ba_{1-x}La_xF_{2+x}$ fluorite structure solid solutions. Recently, the impedance spectra of doped tysonite materials have been investigated by Roos et al. [1984], who again found excellent agreement between data and a circuit which included Cole–Cole branches. However, the activation energies were not consistent with a simple dielectric relaxation, and a model in which conducting species move between inequivalent sites has been developed (Franceschetti and Shipe [1984]). The motion of the defect species may involve several distinguishable jump processes, each governed by a different activation energy. The interpretation of the relaxation branches in the circuit is that they describe a polarization arising from the inequalities in jump probabilities.

The important point to be emphasized here is that, although for a single jump frequency, in the absence of interactions, no particular structure is expected in the conductivity or dielectric constant at that frequency, a more complex model which incorporates several jump frequencies (Fig. 2.1.10c) indicates that frequency dispersion is expected in this range. Although the model of Franceschetti and Shipe was developed for the tysonite structure, a more general treatment for a small number of sublattices has been independently given (Wong and Brodwin [1980]) and confirms the main conclusions. As the number of possible jump frequencies increases, it is apparent that generalization of the model will eventually describe conductivity and diffusion in disordered materials, that is, in materials in which the jump probabilities are described by random variables.

2.1.2.7 Transport in Disordered Media

In a previous section reference was made to the random walk problem (Montroll and Schlesinger [1984], Weiss and Rubin [1983]) and its application to diffusion in solids. Implicit in these methods are the assumptions that particles hop with a fixed jump distance (for example, between neighboring sites on a lattice) and, less obviously, that jumps take place at fixed equal intervals of time (discrete time random walks). In addition, the processes are Markovian, that is, the particles are without memory: the probability of a given jump is independent of the previous history of the particle. These assumptions force normal or Gaussian diffusion. Thus, the diffusion coefficient and conductivity are independent of time.

In recent years more complex types of transport processes have been investigated, and from the point of view of solid state science, considerable interest is attached to the study of transport in disordered materials. In glasses, for example, a distribution of jump distances and activation energies are expected for ionic transport. In crystalline materials, the best ionic conductors are those that exhibit considerable disorder of the mobile ion sublattice. At interfaces, minority carrier diffusion and discharge (for example, electrons and holes) will take place in a random environment of mobile ions. In polycrystalline materials the lattice structure and transport processes are expected to be strongly perturbed near a grain boundary.

In general, the study of transport processes in disordered media has its widest application to electronic materials, such as amorphous semiconductors, and very little attention has been given to its application to ionic conductors. The purpose of this section is to discuss briefly the effect of disorder on diffusion process and to point out the principles involved in some of the newly developing approaches. One of the important conclusions to be drawn is that frequency-dependent transport properties are predicted to be of the form exhibited by the CPE if certain statistical properties of the distribution functions associated with time or distance are fulfilled. If these functions exhibit anomalously long tails, such that certain moments are not finite, then power law frequency dispersion of the transport properties is observed. However, if these moments are finite, then Gaussian diffusion, at least as limiting behavior, is inevitable.

Although the general problem of a random walk on a random lattice is difficult, there has been considerable success in approaching this problem from the point of view of the continuous time random walk, which was first proposed by Montroll and Weiss [1965]. In this approach, the walk is supposed to take place on a regular lattice; disorder is introduced by defining a distribution of waiting or residence times for a particle on a site. In a disordered material there will be a distribution of energy barriers. It therefore seems reasonable that a particle in a deep well will spend more time there than will a particle in a shallow well. The waiting time distribution function $\psi(t)$ therefore describes the probability for an event to happen at a time t after a previous event. The original $\psi(t)$ of Montrol and Weiss was generalized to a position-dependent $\psi(r, t)$ by Scher and Lax [1973a]. The application of the approach to electronic transport in amorphous semiconductors has been discussed in several papers (Scher and Montroll [1975], Pfister and Scher [1978], Scher and Lax [1973b]).

There is a qualitative difference in transport properties depending on the nature of $\psi(t)$. If $\psi(t)$ is such that the time between hops has finite first moment, that is, a mean residence time $\langle t \rangle$ can be defined, then classical diffusion is observed. An example would be

$$\psi(t) = \lambda \exp(-\lambda t) \tag{95}$$

where the mean waiting time is $1/\lambda$. If on the other hand, $\int t\psi \, dt$ diverges, then non-Gaussian or "dispersive" transport is seen. Of particular interest in this respect are $\psi(t)$ with long time tails

$$\psi(t) = \alpha A t^{-1-\alpha}/\Gamma(1 - \alpha), \qquad 0 < \alpha < 1 \tag{96}$$

In other words, the hopping probability is a slowly decaying function of time. Under these conditions, the dispersion of the concentration, $\langle x^2(t) \rangle$, becomes proportional to t^α, and the diffusion coefficient

$$D(t) = (1/2\Delta) \, d\langle x^2 \rangle/dt \tag{97}$$

and conductivity become time-dependent, with a power law dependence on frequency and time. Here Δ is the dimensionality of the system.

The physical origin of a power law distribution function for waiting times might arise from an exponential distribution of activation energies. Suppose, the distribution function of activation energies was of the form

$$K(E^*) = K_0 \exp\left(-E^*/E_0^*\right) \qquad (98)$$

Then, if the waiting time were proportional to the exponential of the activation energy, the distribution of waiting times would have a power law dependence on time, as required by Eq. (96).

The concepts of the continuous time random walk (CTRW) approach have been applied to ionic conductivity in glasses by Abelard and Baumard [1984]. In an alkali silicate glass, it is usually assumed that only a small number of the alkali metal ions are mobile, and the remainder are associated with nonbridging oxygens. These latter form dipoles which may reorient in the presence of an electric field. Interpretation of the complex impedance or dielectric constant of these glasses therefore is usually made in terms of a distribution of these relaxation times, in the manner discussed in the previous section. Abelard and Baumard, however, suggest that a more appropriate approach is to consider all alkali ions mobile, but with a distribution of activation energies associated with the potential wells in which they are situated. There is, therefore, a distribution of waiting times which leads to the observed frequency dependence of the real part of the conductivity.

An interesting extension of the dispersive transport model is its application to dielectric relaxation. As mentioned earlier, Glarum [1960] proposed that frozen-in" polarizations could be relaxed when a defect (e.g., a mobile charge carrier) approached them. Bordewijk [1975] extended the model and showed that in one-dimensional transport a Williams–Watts dielectric relaxation function with $\alpha = \frac{1}{2}$ resulted, but normal Debye relaxation was predicted in three dimensions ($\alpha = 1$). Schlesinger and Montroll [1984] have shown that if diffusion of the mobile defect is restricted to a CTRW with a long-time-tailed $\psi(t)$, then a Williams–Watts dielectric function for the relaxation of the dipoles is expected. The relaxation process is treated as a diffusion controlled chemical reaction.

The essentially different nature of transport processes with $\psi(t) \propto t^{-(1-\alpha)}$ should be stressed. Processes with this type of waiting time distribution function show an absence of scale. They exhibit very sporadic behavior. Long dormancies are followed by bursts of activity. They have been described as *fractal time* processes (Schlesinger [1984]). *Fractal space* processes, in which the absence of scale is present in the spatial aspects of the transport, are considered later in this section.

A different approach to transport in disordered systems has been developed by considering the excitation dynamics of random one-dimensional chains (Alexander et al. [1981]). Such a system may be represented by a master equation of the form

$$dP_n/dt = W_{n,n-1}(P_{n-1} - P_n) + W_{n,n+1}(P_{n+1} - P_n) \qquad (99)$$

where the P's are the amplitude of the excitations (site occupancies) and the W's are the transition probabilities between the nodes or sites, n and so on. This equation is obviously a discrete form of the diffusion equation, with the W's stochastic variables described by a distribution function. The electrical analog to this equation is the random transmission line, described by the equation

$$C_n dP_n/dt = W_{n,n-1} (P_{n-1} - P_n) + W_{n,n+1} (P_{n+1} - P_n) \qquad (100)$$

Here the C_n are the random capacitances, the W_n are the random conductances, and the P_n are the node potentials.

Alexander et al. [1981] have obtained solutions to this type of equation for various types of distribution functions of W for an initial delta function input in P. In particular, they considered distribution functions for which a mean transition rate $\langle W \rangle$ could be defined, and functions which were of the form

$$\rho(W) = \rho_0(T) W^{-\alpha(T)} \qquad (101)$$

where no mean transition rate exists. This is similar to the distribution function of waiting times for a CTRW defined in Eq. (96), and the arguments suggesting its use are essentially the same. The transition rate is an exponential function of activation energies, and the activation energies are supposed to be exponentially distributed, leading to a power law form for the distribution of transition rates. A similar argument can be used if a distribution of jump distances is assumed, that is, configurational disorder rather than randomness in the activation energies is assumed.

As for the case of the CTRW method, qualitatively different solutions are obtained depending on whether a mean transition rate can be defined or not. In the former case, the system behaves as if it were ordered with a single transition rate $[\rho(W) = \delta(W - W_{av})]$, even though the W's are random variables. These systems exhibit a frequency-independent low-frequency conductivity.

For power law distributions, however, the low-frequency conductivity tends to the form

$$\sigma(\omega) \propto (-j\omega)^{\alpha/(2-\alpha)} \qquad (102)$$

as $\omega \to 0$. There may also be a situation in which a crossover between the two distributions occurs as a function of time, in which case the mean square particle displacement is given by

$$\langle x^2 \rangle = 2D_0 t + Bt^{1-s} \qquad (103)$$

and the real part of the conductivity is

$$\sigma'(\omega) = \sigma(0) + A\omega^s \qquad (104)$$

Thus, at low frequencies a constant conductivity would be seen, but at higher frequencies a power law contribution enters. In these equations, D_0 is the limiting diffusion coefficient and A and B are thermally activated constants.

Experimental observations of the frequency dependence of the conductivity in the one-dimensional ionic conductor potassium hollandite (Bernasconi et al. [1979]) show a pronounced power law dependence, as predicted by the model. It was proposed that the transport process in this material was limited by random barrier heights caused by the presence of impurities.

In the previous paragraphs it was pointed out that a discrete time random walk, or a CTRW with a finite first moment for the waiting time distribution, on a lattice with a fixed jump distance led to a Gaussian diffusion process with a probability density given by Eq. (84). The spatial Fourier transform of this equation is

$$\mathbf{n}(q, t) = \exp\left(-Dtq^2\right) \tag{105}$$

Disorder was introduced into this system by postulating a distribution of waiting times. A complementary extension of the theory may be made by considering a distribution of jump distances. It may be shown that, as a consequence of the central limit theorem, provided the single-step probability density function has a finite second moment, Gaussian diffusion is guaranteed. If this condition is not satisfied, however, then Eq. (105) must be replaced by

$$\mathbf{n}(q, t) = \exp\left(-At|q|^{\mu}\right) \tag{106}$$

where μ lies between 0 and 2. This distribution function is known as a Levy or stable distribution. This distribution is a solution of the equation

$$\partial \mathbf{n}(q, t)/\partial t = -A|q|^{\mu} \mathbf{n}(q, t) \tag{107}$$

which is, of course, the Fourier transform of the diffusion equation when $\mu = 2$. A number of authors have considered the type of random walk process, defined by the Levy distribution (Hughes et al. [1981]). A particle executes a walk which may be transient and clustered. In other words, not all regions of space are visited by the walker, and a hierarchy of clusters is developed. The clusters may be self-similar, and Mandelbrot [1983] has stressed the fractal nature of walks with this distribution.

Other workers have also considered transport on a self-similar geometry, through the connection with percolation. Close to a percolation threshold, the conductivity and dielectric constant behave with a power law exponent in the concentration of one of the components, and the percolating cluster at the threshold has been identified as a fractal object. The temporal behavior of the diffusion process close to the percolation threshold has also been considered (Gefen et al. [1983]) and, using scaling arguments, it has been shown that the mean square displacement is

$$\langle r(t)^2 \rangle = at^{\theta} \tag{108}$$

where a and θ are constants, leading to a time- and frequency-dependent diffusion coefficient and conductivity. Experimental verification of this power law dependence has been obtained for two-dimensional percolation in thin gold films (Laibowitz and Gefen [1984]).

An alternative approach, used to describe the properties of ionically conducting glasses, is conceptually closely related to the earlier discussion of transport properties in materials with a small number of sublattices for the conducting species.

It has been recognized for a considerable time that if the translational invariance of the conductivity activation energy barriers is lost, then the dielectric and conductivity properties become frequency-dependent. (See Fig. 2.1.10c). For a material with no dipolar relaxation processes (i.e., ϵ_s is not a function of time), but with a conductivity σ_0, then Eq. (94) becomes

$$Y^* = \sigma_0 + j\omega\epsilon_s\epsilon_0 \tag{109}$$

The equivalent circuit is simply a parallel —RC— combination, and thus

$$\epsilon^* = \epsilon_s - \frac{j\sigma_0}{\omega\epsilon_0} \tag{110}$$

Macedo, Moynihan, and Bose [1972] defined the *conductivity relaxation time*

$$\tau_\sigma = \epsilon_0\epsilon_s/\sigma_0 \tag{111}$$

Hence

$$\epsilon^* = \epsilon_s - \frac{j\epsilon_s}{\omega\tau_\sigma} \tag{112}$$

Macedo and others (Hodge et al. [1975, 1976]) have stressed the electric modulus formalism ($M^* = 1/\epsilon^*$) for dealing with conducting materials, for the reason that it emphasizes bulk properties at the expense of interfacial polarization. Equation (112) transforms to

$$M^* = M_s\frac{j\omega\tau_\sigma}{1 + j\omega\tau_\sigma} \tag{113}$$

where $M_s = 1/\epsilon_s$.

For a material with a single relaxation time τ_σ, a plot of M'' vs. log (f) shows a maximum, in just the same way that ϵ'' shows a maximum for a dielectric relaxation process. Glassy conductors however, often show broad and asymmetric modulus spectra, and, in complete analogy to the discussion of Section 2.1.2.3 Macedo et al. [1972], introduced a distribution function of conductivity relaxation times $G(\tau_\sigma)$ such that

$$M^* = M_s\int_0^\infty G(\tau_\sigma)\frac{j\omega\tau_\sigma}{1 + j\omega\tau_\sigma}\,d\tau_\sigma \tag{114}$$

They were able to fit experimental modulus data for a calcium–potassium nitrate melt and a lithium aluminosilicate glass using a double lognormal distribution function.

The decay function for the electric field after the imposition of a charge on the electrodes

$$\phi'(t) = \mathbf{E}(t)/\mathbf{E}(t = 0) \tag{115}$$

may also be defined, in analogy to the decay of polarization function [see Eq. (63)]. For a single relaxation time,

$$\phi'(t) = \exp{-t/\tau_\sigma} \tag{116}$$

The modulus is related to ϕ' through the expression

$$\frac{M^*}{M_s} = 1 - \left\{ \left(-\frac{d\phi'}{dt} \right) \right\} \tag{117}$$

where, as usual, { } denote a Laplace transform.

Moynihan et al. [1973] used the Williams–Watts form of ϕ',

$$\phi'(t) = \exp{[-(t/\tau_\sigma)^\beta]}, \quad 0 < \beta \le 1 \tag{118}$$

to obtain a much better fit to the same glass data mentioned above. The same function has also been recently used to analyze modulus data for lithium phosphate glasses (Martin and Angell [1986]), where it was found that the parameter β was largely independent of temperature but the distribution broadened with increasing alkali content.

2.1.3 Mass and Charge Transport in the Presence of Concentration Gradients

2.1.3.1 *Diffusion*
In the absence of an electric field and terms in R_i, Eq. (30) reduces to

$$\partial c_i/\partial t = -\nabla \cdot j_i \tag{119}$$

$$= \nabla \cdot (D_i \nabla c_i) \tag{120}$$

In one dimension, for constant D_i,

$$\partial c_i/\partial t = D_i \, \partial^2 c_i/\partial x^2 \tag{121}$$

This equation has been introduced from two points of view. In the macroscopic approach it was assumed that the flux or diffusion current is proportional to a concentration gradient or a chemical potential gradient and also satisfies a continuity equation. The generalizations of and justifications for this approach lie at the basis of nonequilibrium thermodynamics; as such, they are independent of the atomistic nature of the processes involved.

On the other hand, in the atomistic approach, the time-dependent configurations of the system are determined from the probabilities of the elementary atomic process. The random walk approach calculates the probability of finding the system in a certain state after a certain time given an initial distribution of particles. It is then possible to show that the distributions are solutions of the diffusion equation.

It was shown in Section 2.1.2.4 that the general flux equations (e.g., the Nernst–Planck equation) contain, in addition to the diffusion terms, a contribution from migration, that is, the movement of charged particles under the influence of an electric field. Under certain circumstances it is quite possible to carry out experiments in which the field is negligibly small compared to the concentration or activity driving force.

In aqueous electrochemistry this situation is usually achieved by use of a supporting electrolyte. This is an inactive salt that is added to the solution in high concentration to increase the conductivity enough that the migration term in Eq. (78) or Eq. (82) becomes very small. In solid state electrochemistry it is difficult to achieve the same effect in such a simple way. The movement of a minority charge carrier, either electronic or ionic, in a good solid electrolyte is an analogous situation. This result is exploited in the Wagner [1933] asymmetric polarization experiment in which the partial conductivity of electronic species in a solid electrolyte is measured assuming that the driving force for electronic conductivity is an activity gradient rather than an electric field. In the next section, another example, that of chemical diffusion in a majority electronic carrier, is discussed in more detail. It is worth mentioning that local electric fields arising from the coupled motion of two charged species through an approximate electroneutrality condition are not necessarily absent from the equations of this section. This local field may be present and profoundly affect the diffusion of species, without any net average field across the bulk of a sample leading to a migration process. In this section, it is assumed that a field which leads to a migration current is absent, but local fields, such as those present in neutral electrolyte diffusion or ambipolar diffusion, may be present.

The purposes of this section is to discuss the electrical analogs of diffusion processes in the absence of migration and to present suitable electrical equivalent circuits for analysis of data obtained under these circumstances.

From the point of view of impedance spectroscopy, solutions of the diffusion equation are required in the frequency domain. The Laplace transform of Eq. (121) is an ordinary differential equation

$$p\{c\} - c(t = 0) = D \, d^2 \, \{c\} / dx^2 \tag{122}$$

where p is the complex frequency variable

$$p = \sigma + j\omega \tag{123}$$

This transformation of a partial differential equation into an ordinary differential equation illustrates a general advantage of working in the frequency domain. Solutions are of the form

$$\{\Delta c\} = A \exp(-\alpha x) + B \exp(\alpha x) \tag{124}$$

where $\{\Delta c\}$ is the Laplace transform of the excess concentration

$$\Delta c = c(x, t) - c(x, 0) \tag{125}$$

and $\alpha = \sqrt{(p/D)}$. Here A and B are constants to be determined by the boundary conditions. Experimentally, one boundary is usually the interface between the electrode and the electrolyte ($x = 0$). Consider the case of semiinfinite diffusion into the electrode:

$$\Delta c \to 0 \quad \text{as} \quad x \to \infty$$

and therefore $B = 0$.

At $x = 0$ (the electrode–electrolyte interface), the solution is

$$\{\Delta c\}_{x=0} = \{\Delta i\}/zF\sqrt{(pD)} \tag{126}$$

where Δi is the ac current, which is equal to $-zFDd\Delta c/dx$. To calculate the impedance we need a relationship between $\Delta c_{x=0}$ and Δv, the ac component of the voltage. For small perturbations around equilibrium, we may write

$$\Delta v/\Delta c = (dE/dc) \tag{127}$$

where (dE/dc) represents the change in electrode potential with concentration, which may be developed from a model (e.g., ideal solution assumptions) or from a separate thermodynamic measurement. For an ideal solution $dE/dc = RT/zFc$. For small-signal conditions, the perturbation may also be expanded around a steady state dc potential, in which case the surface concentrations due to the dc current must also be calculated from a steady state flux equation.

Taking the Laplace transform of (127) and substituting into (126) gives

$$Z^*(p) = \{\Delta v_{x=0}\}/\{\Delta i\} = (dE/dc)/zF\sqrt{(pD)}. \tag{128}$$

Setting $\sigma = 0$ and separating the real and imaginary parts gives

$$Z^*(j\omega) = (dE/dc)(\omega^{-0.5} - j\omega^{-0.5})/zF\sqrt{(2D)} \tag{129}$$

The complex impedance is therefore inversely proportional to the square root of frequency. In the complex plane it is a straight line inclined at $\pi/4$ to the real axis.

Equation (121) has analogies in both heat conduction and electrical circuit theory. Consider the semiinfinite transmission line composed only of resistors and capacitors (Fig. 2.1.12). If r is the resistance per unit length and c is the capacitance per unit length, then

$$I = -(\partial V/\partial x)/r \tag{130}$$

$$\partial V/\partial t = -(\partial I/\partial x)/c \tag{131}$$

Differentiating (130) and combining the two equations gives

$$\partial V/\partial t = (\partial^2 V/\partial x^2)/rc \tag{132}$$

The analogy with diffusion may be made more specific if we compare the appropriate driving forces and fluxes. The electric potential difference V in the transmission line case is analogous to the electrochemical potential difference in the case of diffusion. Thus,

$$\frac{\partial V}{\partial x} \equiv \frac{RT}{zF} \cdot \frac{1}{c} \cdot \frac{\partial c}{\partial x} \tag{133}$$

for an ideal solution (the case for which Fick's law is least ambiguously valid). The reciprocal of the resistance per unit length is analogous to cDF^2z^2/RT, and the capacitance per unit length is analogous to z^2F^2c/RT.

Thus the reciprocal of the rc product plays the role of the diffusion coefficient. The impedance of the transmission line is

$$Z_R = \sqrt{(r/pc)} \tag{134}$$

which is exactly the same form as Eq. (129) if appropriate substitutions are made. For a nonideal solution, RT/zFc may be replaced by (dE/dc).

So far, only semiinfinite boundary conditions have been considered. For many problems however, thin samples dictate the use of finite-length boundary conditions. A reflective boundary $dc/dx = 0$ has been considered by Ho et al. [1980] and the impedance derived for this case:

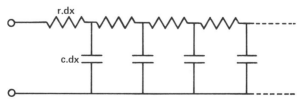

FIGURE 2.1.12. A resistive–capacitive transmission line which describes the behavior of a semiinfinite diffusion process.

$$Z(j\omega) = \frac{dE}{dc} \cdot \frac{1}{zF} \cdot \frac{\coth l\sqrt{(j\omega/D)}}{\sqrt{(j\omega D)}} \tag{135}$$

The equivalent circuit analog of this situation is a finite-length transmission line terminated with an open circuit. A constant activity or concentration is also a common condition for the interface removed from $x = 0$. In this case the finite-length transmission line would be terminated in a resistance, and the impedance is given by the expression

$$Z(j\omega) = \frac{dE}{dc} \cdot \frac{1}{zF} \cdot \frac{\tanh l\sqrt{(j\omega/D)}}{\sqrt{(j\omega D)}} \tag{136}$$

The complex plane representations of these two impedance behaviors are shown in Figure 2.1.13.

In this section the following principal assumptions were made. First, it was assumed that the surface concentrations and potentials were given by their equilibrium or dc steady state values. In other words, there was supposed to be no barrier preventing or slowing down the transfer of matter across the electrode–electrolyte interface. In general, of course, this will not be true, and the impedance associated with the interface forms a very important aspect of impedance spectroscopy as applied to electrochemical situations. This is in contrast to applications where the only interest lies in bulk effects. The interfacial impedance, due to both the storage and the dissipation of energy, will be addressed in Section 2.1.4.

Second, although the form of the diffusion equation was derived from ideal solution theory, it will be seen in the next section that the form of the equation

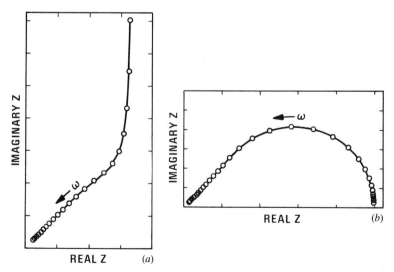

FIGURE 2.1.13. Complex plane representations of the impedance due to a finite-length diffusion process with (*a*) reflective, (*b*) transmissive boundary conditions at $x = 1$.

may be retained, even though the dilute solution assumptions are relaxed, through definition of a chemical diffusion coefficient.

2.1.3.2 Mixed Electronic–Ionic Conductors

It was seen in the previous paragraphs that the presence of a supporting electrolyte leads to a situation in which a charged species may diffuse in an essentially field-free environment. Under these circumstances, Fick's law of diffusion may be solved for the appropriate boundary conditions, and the electrical response of the system may be modeled by a transmission line composed of distributed elements. Although such situations are very common in aqueous electrochemistry, the analogous situation in solids, where a minority ion diffuses in a solid supporting electrolyte, occurs infrequently.

More interesting is the commonly encountered situation where an ion diffuses in a majority electronic conductor. Thus, diffusion in metallic and semiconducting alloys or of inserted species in transition metal oxides and chalcogenides fall into this category. Many electrode reactions are of this type. Lithium diffusion in β-LiAl and other alloys is of interest in negative electrode reactions for advanced lithium batteries; hydrogen and lithium diffusion in oxides (e.g., V_6O_{13}) and sulfides (e.g., TiS_2) are of importance as positive electrode reactions for batteries and electrochromic devices.

In materials of this type the diffusion process may be regarded as involving a neutral atomic species, or as a coupled process in which an ionic and electronic species move together. In the simplest case, where the electronic partial conductivity is much greater than the ionic, the flux equation for a neutral species may be written

$$j = -c^*u \frac{d\mu^*}{dx} = -c^*uRT \frac{d \log a^*}{dx} = -c^*D_k \frac{d \log a^*}{dx} \qquad (137)$$

Here we cannot assume ideal or dilute solution behavior since the mobile species activity may vary widely especially if the electron activity changes rapidly over the stoichiometric domain of the phase of interest. After rearrangement we obtain the equation

$$j = -D_k \frac{d \log a^*}{d \log c^*} \cdot \frac{dc^*}{dx} = -D_k \left[1 + \frac{d \log \gamma^*}{d \log c^*} \right] \frac{dc^*}{dx} \qquad (138)$$

which is equivalent to Fick's law if we write

$$D_c = D_k \left[1 + \frac{d \log \gamma^*}{d \log c^*} \right] \qquad (139)$$

Here u is the mobility, μ the chemical potential, and γ the activity coefficient of the mobile species. The ''*'' denotes that the relations are written for a neutral species. Now D_k is the component diffusion coefficient which, as pointed out in Section 2.1.2.4, obeys the Nernst–Einstein relation

$$D_k = uRT \tag{140}$$

for all situations, irrespective of ideality assumptions.

The term in square brackets in Eq. 138 expresses the variation of activity coefficient of the *neutral* species with concentration. Thus, in addition to the statistical contribution to diffusion, expressed by the familiar gradient-in-concentration term, there is a chemical driving force due to the variation of free energy with composition and hence position. The term in square brackets is known as the thermodynamic enhancement factor and was identified by Darken [1948]. The diffusion coefficient D_c is known as the chemical diffusion coefficient, and its use is appropriate whenever diffusion takes place in an appreciable concentration gradient and when ideal solution laws cannot be applied to the solute. The concept was extended by C. Wagner [1953], and a recent general treatment has been given by Weppner and Huggins [1977]. The more general approach involves elimination of the field term from simultaneous equations of the type

$$j_i = -RTu_i \left[\frac{d \log a_i}{d \log c_i} \frac{dc_i}{dx} + \frac{z_i F c_i}{RT} \frac{d\Phi}{dx} \right] \tag{141}$$

written for ionic and electronic species. The result is a general equation

$$j_i = -D_{k_i} \left[(1 - t_i) \frac{d \log a_i^*}{d \log c_i^*} - \sum_{j \neq i, e, h} t_j \frac{z_i}{z_j} \frac{d \log a_j^*}{d \log c_i^*} \right] \frac{dc_i}{dx} \tag{142}$$

valid for general transference numbers and thermodynamic parameters. For small-signal conditions (constant enhancement factor over the concentration range of the experiment), the diffusion equations are still of the Fick's law type and therefore lead to $\sqrt{\omega}$ dependence of the admittance (Warburg behavior).

In recent years, cells of the type

$$\text{Li} \mid \text{Li}^+ \text{ electrolyte} \mid \text{Li}_y M$$

(where M is a mixed conducting host material for the inserted lithium) have been used to investigate the diffusion of Li in a number of alloys and oxides using ac impedance methods. The boundary condition at the electrolyte–electrode interface is a sinusoidally varying chemical potential of (neutral) lithium. It is important to recognize that the potential difference applied across a cell determines the activity of the electroactive species at the point at which the conductivity of the system changes from being predominantly ionic to being predominantly electronic. In this kind of experiment the thermodynamic enhancement factor is conveniently determined in situ by measuring the dependence of equilibrium cell potential on electrode composition.

For thin samples, the second boundary condition may be modified to include either transmissive or reflective interfaces, as discussed in Section 2.1.3.1.

2.1.3.3 Concentration Polarization

The situation is often encountered where, upon the passage of current through an electrochemical cell, only one of the mobile species is discharged at the electrodes. Examples are (a) the use of a liquid or polymeric electrolyte, where both ions are mobile, and yet where only one is able to participate in the electrode reaction; and (b) a mixed conducting solid in which current is passed by electrons, but in which cations also have a significant transport number.

Consider a system consisting of a binary, unsupported electrolyte between electrodes which are reversible only to the cation. The cell is initially at equilibrium (no net currents are passing). At very short times after the imposition of a potential difference, the concentrations of all species in the bulk of the electrolyte are uniform and the ions move in response to the applied field. The current is determined by the uniform electrolyte conductivity.

$$i = -F^2 \sum c_i z_i^2 u_i \frac{d\Phi}{dx} \tag{143}$$

At long times, on the other hand, the flux of the blocked anion falls to zero, and a constant flux of cations passes through the system. In order to maintain electroneutrality there must also be a gradient in anion concentration and hence in electric potential, which just balances the gradient in anion chemical potential.

$$\frac{d\Phi}{dx} = -\frac{1}{z_- F} \frac{d\mu_-}{dx} \tag{144}$$

Thus, there is effectively a gradient in the concentration of neutral species across the cell, and therefore we must include in the total potential difference a Nernstian term which is equal to the potential difference that would exist immediately after the interruption of current flow but before the reestablishment of uniform concentration profiles. The other contribution to the potential difference, that which is due to the flow of current itself, is a term arising from the gradient in conductivity due to the variations in concentration. This of course arises from differences in the mobility of the two species. It must be distinguished from the ohmic term present at very short times (high frequencies) due to the initially uniform conductivity of the electrolyte. The concentration polarization is therefore the additional polarization which is present due to concentration gradients caused by the current flow; this is compared to the ohmic polarization that would be present if the current flow (and distribution) were the same but the concentration gradients were absent.

Substitution of the condition (144) into the flux equation for cations

$$j_+ = -c_+ u_+ RT \frac{d \log c_+}{dx} - c_+ u_+ F \frac{d\Phi}{dx} \tag{145}$$

(assuming ideal solutions) gives the steady state cation current:

$$i_+ = -2u_+RT\frac{dc^*}{dx} \qquad (146)$$

The ohmic potential difference may then be found by integration of Eq. (144) across the cell using this flux equation. Since the current is constant, the concentration profile must also be uniform. The Nernstian term may be included as the potential of a concentration cell with the same concentration profile. It is possible to show that the ratio of the steady state resistance to the high-frequency resistance depends on the transference numbers of the ions.

At large potentials this model predicts a limiting current density at which point the concentration at one of the electrodes has fallen to zero. For example, in a solid or polymeric electrolyte, with plane parallel electrodes and an initial uniform concentration of c,

$$i_L = \frac{4c^*D_+F}{l} \qquad (147)$$

where l is the thickness of the electrolyte.

It is therefore apparent that in passing from high to low frequency in a system of this kind, there is an additional impedance due to concentration polarization. Macdonald and Hull [1984] considered this effect on the electrical response of this type of system. Under many circumstances, the presence of concentration polarization might be confused with an interface impedance. At different ratios of mobilities of anions and cations, either diffusion-like response (finite-length transmission line behavior) or parallel capacitative–resistive behavior may appear. Ac impedance methods have been used to determine ionic transference numbers in polymeric electrolytes using this principle (Sorensen and Jacobsen [1982]).

2.1.4   Interfaces and Boundary Conditions

2.1.4.1   Reversible and Irreversible Interfaces

Although it is quite reasonable to discuss the bulk properties of homogeneous phases in isolation, it is seldom possible in electrochemical situations to neglect the interfaces, since potentials and fluxes are usually measured or defined at junctions between ionically and electronically conducting phases. In general, two extreme types of interface are recognized.

The first type is an interface which is reversible to the species under consideration. The term *reversible* implies certain thermodynamic and kinetic properties. Thermodynamically, it means that an equilibrium relation of the type

$$^1\Delta^2\eta_i = 0 \qquad (148)$$

may be written for the ith species, which applies to points immediately on each side of the interface in phases 1 and 2. Here, η is the electrochemical potential. Thus, a clean interface between a parent metal M and a binary M^+ conducting solid electrolyte is thermodynamically reversible; the activities of M, M^+, and e^- are all equal across the interface.

Kinetically, the term is less well defined and depends more explicitly on the nature of the experiment. In practice, it means that the exchange current density (the microscopic flux crossing the interface equally in both directions at equilibrium) is very much greater than the net current density crossing the interface during the experiment or the measuring process. At appreciable current densities, however, the net current density may eventually exceed the exchange current density; interface kinetics then become important.

In the electrochemical literature it is useful to refer to a reversible interface or interfacial reaction as one whose potential is determined only by the thermodynamic potentials of the various electroactive species at the electrode surface. In other words, it is only necessary to take into account mass transport to and from the interface, and not the inherent heterogeneous kinetics of the interfacial reaction itself, when discussing the rate of the charge transfer reaction. This nomenclature has two principal disadvantages. First, it neglects the fact that mass transport to the interface, whether migration or diffusion, is inherently an irreversible or dissipative process in a thermodynamic sense. Second, it neglects the time dependence of the system. At short times the rate may be largely determined by interfacial reaction rates; at long times it may be determined by mass transport processes. This is particularly clear when ac experiments are performed; steady states may be achieved in the frequency domain that correspond to transient conditions in the time domain.

An interface may be reversible to one species, but blocking to others. In addition, in multicomponent systems, a reversible electrode may not necessarily define the thermodynamic potentials of all components present at the interface.

An electrode which is reversible to electrons but irreversible to ions is a common situation in both aqueous and solid state electrochemistry. For determinations of *ionic* conductivity in electrolytes, this type of electrode has proved useful, because the concentrations of majority ionic species do not depend critically on the imposition of a well-defined thermodynamic activity of the electroactive neutral species. Measurements with two irreversible electrodes of a nonreactive metal are then permissible; numerous examples are found in the solid–electrolyte literature. Minority electronic transport however, typically depends very strongly on the activity of neutral components, and care must be taken to utilize thermodynamically meaningful experiments to determine minority conductivities. Asymmetric cells using one reversible electrode and one irreversible electrodes are then appropriate, but have actually been little explored using ac impedance methods.

A real electrode with some degree of reversibility will therefore allow a steady state current to pass; in the sense that such a current obeys Faraday's laws, it is termed a *faradic current*. A completely polarizable electrode passes no faradic current. In transient or ac experiment however, a polarizable electrode and a reversible electrode both pass a nonfaradic current, corresponding to charging or discharging of the interface capacitance and perhaps changes in the nature and concentration of any adsorbed species. The distinction between the two types of current is important in developing expressions for the impedance of the electrode-electrolyte interface.

2.1.4.2 Polarizable Electrodes

Within the voltage limits set by the thermodynamic stability range of the electrolyte, foreign metal electrodes may sometimes be regarded as ideally polarizable or blocking. The metal electrodes must not react with the electrolyte, and for the moment adsorption and underpotential deposition will be neglected. From an electrochemical point of view, this is the simplest type of interface and has furnished much of the information we have about the electrified interface.

Depending on the initial positions of the Fermi levels of the electrolyte and electrode, a small amount of charge flows in one direction or the other and a field is created on the electrolyte side of the contact. The mobile charges in the electrolyte distribute themselves over this field; the charge density of the metal is confined to the surface of the electrode. The excess charge density at any point within the electrolyte is given by Boltzmann statistics

$$\rho(x) = \sum z_i F c_i$$

$$= \sum z_i F c_i^0 \exp\left(\frac{-z_i F \Phi}{RT}\right) \tag{149}$$

and the relationship between charge and potential is given by the Poisson equation, and thus

$$\frac{d^2\Phi}{dx^2} = \frac{-1}{\epsilon\epsilon_0} \sum z_i F c_i^0 \exp\left(\frac{-z_i F \Phi}{RT}\right) \tag{150}$$

Solution of this equation leads to the space charge (diffuse double-layer) capacitance. For a symmetrical ($z_+ = z_-$) electrolyte,

$$C_d = \left[\frac{2z^2 F^2 \epsilon\epsilon_0 c_i^0}{RT}\right]^{1/2} \cosh\left(\frac{zF\Phi_0}{2RT}\right) \tag{151}$$

where Φ_0 is the potential at $x = 0$ (measured relative to $\Phi = 0$ at $x = \infty$). The diffuse double layer therefore behaves as a simple parallel plate capacitor. The perturbation in concentration due to the electric field extends into the electrolyte a distance on the order of the Debye length L_D:

$$L_D = \left[\frac{RT\epsilon\epsilon_0}{2z^2 F^2 c_i^0}\right]^{1/2} \tag{152}$$

Thus, the higher the concentration of the electrolyte, the thinner the diffuse double layer.

In addition to the use of Boltzmann statistics, the model has assumed:

a. Point charges, and hence no limit on the distance of closest approach to the interface.

b. A uniform dielectric constant.

c. A sharp boundary between the metal and the electrolyte; that is, the electronic wave functions do not extend beyond the geometrical plane of the interface.

d. No screening effects such as those found in the Debye–Huckel theory of electrolytes.

At some potential, the situation will occur where there is no excess charge on either side of the interface and the concentration profiles are flat. This point is known as the potential of zero charge.

Up to now no real distinction has been made between solid and liquid electrolytes. In an extrinsically conducting solid, the complementary charge carrier will be absent; therefore, it will not be included in the distribution. However, the theory is basically equally applicable to solids, molten salts, and polar solvent electrolytes.

In the presence of a polar solvent molecule such as water, considerable attention has been focused on the role of the solvent. Since the dipole moment is free to rotate in the presence of an electric field, it is reasonable that in a layer of water close to the interface there will be a net dipolar orientation and the water will not exhibit its normal dielectric constant. In addition, hydrated ions will not be able to approach indefinitely close to the interface. Thus, up to a monolayer of charge will exist at a distance of closest approach to the electrode; this distance is determined by the size of the (hydrated) ion. Beyond this inner layer, the diffuse layer will extend back into the solution. The interface therefore behaves as two capacitors in series: an inner (Stern) layer and an outer (Gouy–Chapman) layer. The model takes into account both the dipolar nature of the solvent and some of the finite-size effects.

In a solid, of course, there is no solvent. However, we still expect an inner-layer capacitance, since there is still a finite distance of closest approach of the ions to the interface. Therefore,

$$C_i = \epsilon\epsilon_0/d \tag{153}$$

where C_i is the capacitance per unit area. If d is of the order of a few angstroms and ϵ is of order 1, then C_i should lie between 1 and 10 μF/cm^2. We also expect it to be independent, or a slowly varying function, of interfacial potential difference. Since C_d, however, depends exponentially on voltage, we expect it to become large quickly, and therefore C_i will dominate the interface capacitance except when it is close to the potential of zero charge.

There are few experimental studies of the solid electrolyte–solid electrode interface carried out in such a way that meaningful potential capacitance data can be obtained. This would involve the asymmetric cell type of arrangement; for example,

$$M/M^+ \text{ electrolyte}/\text{inert metal}$$

where the thermodynamic quantities of the electrolyte are fixed at the inert metal-electrolyte interface by the application of a potential difference. Some studies of this type have been performed, but there seems to be few instances of the observation of a well-defined diffuse double-layer capacitance. Most of the experiments were, however, performed on highly conducting materials, which would be expected to have very thin diffuse double layers.

A possible exception to this is the study of the graphite–AgBr interface (Kimura et al. [1975]), where a broad minimum in capacitance was found at potentials somewhat positive of the Ag–AgBr electrode potential at temperatures between 219 and 395°C. The minimum was somewhat broader than expected from the theoretical model, but of the correct order of magnitude. On the other hand, a similar experiment by Armstrong and Mason [1973] showed no particular minimum in capacitance at a similar temperature.

More often, the double-layer capacitance for the silver conductors seems to show a small potential dependence, more easily interpreted, at least qualitatively, in terms of an inner-layer phenomenon.

The experimental study of the solid–solid interface is complicated by a further problem. It is often (perhaps usually) observed that, instead of a purely capacitive behavior, the interface shows significant frequency dispersion. Several authors have found excellent agreement of this behavior with the dispersion shown by the constant-phase element (Bottelberghs and Broers [1976], Raistrick et al. [1977]). Although the amount of frequency dispersion is influenced by electrode roughness and other aspects of the quality of the interface (i.e., nonuniform current distribution), these are evidently not the only contributions to the observation of CPE behavior. Although no well-defined microscopic theory of the CPE has emerged, this empirically important aspect of both interface and bulk behavior is discussed further in this section and in Section 2.2. As is true for the case of the potential dependence of the capacitance, there have been too few studies of the frequency dispersion of the interface. It should be mentioned that the microscopically smooth liquid metal–aqueous electrolyte interfaces apparently do not show frequency dispersion of the capacitance if the systems are quite pure.

2.1.4.3 Adsorption at the Electrode–Electrolyte Interface

In the previous section the distance of closest approach of ions to a planar electrode–electrolyte interface was discussed. In solid–electrolyte systems, this distance is assumed to be approximately the radius of the mobile ion. In the presence of a polar solvent the hydration sheath of the ion and the solvent layer adjacent to the metal are also important. The only forces acting on the interface have been assumed to be electrostatic in origin. These forces orient the solvent dipoles and determine the distribution of ions with distance from the interface.

It is possible however, that an ion can interact chemically with the electrode material. If this happens the ion may break through the solvent layers or, as in the case of the solid, become displaced from a normal lattice site. This possibility is known as specific adsorption. In aqueous electrochemistry the locus of the centers of the specifically adsorbed ions is known as the inner Helmholtz plane. Neutral molecules may also adsorb and hence affect the faradic current, for example, by blockage of the reaction sites. Neutral molecule effects have not been studied in the case of solid systems and will therefore not be considered further.

In order to include adsorption in a discussion of the electrical response, it is necessary to know the relationship between the surface concentration of the adsorbed species and the concentration in the electrolyte just outside the double layer. This last concentration can then be related to the bulk or average concentration through appropriate diffusion equations.

For a neutral molecule, potential dependence will still be expected, since at large potential differences the force acting on the dipole of a polar solvent will be sufficient to compete with all but the strongest adsorption bonds.

A simple isotherm, due originally to Langmuir, assumes that the free energy of adsorption ΔG_i^0 is the same all over the surface and that interactions between adsorbed species are neglected. Under these conditions, the surface concentration Γ_i is related to the surface concentration at full coverage Γ_0 by the expression

$$\frac{\Gamma}{\Gamma_0 - \Gamma} = \frac{\theta}{1 - \theta} = a_i^b \exp\left(\frac{-\Delta G_i^0}{RT}\right) \exp\left(\frac{-\Phi z_i F}{RT}\right) \quad (154)$$

where a_i^b is the bulk activity of i.

The capacitance associated with the adsorption can be obtained by differentiation of the charge due to the adsorbed species:

$$q = \theta q_i \quad (155)$$

where q_i is the charge corresponding to one monolayer:

$$C = dq/d\Phi = (dq/d\theta)(d\theta/d\Phi) = q_i (z_i F/RT) \theta(1 - \theta) \quad (156)$$

Various attempts have been made to include interactions between adsorbed species. As pointed out by Conway et al. (1984), the correct way to handle interactions is to include the appropriate pairwise or long-range interaction term into the partition function, which allows calculation of the Helmholtz free energy and the chemical potential. These quantities are a function of θ due to (a) the configurational term, as included in the Langmuir case; and (b) the interaction or deviation from ideality.

As an example, Frumkin's isotherm may be derived by assuming a pairwise interaction of the form

$$U(\theta) = r\theta^2/2 \tag{157}$$

where r is positive for a repulsive interaction and negative for an attractive force. This leads to

$$\frac{\theta}{1-\theta} = a_i^b \exp\left(\frac{-\Delta G_i^0 - r\theta}{RT}\right) \exp\left(\frac{-\Phi zF}{RT}\right) \tag{158}$$

This yields a capacitance of the form

$$C(\theta) = q_i \left(\frac{z_i F}{RT}\right) \cdot \frac{\theta(1-\theta)}{1 + r\theta(1-\theta)} \tag{159}$$

Comparing this expression with Eq. (156) indicates that the new capacitance expression can be expressed as a "Langmuir" capacitance in series with an "interaction" capacitance:

$$C(\theta) = (C_L^{-1} + C_I^{-1})^{-1} \tag{160}$$

Other expressions for different forms of the interaction term have been given by Conway et al. (1984).

The Temkin isotherm attempts to account for heterogeneity of the electrode surface by making the energy of adsorption vary linearly with coverage, which gives

$$\exp(r\theta) = Ka_i^b \exp(z_i F\Phi/RT) \tag{161}$$

and

$$C(\theta) = q_1 (z_i F/RT) \cdot 1/r \tag{162}$$

The rates of adsorption are usually rapid and hence the kinetics are determined by other electrochemical or chemical steps and mass transport. Armstrong has pointed out that in solid–electrolyte systems, where the interfacial potential difference cannot be varied independently of the concentration of the mobile species, the adsorption of that species cannot be controlled by a diffusional process.

Raleigh [1976] has put forward a model of competitive chemisorption of anions and cations in silver halides that leads to a broad maximum in capacitance at the potential of zero charge, in agreement with observations on some of these compounds. This approach is greatly extended in Macdonald et al. [1980].

The kinetics of complex electrochemical reactions in the presence of adsorbed intermediates and its effect on the impedance of the interface is discussed in Section 2.1.4.4.

2.1.4.4 *Charge Transfer at the Electrode–Electrolyte Interface*

The rate of heterogeneous charge transfer reaction

$$O + ne = R \tag{163}$$

is given by the expression

$$-i_F = nF[k_f c_O - k_b c_R] \tag{164}$$

where i_F is the faradic current density, $k_{f,b}$ are the forward and reverse rate constants, and $C_{O,R}$ are the concentrations of the reactants and products at the interface at time t.

The current, in general, is composed of a steady state or dc part determined by the mean dc potential E and the mean dc concentrations at the interface, c_O and c_R, and an ac part, Δi_F, determined by the ac perturbing potential ΔE and the fluctuating concentrations Δc_i. The faradic impedance is given by the ratio of the Laplace transforms of the ac parts of the voltage and current

$$Z_F = \{\Delta E\}/\{\Delta i_F\} \tag{165}$$

Because charge transfer is involved, the presence of an electric field at the interface affects the energies of the various species differently as they approach the interfacial region. In other words, the activation energy barrier for the reaction depends on the potential difference across the interface. It is convenient to express the potential dependence of the rate constants in the following manner:

$$k_f = k_0 \exp - \alpha(E - E^0)nF/RT \tag{166}$$

$$k_b = k_0 \exp (1 - \alpha)(E - E^0)nF/RT \tag{167}$$

where k_0 is the rate constant at the formal electrode potential E^0 and α is the apparent cathodic transfer coefficient. Hence

$$\frac{k_f}{k_b} = \exp \frac{nF}{RT}(E - E^0) \tag{168}$$

Generally, we can express Δi_F as an expansion of the ac parts of the concentrations and electrode potential,

$$\Delta i_F = \Sigma \left(\frac{\partial i_F}{\partial c_i}\right) \Delta c_i + \left(\frac{\partial i_F}{\partial E}\right) \Delta E + \text{higher-order terms} \tag{169}$$

Neglecting all but the first-order terms (linearization) and solving for ΔE,

$$-\Delta E = \frac{1}{(\partial i_F / \partial E)} \left[\Sigma \left(\frac{\partial i_F}{\partial c_i} \right) \Delta c_i - \Delta i_F \right] \qquad (170)$$

and hence

$$Z_F = \frac{1}{(\partial i_F / \partial E)} \left[1 - \Sigma \left(\frac{\partial i_F}{\partial c_i} \right) \frac{\{\Delta c_i\}}{\{\Delta i_F\}} \right] \qquad (171)$$

The first term is the *charge transfer resistance*; the second term contains the influence of the ac part of the mass transport on the impedance. Here $\{\Delta c_i\}/\{\Delta i_F\}$ can be expressed as a solution of the diffusion equation. For example, for semi-infinite diffusion to a plane, we can use Eq. (126):

$$\frac{\{\Delta c_i\}}{\{\Delta i_F\}} = \frac{1}{nF \sqrt{(pD_i)}} \qquad (172)$$

The coefficients in parentheses may be evaluated from the rate expressions discussed above,

$$\left(\frac{\partial i_F}{\partial E} \right) = k_f \frac{n^2 F^2}{RT} \left[\alpha c_O + (1 - \alpha) c_R \exp \frac{nF}{RT} (E - E^0) \right] \qquad (173)$$

and

$$-\left(\frac{\partial i_F}{\partial c_O} \right) = nFk_f; \qquad -\left(\frac{\partial i_F}{\partial c_R} \right) = -nFk_f \exp \frac{nF}{RT} (E - E^0) \qquad (174)$$

Here c_O and c_R are determined by the solution of the appropriate dc mass transport equations. The coefficients may then be substituted into Eq. (166) to give the overall faradic impedance.

At the equilibrium potential E_r, the net current is zero; therefore, c_O and c_R are equal to their bulk values c_O^* and c_R^* and are related through the Nernst equation

$$\frac{c_O^*}{c_R^*} = \exp \frac{nF}{RT} (E_r - E^0) \qquad (175)$$

Under these circumstances, the coefficient $(\partial i_F / \partial E)$ simplifies to give

$$\frac{\partial i_F}{\partial E} = \frac{n^2 F^2}{RT} k^0 \exp \left[- \frac{\alpha nF}{RT} (E_r - E^0) c_O^* \right] \qquad (176)$$

and the charge transfer resistance is

$$r_{ct} = RT/nFi_0 \tag{177}$$

where the exchange current density is

$$i_0 = nFk^0 c_O^* \exp\left[-\alpha(nF/RT)(E_r - E^0)\right] \tag{178}$$
$$= nFk^0 c_O^{*(1-\alpha)} c_R^{*\alpha}$$

When mass transport to the electrode is unimportant, substitution of Eq. (178) into Eq. (164) gives the Butler–Volmer equation

$$-i_F = i_0\left[\exp -\frac{nF}{RT}\alpha(E - E_r) - \exp\frac{nF}{RT}(1 - \alpha)(E - E_r)\right] \tag{179}$$

When E is sufficiently far removed from E_r, the current in one direction may be neglected, leading to the Tafel relation

$$(E - E_r) = a + b\log i \tag{180}$$

It should be emphasized that the development of the expressions for charge transfer kinetics given here is not completely general. It rests on the assumptions of absolute rate theory. More general treatments have been given in the literature where no a priori assumption of the form of the dependence of the rate constants on potential is made (Birke [1971], Holub et al. [1967]). A point which arises from these more general treatments is worth pursuing here. For the case of semiinfinite diffusion to a planar interface, the faradic impedance may be written in the form

$$Z_F = r_{ct} + (\sigma_O + \sigma_R)(1/\sqrt{\omega})(1 - j) \tag{181}$$

where r_{ct} is given by the inverse of Eq. (168), and σ are of the form

$$\sigma_O = \frac{RT}{n^2F^2\sqrt{2}} \cdot \frac{1/\sqrt{D_O}}{\alpha c_O + (1 - \alpha)c_R \exp(nF/RT)(E - E^0)} \tag{182}$$

and

$$\sigma_R = \frac{RT}{n^2F^2\sqrt{2}} \cdot \frac{(1/\sqrt{D_R})\exp(nF/RT)(E - E^0)}{\alpha c_o + (1 - \alpha)c_R \exp(nF/RT)(E - E^0)} \tag{183}$$

The terms in $\sigma_i/\sqrt{\omega}$ correspond to the normal Warburg impedance; they do not contain the heterogeneous rate constants. The more general treatments, however, indicate that the Warburg impedance does in general contain coefficients that depend on the rate constants and their potential dependence. It is only on the basis of absolute rate theory that these coefficients cancel out of the final expression.

The complete equivalent circuit for a single-step charge transfer reaction in the presence of diffusion is given in Figure 2.1.14. The electrolyte resistance and double-layer capacitance have also been added to this figure.

A second aspect of the theory developed in this section is the assumption that the faradic current is decoupled from the nonfaradic current. In other words, the impedance due to the double-layer capacitance is included afterward and placed in parallel to the faradic impedance, since

$$i = i_F + i_{NF} \tag{184}$$

In general, however,

$$i_{NF} = \left(\frac{dq}{dt}\right) = \left(\frac{\partial q}{\partial E}\right)\left(\frac{\partial E}{\partial t}\right) + \Sigma \left(\frac{\partial q}{\partial c_i}\right)\frac{\partial \Delta c_i}{\partial t} \tag{185}$$

where the summation extends over all species, including O and R. Thus, the nonfaradic component is coupled to the faradic current unless experimental steps are taken to decouple them. This is usually achieved by making the concentration of electroactive ions very small compared to the inactive charge carriers, which do most of the double-layer charging.

In solid electrolytes, however, the unsupported electroactive species is often the sole charge carrier. It is thus impossible to change the interfacial potential difference without changing the concentration of ions in the double-layer region. This means of course, that the normal Warburg impedance is not seen, but it also means that there is a coupling between the faradic current and the double-layer charging.

This has been recognized by Armstrong [1974], who has proposed the rate equation

$$i = i_0 \left[1 + \frac{C_{dl}(E - E_r)}{|q_-| - \Delta E C_{dl}} \right] \exp \frac{nF}{RT} \alpha(E - E_r) \tag{186}$$

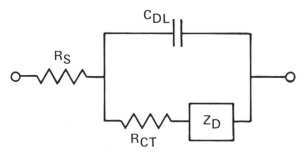

FIGURE 2.1.14. The Randles equivalent circuit, which describes the response of a single-step charge-transfer process with diffusion of reactants and/or products to the interface.

for metal deposition from a solid electrolyte. Here, ΔE is the difference between E and the potential of zero charge, q_- is the charge density of anions in the inner layer, and C_{dl} is taken independent of voltage.

The treatment given above for a single-step charge-transfer reaction may be readily extended to more complex reaction schemes. For a multistep reaction, the partial currents of the individual steps must be appropriately coupled and the mass transport relations defined for each step.

For example, for a surface-adsorbed species, intermediate in a two-step reaction, there will be an additional relationship of the type

$$d\Gamma/dt = \Delta i_F^1/n_1 F - \Delta i_F^2/n_2 F \tag{187}$$

where Γ is the surface concentration of the adsorbed intermediate produced by reaction 1 and removed by reaction 2. This case of considerable importance in aqueous electrochemistry. The form of the impedance and the expected equivalent circuits have been discussed by Gerischer and Mehl [1955], Armstrong and Henderson [1972], Grahame [1952], and Epelboin and Keddam [1970]. In the absence of mass transport control, the equivalent circuit is of the form shown in Figure 2.1.15. Here R_{CT} is a charge transfer resistance, and R_{ADS} and C_{ADS} are components which contain the contribution of the surface concentration (coverage) of the adsorbed intermediate and the rate of adsorption or desorption, respectively. Under certain circumstances R_{ADS} and C_{ADS} can become negative, leading to the appearance of inductive behavior in the impedance spectrum.

As applications of impedance spectroscopy to very complex reactions in the solid state have not yet been made, further development of the theory of the faradic impedance seems unwarranted here. The linear operator approach to calculating the impedance of the systems is due to Rangarajan [1974] and is well described in the recent review of Sluyters-Rehbach and Sluyters [1984].

2.1.5 Grain Boundary Effects

It was suggested earlier that the electrical analog of an isotropic, homogeneous, ionically conducting solid is a pure resistance in parallel with a high-frequency

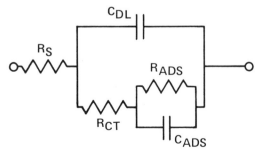

FIGURE 2.1.15. An equivalent circuit which describes the electrical response of an electrochemical reaction with a strongly adsorbed intermediate.

ideal capacitor. This model assumes the absence of electrode polarization and of relaxation processes within the crystal that would lead to additional parallel branches in the equivalent circuit. This model is generally accepted, and several studies of single-crystal materials have demonstrated its validity.

Many solids are, however, studied in polycrystalline form, either because they are only available as such or because this is the manner in which they will be utilized. Polycrystalline materials usually have less than theoretical density (void-age) and misorientated grains (important in anisotropic materials). In the simplest case, these effects would lead to purely geometric reductions in the conductivity with respect to the single crystal. In addition, impurities may be present as a second phase at the grain boundaries. Because of the importance of ac impedance spectroscopy as a tool for measuring ionic conductivity, there have been several studies of the effect of polycrystallinity on the impedance of solid electrolytes.

The problem was first attacked in a modern manner by Bauerle [1969], whose paper was the first application of impedance spectroscopy to solid electrolytes. He studied both high-purity and "impure" polycrystalline zirconia. Bauerle found that the presence of a second phase at the grain boundaries in dense material led to the introduction of a second time constant in the equivalent circuit. This additional impedance was absent in the very-high-purity material. Bauerle envisaged the ionically insulating second phase as introducing a constriction in the area of contact between the grains of the highly conducting phase. Beekmans and Heyne [1976] found similar behavior in calcia-stabilized zirconia and suggested as well that a distribution of time constants for the grain boundary behavior was appropriate, rather than the single —RC— time constant as was suggested by Bauerle.

Later, it became apparent that a second phase need not be present in polycrystalline materials for the grain boundaries to make a contribution to the impedance of the system. There have been several studies of the impedance of polycrystalline sodium β-alumina, a very nonisotropic solid electrolyte. Hooper [1977] systematically studied the relationship between single-crystal and polycrystalline material and showed that it was possible to extract "true" bulk values from polycrystalline samples. This intragrain conductivity had the same activation energy as the single-crystal material; there still was, however, a relatively small difference between the absolute conductivity values, probably mostly due to the geometric effects introduced by the anisotropy of the material, and preferential orientation in the pressed samples. Grain boundary (intergrain) conductivity had a greater activation energy and disappeared at high temperatures.

There have also been several studies of the more isotropically conducting materials based on the Li_4SiO_4 and γ-II Li_3PO_4 ("LISICON") structures (Ho [1980] and Bruce and West, [1983]). Ho varied the density of polycrystalline $Li_{4+y}Si_{1-y}Al_yO_4$ from nearly the theoretical density down to about 60% of theoretical density. At the highest densities, only a single circular arc was seen in the impedance plots, but at lower densities two arcs became apparent. The resistance associated with the lower-frequency arc exhibited a higher activation energy than that associated with the higher-frequency arc, which was attributed to intergrain impedance. As with Hooper's study of β-alumina, this contribution disappeared at higher temperatures. Ho's study and the 1976 work of Raistrick et al. [1976] on other polycrystalline alumino silicates noted, however, that except for the very densest of materials, polycrystalline samples always showed some anomalous frequency dispersion. The circuit element now often known as a constant-phase element (CPE) was introduced to fit the data

$$Y^*_{CPE} = A(j\omega)^\alpha \tag{188}$$

Each of the circular arcs found in the polycrystalline materials was of the form shown in Figure 2.1.16. It seems that polycrystallinity introduces anomalous frequency dispersion into the bulk impedance behavior before a second and separate contribution from an intergrain impedance appears. A recent study by Bruce and West [1983] of polycrystalline LISICON essentially reached the same conclusions. Unlike Ho, however, the same activation energy was found for both the inter- and intragrain resistances. This suggests that essentially the same physical processes are involved and that the authors attributed the intergrain impedance to a constriction effect not unlike that proposed by Bauerle. Bruce and West, however, attributed the constriction to the smaller area of contact between grains rather than to the presence of an ionically insulating second phase. A recent investigation of the effect of a single-grain boundary on the response of an otherwise single-crystal CaO-doped CeO_2 specimen also found two semicircles in the impedance plane (El Adham and Hammou [1983]). Depending on the presence or absence of CaO enrichment at the grain boundary, a different activation energy was present for the resistance associated with the boundary. A detailed modeling of the properties of a constriction resistance of this type should be possible, but does not yet appear to have been carried out. It is also possible that close to a grain boundary, the transport properties of the crystal are controlled by imperfections, expected to be present there in higher concentration than in the center of a grain, leading to an additional contribution to the intergrain impedance. This idea is very close to the observation of conductivity enhancement due to heterogeneous doping of solid

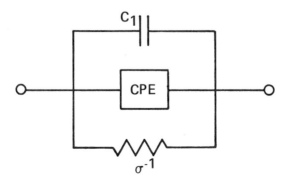

FIGURE 2.1.16. An equivalent circuit which describes the electrical response of polycrystalline solid electrolytes.

electrolytes. Here, a second insoluble, nonconducting phase is introduced into the solid electrolyte as finely dispersed particles. The internal space charge created at the phase boundaries may lead to a significant increase in the concentration of mobile defects. A detailed impedance study of such a system would be of considerable interest, but does not yet appear to have been carried out.

2.1.6 Current Distribution, Porous and Rough Electrodes—the Effect of Geometry

2.1.6.1 *Current Distribution Problems (Ibl [1983b])*
It was pointed out in Section 2.1.2.4 that in most electrochemical systems, including those situations in which there is a concentration gradient but in which regions close to interfaces between different phases are excluded, electroneutrality is a reasonable assumption. Under these circumstances, the potential variation is given by Laplace's equation

$$\nabla^2 \Phi = 0 \tag{189}$$

In principle, this equation may be solved subject to the following boundary conditions:

$$\Phi = \text{constant (conducting boundary)}$$

$$\partial \Phi / \partial n = \text{constant (insulating boundary)}$$

where n is the normal to the boundary.

The potential and current distributions derived from these boundary conditions are called the *primary distribution*. They depend only on the geometry of the system. For very simple geometries, analytical solutions to Laplace's equation have been found. In recent years, numerical solutions have often become the preferred method, and both finite-difference and finite-element methods, as well as techniques based on Green's function methods, are valuable. Equation (189) corresponds to steady temperature in heat conduction problems; a useful discussion is found in Carslaw and Jaeger [1959].

Often, in solid state experiments, the most common experimental arrangement is the most satisfactory from a primary current distribution point of view. The electrodes completely cover the ends of the electrolyte and there is no spreading of the current lines. The current distribution should, however, be considered whenever more complex geometries are involved or the placement of a reference electrode is in question.

The real importance of current distribution problems in impedance measurements, however, lies in the fact that the distribution is frequency-dependent. This arises because of the influence of interfacial polarization combined with the geometrical aspects of the arrangement.

The electrode–electrolyte interface is not an equipotential surface. This is because the interfacial potential difference is typically a function of the local current density. Thus, even for a purely resistive interface impedance—independent of local current density—there is a smoothing effect on the current distribution in the system. The greater the current density at a particular point, the greater the potential drop across the interface, which in turn tends to lower the local current density. The tendency is therefore to make a more uniform current distribution. The magnitude of the smoothing depends on the relative magnitudes of the interface and bulk impedances, as well as the geometry of the system. The current distribution in the presence of interfacial polarization (but neglecting mass transport effects) is called the *secondary distribution*.

In general, interfacial impedance is partly capacitative as well as resistive in nature. At high frequencies, the capacitance short-circuits the interface, and the primary distribution is observed for the ac part of the current. As the frequency is lowered, the interface impedance increases, causing a changeover to the secondary distribution. Of necessity, this effect leads to a frequency dependence of the equivalent circuit parameters which describe the system. Of course, if the primary distribution is uniform, there will be no frequency dispersion arising from this source.

The question of the frequency dependence of the current distribution and its effect on the measured impedance of a solid state electrochemical system, has been hardly considered, although it is important in discussing the impedance of, for example, porous gas electrodes on anion conductors, of rough electrodes (discussed below), and also perhaps of polycrystalline materials. In aqueous electrochemical situations the effect has been considered with respect to the rotating disk electrode, where there may be severe current distribution problems.

2.1.6.2 *Rough and Porous Electrodes*
It is recognized that porosity or roughness of the electrode surface could be expected to lead to a frequency dispersion of the interfacial impedance even in the absence of detailed considerations of the current distribution problems as outlined above.

A simple approach to the problem of porous and rough interfaces is based on the use of transmission line analogies (de Levie [1967]). Consider a cylindrical pore in a conducting electrode. If the series resistance of the electrolyte per unit length is r, and the interfacial capacitance per unit length is c, then the pore behaves as a transmission line and has an impedance given by equation 2.134. This approach can be extended to more complex situations which include pores of finite depth, non-uniform pores, and situations where the interfacial capacitance is replaced by a complex faradic admittance, corresponding to an electrochemical reaction taking place down the depth of the pore. It is also possible to include finite electrode resistance.

In order to describe a rough electrode, de Levie [1965a] suggested a model based on the V-shaped groove shown in Figure 2.1.17. It was assumed that the double-layer capacitance was uniform over the true surface of the electrode and that the current lines were perpendicular to the *macroscopic* surface of the electrode. The impedance of the groove is then analogous to a transmission line in

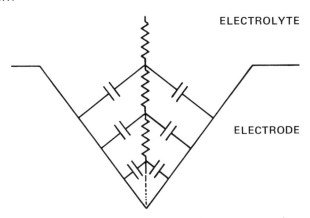

ELECTROLYTE

ELECTRODE

FIGURE 2.1.17. Transmission line model for a V-shaped groove in an electrode surface.

which the components are a function of the distance into the groove. The transmission line equation may be readily modified for the present case, where r and c are functions of x, which is the distance down the line:

$$I = -[1/r(x)]\partial V/\partial x \qquad (190)$$

$$\partial V/\partial t = -[1/c(x)]\partial I/\partial x \qquad (191)$$

On combining these equations, we obtain

$$-c(x)\frac{\partial V}{\partial t} = \frac{1}{r(x)^2} \cdot \left[\frac{\partial V}{\partial x}\frac{\partial r(x)}{\partial x} - r(x)\frac{\partial^2 V}{\partial x^2}\right] \qquad (192)$$

The Laplace transform of this equation is, like the diffusion equation, an ordinary differential equation

$$r^2\frac{d^2\{V\}}{dx^2} - r\frac{dr}{dx}\cdot\frac{d\{V\}}{dx} - r^3cp\{V\} = 0 \qquad (193)$$

For r and c simple functions of x, the equation becomes a modified Bessel equation and may be solved analytically. For example, in the case of the groove geometry, as considered by de Levie,

$$Z = (\rho/\tan\beta)I_0(\lambda)/\lambda I_1(\lambda) \qquad (194)$$

where Z is the impedance, 2β is the angle at the apex of the groove, ρ is the specific solution resistivity, and λ is $2\sqrt{(\rho l p \kappa/\sin\beta)}$; κ is the interface capacitance per unit area. In the limit of low and high frequency, the phase angle of the impedance changes from $\pi/2$ to $\pi/4$.

Using this approach, a number of different geometries can be analyzed (Keiser [1976]). The interface impedance, considered in the example above to be a pure capacitance, could be generalized to include both real and imaginary components, for example, a Warburg impedance. In general, however, ρ and κ would become functions of distance into the groove if significant diffusional effects were included in the calculation, and the diffusion layer thickness, relative to the thickness of the surface features, becomes important.

As pointed out by de Levie, however, the most important weakness in the model is the assumption that the current distribution is normal to the macroscopic surface, that is, a neglect of the true current distribution. For a rough surface, the lines of electric force do not converge evenly on the surface. The double layer will therefore be charged unevenly, and the admittance will be time and frequency dependence.

The tangential components of the interface charging were recognized and included in a qualitative model by Scheider [1975], who suggested the use of branched transmission lines to model the effects of uneven surface topology. The suggestion was significant in that transmission lines of this type do represent a circuit which, unlike the unbranched transmission line of de Levie, agrees with experimental observations of the impedance at rough electrodes.

The basic type of line suggested by Scheider is shown in Figure 2.1.18 for a single type of branching. In general, both the series (as shown) and the parallel components of a simple —RC— line may be replaced by other transmission lines, which may themselves be branched. The degree of branching may be unlimited. Let the series impedance per unit length be z and the parallel admittance per unit length be y (Fig. 2.1.19). In addition, let all the z and all the y be independent of distance down the line. Using the continued-fraction approach to write the total impedance Z_T,

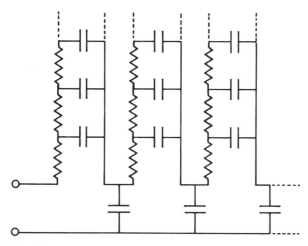

FIGURE 2.1.18. A branched transmission line circuit which shows CPE behavior.

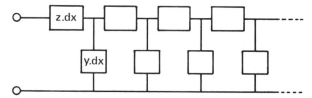

FIGURE 2.1.19. A generalized transmission line where z and y are respectively the series impedance and interfacial admittance per unit length.

$$Z_T = \cfrac{1}{y + \cfrac{1}{z + \cfrac{1}{y + \cdots}}} \tag{195}$$

For an infinite line,

$$Z_T = \cfrac{1}{y + \cfrac{1}{z + Z_T}} \tag{196}$$

or

$$Z_T = -\frac{z}{2} + \sqrt{\frac{z^2}{4} + \frac{z}{y}} \tag{197}$$

In the limit of $dx \to 0$, z, $y \to 0$ but z/y is finite. Therefore,

$$Z_T = \sqrt{\frac{z}{y}} \tag{198}$$

Suppose, if z represents a transmission line $z = A(j\omega)^{-1/2}$, and y is a pure capacitance $y = j\omega C$, then

$$Z_T = \sqrt{\left(\frac{A}{C}\right)(j\omega)^{-3/4}} \tag{199}$$

Evidently, a first-order branching of the series component leads to an impedance with a phase angle of $(3/4)(\pi/2)$.

In fact, the value of the frequency exponent is limited only by the degree of branching of the circuit. The interval between frequency exponents is

$$[1/2]^{\theta+1}$$

where θ is the order of the branching. The frequency dependence is determined by the branching type and not by the magnitude of the components.

The important feature of these lines is that they produce a constant phase angle, like a Warburg impedance, but with the phase angle not restricted to $\pi/4$. This is exactly the behavior often found at the electrode–electrolyte interface and has been termed a *constant-phase element* (CPE). It appears to be true that roughness is an important contributing factor to the observed frequency dispersion. Scheider's model, however, remains qualitative, and the microscopic link between the topology and the circuit is absent.

In general, a transmission line with nonuniform components, such as that described by Eq. (192), does not lead to CPE behavior. Schrama [1957] has shown that for lines with a particular type of nonuniformity, CPE behavior is predicted. This relationship is, for a discrete —RC— line,

$$R_k = \frac{2\Gamma(1 - \alpha)}{\Gamma(\alpha)} \cdot \frac{\Gamma(\alpha + k)}{\Gamma(1 - \alpha + k)} \cdot h^\alpha \tag{200}$$

$$C_k = (2k + 1)\frac{\Gamma(\alpha)}{\Gamma(1 - \alpha)} \cdot \frac{\Gamma(1 - \alpha + k)}{\Gamma(1 + \alpha + k)} \cdot h^{1 - \alpha} \tag{201}$$

where h is a positive real number. Schrama suggested an interpretation in terms of nonuniform diffusion coefficients and driving forces.

Recently it has been shown (Liu [1985]) that at a fractal interface, a nonuniform transmission line will model the electrical response. The fractal geometry assumed was that of the triadic bar of Cantor, illustrated schematically in Figure 2.1.20a. The equivalent circuit corresponding to such an interface is shown in Figure

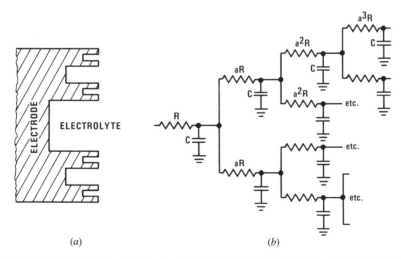

(a)   (b)

FIGURE 2.1.20. (a) Formal model for a fractal electrode–electrolyte interface, and (b) an equivalent circuit which models the double-layer charging behavior.

2.1.20*b*. It is evident that the circuit may be folded over into an —RC— transmission line with the resistance and capacitance per unit length related to one another in a specific manner.

The possible relationship between CPE behavior and a fractal interface geometry have also been emphasized by Le Mehaute [1984] and Le Mehaute and Crepy [1983].

This section was written by Ian D. Raistrick. The U.S. government retains a royalty-free license to reproduce or reuse this material for in-house or governmental purposes.

2.2 PHYSICAL AND ELECTROCHEMICAL MODELS

2.2.1 The Modeling of Electrochemical Systems

Numerous theoretical models have been developed to explain and predict the behavior of electrochemical systems and to guide the design of systems with desired characteristics. The models which have been developed thus far fall generally into three broad categories, corresponding to three different levels of description of the system. From a practical standpoint, no one group of models is more important than another: the models which may be helpful to a materials scientist in fabricating a new solid electrolyte may provide no assistance at all to the engineer attempting to design a portable power source which meets rigid cost and performance specifications.

At the most fundamental level of description are atomistic or microscopic models which attempt to provide an accurate description of the motions of individual charge-carrying particles in the system. At the least detailed level are the equivalent circuit models, in which hypothetical electrical circuits, consisting of elements with well-defined electrical properties, are used to describe the response of the system to a range of possible signals. Such models are of special interest in impedance spectroscopy, since the frequency response behavior of linear electrical circuits is now extremely well understood. An introduction to equivalent circuit models is provided in Section 2.2.2 below. The level intermediate between equivalent circuits and microscopic models might be termed the *continuum level*, as the bulk regions of the electrodes and electrolyte are regarded as continuous media. The transport of mobile species is thus assumed to be governed by differential equations, and the transfer of charge across interfaces obeys rate laws which serve as boundary conditions for these equations. The parameters of the continuum model can be regarded as averages of the parameters appearing in an atomistic model. As can be seen in Section 2.2.3, the continuum parameters can usually be related to the parameters of an equivalent circuit model, and the analysis of the continuum model can thus guide in the construction of an appropriate equivalent circuit.

2.2.2 Equivalent Circuits

2.2.2.1 Unification of Immittance Responses

For a long time dimensionless normalization has been used in the dielectric constant measurement area of IS. As in Eq. (1) in Section 1.3, one supposes that there exists a low-frequency limiting value of the dielectric constant, ϵ_s, and a high-frequency limiting value ϵ_∞. In the latter case there may be even higher-frequency dispersions than that whose limit is ϵ_∞, but in ordinary IS it is usually sufficient to establish (or assume!) that $\epsilon = \epsilon_\infty$ over a wide range of high frequencies, and primary attention is directed to the response in the frequency region between $\epsilon = \epsilon_s$ and ϵ_∞. Then the normalized response may be written

$$\kappa \equiv \frac{\epsilon - \epsilon_\infty}{\epsilon_s - \epsilon_\infty} \tag{1}$$

where $\kappa \to 1$ as $\nu \to 0$ and $\kappa \to 0$ as $\nu \to \infty$ (or to the region where $\epsilon \cong \epsilon_\infty$). The function κ is a normalized immittance, defined at the dielectric constant level.

For solid electrolytes one usually is concerned with intrinsically conducting systems rather than with intrinsically nonconducting (dielectric) ones. It is then appropriate and usual to consider basic system response at the impedance rather than the complex dielectric constant level. Then if one assumes that the overall impedance of the system, Z_Z, approaches R_0 at sufficiently low frequencies and R_∞ at sufficiently high ones, one can form the normalized dimensionless quantity

$$I_Z \equiv \frac{Z_Z - R_\infty}{R_0 - R_\infty} \tag{2}$$

in analogy to Eq. (1). We have given a specific expression for I_Z in Eq. (6) of Section 1.3. Again, $I_Z \to 1$ as $\nu \to 0$ and $I_Z \to 0$ as $\nu \to \infty$.

Since Eqs. (1) and (2) are similar in form, we can combine them in the single expression

$$I_k = \frac{U_k - U_{k\infty}}{U_{k0} - U_{k\infty}} \tag{3}$$

where $k = \epsilon$ or Z, $U_\epsilon = \epsilon^* = \epsilon' + j\epsilon''$, and $U_Z = Z = Z' + jZ''$. As before, U_{k0} and $U_{k\infty}$ are, respectively, the low- and high-frequency limiting real values of U_k. We actually use ϵ^* rather than ϵ here so that the imaginary parts of I_ϵ and I_Z may be defined with the same sign.

Now as already mentioned in Section 1.3, the I_Z of that section's Eq. (6) is of just the same form as the well-known Cole–Cole dielectric dispersion response function (Cole and Cole [1941]). In its normalized form, the same I_k function can thus apply at either the impedance or the complex dielectric constant level. We may generalize this result (J. R. Macdonald [1985a,c,d]) by asserting that *any* IS response function that can be normalized as in Eq. (3) may be used at either the complex dielectric constant level or at the impedance level. It is very important to note that when the same function (with possibly different parameter values) is applied at both the complex dielectric constant and the impedance levels, it defines different systems at these levels. This matter is discussed in more detail in J. R. Macdonald [1985c]. Thus a theoretical derivation of response at one level automatically yields response of the same kind for the other level but applying to a different type of system.

The above results allow us to use a single general $I(\omega)$ function to represent normalized response at either the ϵ or Z system levels. When the k subscript of I_k is omitted, it will be understood to be general in this sense. The use of the general normalized immittance response function I allows one to subsume two kinds of systems and response with a single function and will be so employed in the next section. Table 2.2.1 shows how I_k, for $k = Z$ or ϵ, is related for the various immittance levels to the specific conductive and dielectric systems functions. Here, as before, $\mu \equiv j\omega C_c$. Alternatively, to maintain dimensionless quantities at all levels, one might replace μ by $j\omega\tau \equiv js$, where τ is a specific relaxation time and $s \equiv \omega\tau$. Note that all functions are simply related to I. Let us illustrate these relations with a specific example. Take, for concreteness, $I_k = F_k(\psi_{kC}, s)$, the specific response function of Eq. (6) in Section 1.3. Then we can write

$$\epsilon_\epsilon = \epsilon_\infty + (\epsilon_s - \epsilon_\infty) F_\epsilon(\psi_{\epsilon C}, s) \tag{4}$$

$$Y_\epsilon = i\omega C_c \epsilon_\epsilon \tag{5}$$

$$Z_\epsilon = \left\{ i\omega C_c \left[\epsilon_\infty + (\epsilon_s - \epsilon_\infty) F_\epsilon(\psi_{\epsilon C}, s) \right] \right\}^{-1} \tag{6}$$

and

$$Z_Z = R_\infty + (R_0 - R_\infty) F_Z(\psi_{ZC}, s) \tag{7}$$

$$Y_Z = Z_Z^{-1} \tag{8}$$

$$\epsilon_Z = \left\{ i\omega C_c \left[R_\infty + (R_0 - R_\infty) F_Z(\psi_{ZC}, s) \right] \right\}^{-1} \tag{9}$$

TABLE 2.2.1. Relations Between the General, Unified Immittance Function I and Specific System Functions

Conductive System ($k = Z$)	General System Normalized	Dielectric System ($k = \epsilon$)
M	μI_k	Y
Z	I_k	ϵ
Y	I_k^{-1}	M
ϵ	$(\mu I_k)^{-1}$	Z

Note: Here $\mu \equiv j\omega C_c$.

We have used specific subscripts here to designate which type of system is involved. Now when one compares, say, Eqs. (6) and (7), both at the impedance level but for different types of systems, or Eqs. (4) and (9), both at the complex dielectric constant level, one sees that although the normalized expressions for I_Z and I_ϵ are of exactly the same form, Z_ϵ and Z_Z as well as ϵ_ϵ and ϵ_Z yield very different frequency response. The main unification produced by the introduction of the general $I \equiv I_k$ is that of allowing a single function to represent typical normalized response of either a conductive ($k = Z$) or a dielectric ($k = \epsilon$) system.

2.2.2.2 Distributed Circuit Elements

Diffusion-Related Elements. Although we usually employ ideal resistors, capacitors, and inductances in an equivalent circuit, actual real elements only approximate ideality over a limited frequency range. Thus an actual resistor always exhibits some capacitance and inductance as well and, in fact, acts somewhat like a transmission line, so that its response to an electrical stimulus (output) is always delayed compared to its input. All real elements are actually distributed because they extend over a finite region of space rather than being localized at a point. Nevertheless, for equivalent circuits which are not applied at very high frequencies (say over 10^7 or 10^8 Hz), it will usually be an adequate approximation to incorporate some ideal, lumped-constant resistors, capacitors, and possibly inductances.

But an electrolytic cell or dielectric test sample is always finite in extent, and its electrical response often exhibits two generic types of distributed response, requiring the appearance of distributed elements in the equivalent circuit used to fit IS data. The first type, that discussed above, appears just because of the finite extent of the system, even when all system properties are homogeneous and space-invariant. Diffusion can lead to a distributed circuit element (the analog of a finite-length transmission line) of this type. When a circuit element is distributed, it is found that its impedance cannot be exactly expressed as the combination of a finite number of ideal circuit elements, except possibly in certain limiting cases.

The second generic type of distributed response is quite different from the first, although it is also associated with finite extension in space. In all ordinary IS experiments one uses electrodes of macroscopic dimensions. Therefore, the total macroscopic current flowing in response to an applied static potential difference is the sum of a very large number of microscopic current filaments originating and ending at the electrodes. If the electrodes are rough and/or the bulk properties of the material are inhomogeneous, the individual contributions to the total current will all be different, and the distribution in electrode surface or bulk properties will lead to a distributed resistance (many different elemental resistances) or conductance.

The situation is even more complicated when small-signal frequency and time dependence is considered. Consider a material involving ion-hopping conduction. The immediate microscopic surroundings of different ions may be different at a given instant either because of inhomogeneous material properties or because the dynamic relaxation of the positions of atoms surrounding an ion has progressed a different amount for different ions. The result may be described in terms of a distribution of relaxation times, which, for example, might be associated with a distribution of hopping-barrier-height activation energies. Such a distribution of relaxation times will lead to frequency-dependent effects which may, at least approximately, often be described through the use of certain simple distributed circuit examples.

The first distributed element introduced into electrochemistry was the infinite-length Warburg [1899] impedance, often termed *the* Warburg impedance by those possibly unaware of the more general finite-length Warburg solution. The infinite-length Warburg impedance is obtained from the solution of Fick's second law, the diffusion equation, for one-dimensional diffusion of a particle in a semiinfinite space, a situation mathematically analogous to wave transmission on a semiinfinite distributed RC transmission line (see, e.g., Franceschetti and Macdonald [1979c]). Diffusion of atomic oxygen in an infinitely thick electrode might be described by this impedance, an impedance which we shall designate $Z_{W\infty}$ (see below). But real physical situations never involve infinite lengths (although this limit may sometimes be a useful one to consider). The solution for the diffusion of particles in a finite-length region (equivalent to a finite-length, shorted transmission line) appears first to have been presented by Llopis and Colon [1958], for the *supported* situation, where the finite length considered was the thickness of the Nernst diffusion layer, appropriate for a stirred electrolyte or a rotating electrode. But particles diffusing in an electrode of thickness l_e or in an electrolytic cell of unstirred liquid or in solid material are free to move through the entire available region l_e or l. Thus it is reasonable to take the finite-length region where diffusion occurs as l or l_e in cases of present interest. General Warburg response for charge motion in a finite-length region of an *unsupported* electrolyte appears in the first exact solution of this problem (Macdonald [1953]). It was identified and discussed in later work (Macdonald [1971a,b, 1974a,b] and Macdonald and Franceschetti [1978]). These results, particularly appropriate for solid electrolytes, will be discussed later in Section 2.2.3.3. Here it is sufficient to give the expression for Z_W for an uncharged particle diffusing in a finite-length region of length l_e, which might be the thickness of an electrode (Franceschetti and Macdonald [1979c]), and show how it reduces to $Z_{W\infty}$ as $l_e \to \infty$. The result may be written

$$Z_W \equiv R_{D0} \left[\tanh (\sqrt{js}) / \sqrt{js} \right] \qquad (10)$$

where $s \equiv l_e^2(\omega/D)$ and D is the diffusion coefficient of the diffusing particle. Here the diffusion resistance R_{D0} is the $\omega \to 0$ limit of $Z_W(\omega)$. It may be expressed in a form involving various rate constants if so desired (Macdonald and Franceschetti [1978], Franceschetti [1981]). Series expansion readily shows that when $s \ll 3$, Z_W is well approximated by R_{D0} in parallel with a capacitance C_{D0}, where

$$C_{D0} \equiv l_e^2 / 3DR_{D0} \qquad (11)$$

When plotted in the complex plane, Z_W leads to an initial straight-line region with $\theta = 45°$: it reaches a peak value of $-Z''_w \simeq 0.417 R_{D0}$ at $s \simeq 2.53$ and then begins to decrease toward the real axis, finally approaching it vertically, as required by the limiting R_{D0} and C_{D0} in parallel.

When $s \gg 3$, the tanh term approaches unity and Z_W approaches $Z_{W\infty}$, given by

$$Z_{W\infty} = R_{D0}/\sqrt{js} = (R_{D0}/l_e)(2\omega/D)^{-1/2}(1-j) \tag{12}$$

clearly showing the 45° response of $Z_{W\infty}$. Let us define the frequency-dependent diffusion length as $l_D \equiv \sqrt{D/\omega}$; then $s \equiv (l_e/l_D)^2$. It is obvious that when $l_D \ll l_e$, for example, at high frequencies, $Z_{W\infty}$ response is found: the diffusion length is then much less than the entire region available for diffusion. But when l_D begins to approach l_e, $Z_{W\infty}$ response is no longer appropriate since diffusion begins to be limited, and $Z_{W\infty}$ must then be replaced with Z_W. In fact, it is always reasonable and appropriate to use Z_W. The quantity Z_W will always be referred as the Warburg or diffusion impedance in this work. Finally, the resistance R_{D0} is proportional to l_e and is thus extensive. Equation (11) also shows that C_{D0} is also extensive (but proportional to l_e rather than to l_e^{-1} as in an ordinary plane parallel capacitance). Thus Warburg response becomes extensive and depends on electrode separation at sufficiently low frequencies. But as Eq. (12) shows, $Z_{W\infty}$ is entirely intensive since (R_{D0}/l_e) is itself intensive. One way of identifying Warburg response is to validate the transition from intensive to extensive behavior using measurements with two or more different values of l_e.

The Warburg impedance Z_W is the diffusion analog of the impedance of a finite-length, uniformly distributed RC transmission line (see Fig. 2.2.1) with a short at the far end, equivalent in the diffusion case to unhindered disappearance of the diffusing particles at $x = l_e$. But this special situation, while common, by no means includes all cases of interest. An expression for the impedance Z_D of the finite-length diffusion problem with more general conditions at the far end has been presented by Franceschetti and Macdonald [1979c]. It is recommended that the

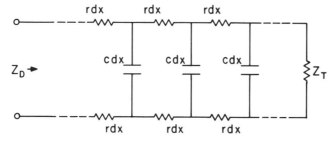

FIGURE 2.2.1. Uniform continuous transmission line involving series resistance of $2r$ per unit length and shunt capacitance of c per unit length, terminated by an impedance Z_T. When $Z_T = 0$, $Z_D = Z_W$, and when $Z_T = \infty$, $Z_D = Z_{D0C}$.

general Z_D be used initially in an equivalent circuit representation and for fitting unless and until it can be established that $Z_D \cong Z_W$.

Although we shall not discuss the general $Z_D(\omega)$ further here, there is one additional specific case which follows from it and deserves mention. Suppose that the finite-length transmission line analog is open-circuited (see Franceschetti and Macdonald [1979c]). Then no direct current can flow in the actual system, as it could with Z_W (but not $Z_{W\infty}$), and the concentration of the diffusing particle increases at the far end of the diffusion region where complete blocking occurs. The final low-frequency behavior of the open-circuit impedance, Z_{D0C} is thus capacitative. Its representation at the dielectric constant level is

$$\epsilon \equiv Y/j\omega C_c = (C_{D0C}/C_c)[\tanh \sqrt{js}/\sqrt{js}] \qquad (13)$$

Thus, the process leads to the limiting capacitance C_{D0C} as $\omega \rightarrow 0$: further, the frequency response at this level is exactly the same as that for Z_W at the impedance level [see Eq. (10)] and thus involves an initial straight line at $\theta = 45°$. At the impedance level, $Z_{D0C} \equiv (j\omega C_c \epsilon)^{-1}$, is given by

$$Z_{D0C} = (\tau_{D0C}/C_{D0C})[\text{ctnh} (\sqrt{js})/\sqrt{js}] \qquad (14)$$

where we have written $s \equiv \omega\tau_{D0C}$ so that $\tau_{D0C} \equiv l_e^2/D$. The appearance of response of the present type has been found in electrochromic thin films (see Glarum and Marshall [1980], Ho et al. [1980], and Franceschetti and Macdonald [1982]). Equations (10) and (13) show immediately that the general I_k function associated with shorted or open-circuited diffusion is just

$$I_k = \tanh \sqrt{js}/\sqrt{js} \qquad (15)$$

The Constant Phase Element and Its Simple Combinations. Although Warburg and open-ended diffusion effects frequently appear in supported situations and sometimes in unsupported ones and exhibit characteristic $\theta = 45°$ lines in the Z^* or ϵ plane, one often finds approximate straight-line behavior over a limited frequency range with $\theta \neq 45°$ (e.g., McCann and Badwal [1982]). Then the frequency response of Z' and Z'' is no longer proportional to $\omega^{-1/2}$ but to some other power of ω. To describe such response it is convenient to write, as in Eq. (7) in Section 1.3, at the admittance level,

$$Y_{CPE} = A_0(j\omega)^\psi = A_0\omega^\psi[\cos(\psi\pi/2) + j\sin(\psi\pi/2)] \qquad (16)$$

where A_0 and ψ are frequency-independent parameters which usually depend on temperature, and $0 \leqslant \psi \leqslant 1$. This admittance has been designated the constant-phase element (CPE) admittance because its characteristic feature, and that of Z_{CPE} as well, is a phase angle independent of frequency. Although a slightly more general form $a_0(j\omega\tau)^\psi$ may be written, the combination $(a_0\tau^\psi)$ cannot be resolved into its parts using single-temperature frequency response measurements and fitting. The parameter A_0 will be intensive for interface processes and may be extensive for bulk ones. Unlike the finite-length Warburg impedance, the CPE exhibits no transition from intensive to extensive behavior as the frequency decreases. Note that a resistance R_∞ in series with $Z_{CPE} = Y_{CPE}^{-1}$ yields an inclined spur (straight line) in the Z plane, with an $\omega \rightarrow \infty$ intercept of R_∞.

The importance of constant phase response was probably first emphasized by Fricke [1932]; the CPE was explicitly mentioned by Cole and Cole [1941], and its importance and ubiquity have been independently emphasized in recent times by Jonscher [1974, 1975a, b, 1980, 1983]. Some discussion of its history, relation to physical processes, and applicability has recently been given (Macdonald [1984]). Note that it describes an ideal capacitor for $\psi = 1$ and an ideal resistor for $\psi = 0$. It is generally thought to arise, when $\psi \neq 0$ or 1, from the presence of inhomogeneities in the electrode–material system, and it can be described in terms of a (nonnormalizable) distribution of relaxation times (Macdonald and Brachman [1956]), or it may arise from nonuniform diffusion whose electrical analog is an inhomogeneously distributed RC transmission line (Schrama [1957]).

Although CPE-like response appears in the majority of experimental data on solid and liquid electrolytes, it is always well approximated only over a finite range of frequency. In fact, the CPE cannot be applied for all frequencies and becomes physically unrealizable for sufficiently low or high frequencies (Macdonald [1984, 1985b, c, d]). Although many response theories lead to the CPE type of response for a finite frequency range, they must deviate from such response at the frequency extremes in order to yield realistic, physically realizable response. Because of the lack of full physical realizability, the CPE, as in Eq. (16), cannot be normalized in the usual I_k fashion. For example, the $\omega \rightarrow 0$ limits of neither ϵ_{CPE} nor Z_{CPE} exist. With this understood, we shall nevertheless write a unified expression for the CPE, taking it to represent just a dimensionless form of either ϵ or Z. Then we have

$$I_{CPEk} = B_k(j\omega)^{-\psi_k} \equiv (js)^{-\psi_k} \qquad (17)$$

where B_k is a frequency-independent constant and the second form is less general than the first (Macdonald [1984]). For $k = Z$, one usually sets $\psi_Z = n$ and for $k = \epsilon$, $\psi_\epsilon = 1 - n$, where as usual $0 \leqslant n \leqslant 1$. These choices ensure that at the admittance level, for either $k = \epsilon$ or $k = Z$, the fractional exponent ψ in Eq. (16) is just n.

There are three important subcircuits, shown in Figure 2.2.2, which consist of the CPE in conjunction with other circuit elements. Although they can always be treated by considering the CPE contribution separately, their wide use as combined elements and their historical importance justify their separate discussion as compound circuit elements and our assignment of specific designations to them. Further, it has not usually been recognized how such compound expressions can involve individual CPE's. The first subcircuit, that in Figure 2.2.2a, the ϵARC, yields a depressed symmetrical semicircular arc in the complex ϵ plane; the second, shown in Figure 2.2.2b, the ZARC, yields an arc in the Z plane; and the

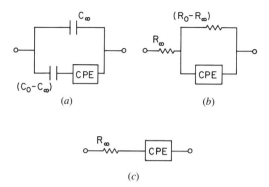

FIGURE 2.2.2. Three compound circuits involving the CPE: (a) the ϵARC, (b) the ZARC, and (c) the YARC.

third, the YARC, leads to an arc in the Y plane. The depression of the arc of course depends on the CPE parameter ψ or n. We have already discussed in Section 1.3 the ZARC and simple ways of analyzing data exhibiting it. The same methods apply to depressed arcs in any complex plane.

The circuit shown in Figure 2.2.2a is just that originally proposed by Cole and Cole [1941] for dielectric systems. It yields capacitances C_0 in the $\omega \rightarrow 0$ limit and C_∞ in the $\omega \rightarrow \infty$ limit. When one uses the Eq. (16) definition of the CPE admittance, it leads to the following expression for ϵ:

$$\epsilon = \epsilon_\epsilon = \epsilon_\infty + \frac{\epsilon_s - \epsilon_\infty}{1 + (j\omega\tau)^{\psi_\epsilon}} \tag{18}$$

where $\epsilon_\infty \equiv C_\infty/C_c$, $\epsilon_s \equiv C_0/C_c$, $\tau \equiv [(C_0 - C_\infty)/A_0]^{1/\psi_\epsilon}$, and $\psi_\epsilon \equiv 1 - \psi$. Cole and Cole related Eq. (18) to a particular distribution of relaxation times, and it has been widely used with $\psi_\epsilon \equiv 1 - \alpha$ for the interpretation of IS results for dielectric and low-conductivity liquid and solid materials. Clearly when $0 < \psi_\epsilon < 1$, Eq. (18) leads to a depressed arc in the complex ϵ plane. For $\psi_\epsilon = 1$, it yields a single-relaxation-constant Debye curve, a full semicircle.

In terms of the original CPE parameters of Eq. (16), the ϵARC admittance of Figure 2.2.2a may be written as

$$Y_\epsilon = Y_{\epsilon ARC} = j\omega \left[C_\infty + \frac{C_0 - C_\infty}{1 + [(C_0 - C_\infty)/A_0](j\omega)^{1-\psi}} \right] \tag{19}$$

For simplicity we shall usually ignore C_∞ in the definition of ϵARC; it may readily be included when needed. The ϵARC circuit element will, in fact, often appear in solid electrolyte equivalent circuits with C_∞ zero (or appearing elsewhere in the overall circuit). Then it is clear that the ϵARC function of Eq. (19) may be considered to represent a distributed (complex) capacitor. When $\psi = 0$, it involves an ordinary capacitor C_0 and resistor A_0^{-1} in series, and when $\psi = 1$, it involves

C_0 in series with the capacitor A_0. These results are consistent with the behavior of the CPE alone. The CPE reduces to an ideal capacitor for $\psi = 1$ and to a resistor for $\psi = 0$. A recent example of the use of the ϵARC function for the analysis of a polycrystalline sample is provided by the work of Casciola and Fabiani [1983].

It was independently suggested some time ago (Ravaine and Souquet [1973], Sandifer and Buck [1974], and Macdonald [1976b]) that the following impedance form might be used to describe the depressed arcs which often appear when impedance data on solids is plotted in the Z^* plane, namely,

$$Z_Z = Z_{ZARC} = \frac{R_0}{1 + (j\omega\tau)^{1-\alpha}} \tag{20}$$

where no R_∞ is included and $0 \leqslant \alpha \leqslant 1$. When R_∞ is included, one can write

$$Z_Z = R_\infty + \frac{R_0 - R_\infty}{1 + (j\omega\tau)^{\psi_Z}} \tag{21}$$

which is just the impedance of the Figure 2.2.2b circuit with $\psi = \psi_Z$. This is an exact analog, at the impedance level, of the Cole–Cole complex dielectric constant expression of Eq. (18). Although the two forms may be described in terms of the same formal distribution of relaxation times, this distribution applies at different response levels for the two cases and thus describes quite different system behavior (Macdonald and Brachman [1956]), Macdonald [1985a–c]). Now it is clear that Eq. (20), which applies when $R_\infty = 0$ or is neglected, may be rewritten as

$$Z_Z = \frac{R_0}{1 + A(j\omega)^{\psi_Z}} \tag{22}$$

involving the parameter $A \equiv \tau^{\psi_Z}$. Now the parallel combination of a resistance R_0 and a CPE with parameter A_0, as in Figure 2.2.2b, yields just

$$Z_Z = \frac{R_0}{1 + A_0 R_0 (j\omega)^{\psi_Z}} \tag{23}$$

equivalent to the result in Eq. (22) if $A = A_0 R_0$ or $\tau = (A_0 R_0)^{1/\psi_Z}$.

The foregoing results and the definition of I_k of Eq. (3) lead immediately to

$$I_k = \left[1 + (js)^{\psi_k}\right]^{-1} \tag{24}$$

where, as usual, $s \equiv \omega\tau$ (see the discussion in the last section). Since when $k = Z$ one obtains the ZARC function and, when $k = \epsilon$, the Cole–Cole equation, we suggest that the general normalized response function of Eq. (24) be designated the ZC function. Again it is most appropriate to take $\psi_Z = \psi = \alpha$ and $\psi_\epsilon = 1 - \alpha$. Although CNLS fitting of data with either a CPE and R in parallel, as in Eq.

(23), or with the unified expression of Eq. (24), involving τ, will yield exactly the same fit, the two approaches involve different parameterizations (R_0, A_0, and ψ_Z or R_0, τ, and ψ_Z). One or the other will generally yield smaller estimated standard errors for A_0 or τ and less correlation of one of these quantities with the other parameters. That choice should be used. Recent analysis of Na β-alumina data (Macdonald and Cook [1985]) gave better results, for example, with the τ parameterization.

It has already been mentioned that the series combination of a resistance R_∞ and a CPE as in Figure 2.2.2c leads to a depressed arc in the Y plane. Since such arcs are also often encountered experimentally, it is reasonable to define them as YARCs, for which the admittance may be written as

$$Y_{\text{ZARC}} = \frac{G_\infty}{1 + (R_\infty A_0)^{-1}(j\omega)^{-\psi}} \tag{25}$$

where $G_\infty \equiv 1/R_\infty$. The similarity to Eq. (23) is obvious, although frequency increases along the ZARC and the YARC in opposite directions, as usual for Z- and Y-plane plots. Although it is possible to define a dimensionless function like I_k which can represent either Y-system or M-system response, just as I_k represents either conductive or dielectric system response, the matter will not be pursued here. In fact, a single dimensionless function with superscript $k = \epsilon, Y, Z$, or M may be used to represent response of any of the four different immittance-level systems if the normalization is properly defined at each level.

It is worth mentioning that although Eqs. (18)–(25) may be interpreted as involving nonuniform diffusion (Schrama [1957]) either in bulk or at an interface, another allied but somewhat different approach which also leads to Eq. (20) has recently been proposed (Le Mehaute and Crepy [1983]) without reference to its earlier history and use. This theory involves mass transfer at a fractal interface, one with apparent fractal dimensionality d, with $d = \psi^{-1} = (1 - \alpha)^{-1}$. A more solidly based treatment of a fractal interface has recently been published by Liu [1985]. Unfortunately, neither of these approaches provides a quantitative interpretation in terms of microscopic parameters of why ψ, determined from data fitting on solids or liquids, often depends appreciably on temperature.

The above results show that the $\psi_\epsilon = 1 - \alpha$ parameter which appears in the ϵARC Cole–Cole function, Eq. (18), associated with a CPE and ideal capacitor in series, and the ψ's appearing in the ZARC and YARC functions, Eqs. (23) and (25), associated with a CPE and resistor in parallel or in series, may all be interpreted as the ψ of a CPE. The ψ values estimated from fitting with these forms are thus comparable. Although the CPE has sometimes been found in equivalent circuit data fitting to appear separately and not directly in any of the above compound forms (e.g., Macdonald, Hooper, and Lehnen [1982]), its presence as a direct part of the ϵARC, ZARC, and YARC functions, ones which have long been used in the interpretation of a wide variety of IS data on dielectric and conduction materials, underlines the wide usefulness of the CPE.

As already mentioned, the lack of any physically based relation for the temperature dependence of the CPE and CPE-like fractional exponent ψ (or n or α) is an important weakness in the theories which lead to frequency response with such exponents. A new theory which involves a distribution of activation energies and does predict temperature dependence for ψ often in agreement with experiment (Macdonald [1985a,c,d]) will be discussed in Section 2.2.3.4, along with some empirical frequency response relations suggested by Jonscher. We summarize the various distributed elements (simple and compound) discussed in this section in Table 2.2.2.

2.2.2.3 Ambiguous Circuits

Let us now further consider the inherent ambiguity of equivalent circuit fitting. One example of two different equivalent circuits having the same overall impedance at all frequencies has already been presented in Figure 1.2.2. Incidentally, if we change all the resistors in both circuits to capacitors, we have another instance of the same kind of ambiguity. Another series of circuits which may all have the same impedance is shown in Figure 2.2.3 (see Franceschetti and Macdonald [1977]). Here we have again given the actual relations between the various components. Some adsorption models (see later) yield inductive-type behavior and a resulting arc which falls below the real axis in the Z^* plane. Sometimes the apparent inductance can be very large. But it is only an apparent inductance since real inductance requires storage of energy in a magnetic field and there is no appreciable ac magnetic field energy present in low-current IS measurements. The actual situation involves an inductive type of phase shift, but rather than represent it by the inductive circuits of Figure 2.2.3a and b, which give a somewhat misleading picture of the process, we recommend following earlier work (e.g., Franceschetti and Macdonald [1977]) and using circuit (c), which involves both a neg-

TABLE 2.2.2. Some Distributed Elements and Their Descriptions at the Impedance Level

Symbol	Name and Description	Defining Equation
Z_W	Finite-length Warburg diffusion	(10)
$Z_{W\infty}$	Infinite-length Warburg diffusion	(12)
Z_D	Diffusion with general boundary conditions	—
Z_{DOC}	Open-circuit (blocked) diffusion	(14)
Z_{CPE}	Constant-phase element	(16), (17)
$Z_{\epsilon ARC}$	Depressed semicircle in complex dielectric constant plane (see Fig. 2.2.2a)	(18), (19)
Z_{ZARC}	Depressed semicircle in impedance plane (see Fig. 2.2.2b)	(20), (21)
Z_{ZC}	ZC element, general form of ϵARC and ZARC	(24)
Z_{YARC}	Depressed semicircle in admittance plane (see Fig. 2.2.2c)	(25)

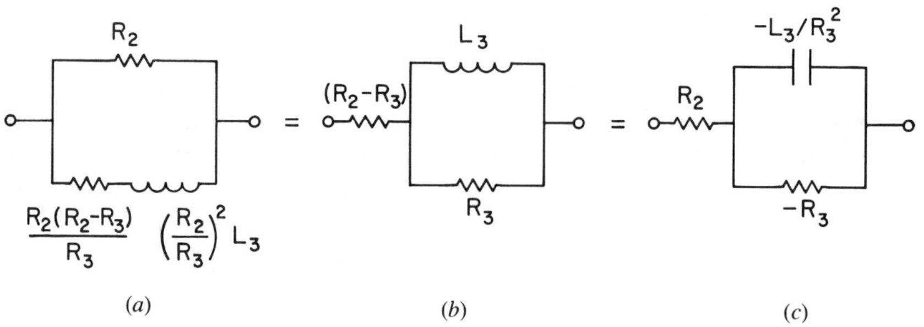

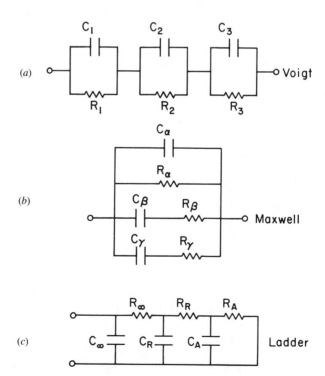

FIGURE 2.2.3. Three circuits having the same impedance at all frequencies.

ative differential capacitor and a negative differential resistor. Since adsorption often can be represented electrically by a positive resistor and capacitor in the (c) type of circuit, continuity is served by allowing both these elements to become negative when appropriate. It is then unnecessary to pass from an ordinary RC circuit to a LC one as adsorption changes; instead the R and C can just go from positive values to negative ones.

Figure 2.2.4 presents three more electrical circuits often encountered in IS work. They exhibit three time constants ($N = 3$) and can all yield the same impedance

FIGURE 2.2.4. Three further circuits which can have the same impedance at all frequencies when the parameters of the circuit are properly interrelated.

for all frequencies when their elements are properly related. All three circuits yield three distinct arcs in the Z^* plane when the three time constants are well separated. Starting from the Voigt circuit with only two time constants, one can find relatively simple algebraic formulas yielding expressions for the elements in the other two $N = 2$ circuits which ensure that the impedance is the same (Novoseleskii et al. [1972]), but such simple relations do not always exist when one starts with another of the circuits or when there are three or more time constants present. In practice, however, the detailed relations between the elements are not particularly important when CNLS fitting procedures are available. First, parameter estimates for any of the circuits may be obtained by such fitting and those for the different circuits compared. Incidentally, the degree of fit is completely independent of which of the three circuits is employed. Second, when the time constants are separated by factors of 100 or more, the R's and C's of, say, the top and bottom circuits closely approach each other. It is particularly when this condition is not satisfied, however, that CNLS fitting is necessary to resolve the overlapping arcs in the Z^* plane.

But CNLS fitting is not always available or may not be justified for preliminary fits. When $N = 2$ in the circuits of Figure 2.2.4, several of the relations between the circuit elements of the three types of circuits can prove very useful in graphical fits. These relationships are summarized in the appendix at the end of this section. Once the parameters of a particular $N = 2$ circuit have been graphically estimated from impedance spectra, estimates of the parameters of the other two circuits may be obtained using these relations, avoiding the need for graphical fitting of the other circuits. Indeed, the equivalence relations are the easiest way to obtain the parameters of the ladder network (Figure 2.2.4c), which cannot be well estimated from either an impedance or admittance plot but require a laborious process involving more than one type of plot (de Levie and Vukadin [1975]). It is simple to fit to the $N = 2$ circuit of Figure 2.2.4a or b and transform to the ladder representation. Further, the equivalence relations may be useful when no IS data are available but circuit element estimates are, as in published work of others.

Since all three of the Figure 2.2.4 circuits are equivalent as far as fitting is concerned, how does one choose between them, particularly in cases when element estimates for the different circuits are quite different? First, one may use continuity and knowledge of the physical processes involved, as in the above brief discussion of inductive-like effects in adsorption. Second, one may be able to compare the circuits with the predictions of a physical model—one which yields simpler expressions for the elements of one of the circuits than for the others. This has been done for the unsupported conduction case (Franceschetti and Macdonald [1977]) and the work showed that in the case of charge of a single-sign mobile the ladder circuit was much superior to the others (see Section 2.2.2.3).

Finally, one should apply the criterion of simplicity. Given equally good fits, the circuit with the smallest number of elements should be used. Second, when electrode separation l, temperature T, or possibly oxygen partial pressure $p(O_2)$ (see, e.g., Verkerk and Burggraaf [1983], Badwal [1984]) is changed, one expects some or all of the fitting parameters to change. But that circuit in which the changes are least, simplest, and/or closest to theoretical expectations should certainly generally be preferred. By carrying out CNLS fitting with several different but plausible circuits, such as those in Figure 2.2.4, for various l, T, and/or $p(O_2)$ conditions, one can often reach an unambiguous choice of the "best" fitting circuit (out of those considered) to use. Since various processes occur in various, often widely separated different frequency regions, it should be emphasized that IS measurements must include such regions to allow identification and analysis of the individual processes present. Generally, then, as wide a frequency span as possible should be covered by the experimental measurements.

Appendix. This appendix summarizes the relations between the elements of the three circuits of Figure 2.2.4 when $N = 2$. Here the subscripts a and b are used in place of the α and β of Fig. 2.2.4.
(a) Voigt → Maxwell

$$C_a = \frac{C_1 C_2}{C_1 + C_2} \tag{A1}$$

$$C_b = \frac{(R_1 C_1 - R_2 C_2)^2}{(C_1 + C_2)(R_1 + R_2)^2} \tag{A2}$$

$$R_a = R_1 + R_2 \tag{A3}$$

$$R_b = \frac{R_1 R_2 (R_1 + R_2)(C_1 + C_2)^2}{(R_1 C_1 - R_2 C_2)^2} \tag{A4}$$

(b) Maxwell → Voigt

$$C_{1,2} = 2C_a \left(1 \mp \frac{R_b/R_a - C_a/C_b + 1}{k^{1/2}} \right)^{-1} \tag{A5}$$

$$R_{1,2} = \frac{R_a}{2} \left(1 \pm \frac{C_a/C_b - R_b/R_a + 1}{k^{1/2}} \right) \tag{A6}$$

where

$$k = \left(\frac{C_a}{C_b} + \frac{R_b}{R_a} + 1 \right)^2 - 4 \frac{C_a R_b}{C_b R_a} \tag{A7}$$

Here $R_{1,2}$ and $C_{1,2}$ are defined such that $R_1 C_1 > R_2 C_2$.
(c) Maxwell → Ladder

$$C_\infty = C_a \tag{A8}$$

$$R_\infty = \frac{R_a R_b}{R_a + R_b} \tag{A9}$$

$$C_R = C_b \left(\frac{R_a + R_b}{R_a} \right)^2 \tag{A10}$$

$$R_R = \frac{R_a^2}{R_a + R_b} \tag{A11}$$

(d) Ladder → Maxwell

$$C_a = C_\infty \tag{A12}$$

$$R_a = R_\infty + R_R \tag{A13}$$

$$C_b = \left(\frac{R_R}{R_\infty + R_R}\right)^2 C_R \tag{A14}$$

$$R_b = \frac{R_\infty}{R_R}(R_\infty + R_R) \tag{A15}$$

2.2.3 Modeling Results

2.2.3.1 Introduction

In any modeling situation one must first specify the physical conditions considered. Is the system in equilibrium or in a steady state? What species of mobile and immobile charges are present? Is the material between the electrodes homogeneous (liquid or single-crystal) or not (amorphous or polycrystalline)? In the polycrystalline case, what boundary conditions should be used at the interfaces between crystallites? In all cases, what kind of electrodes are assumed and thus what are the appropriate boundary conditions at the electrode–material interfaces?

Perhaps the most general problem one would like to solve in the present area, a sufficiently general situation that would include almost all simpler ones of interest, is the following: a biased situation with applied dc p.d. of arbitrary size and a small-signal ac p.d. also applied; an arbitrary number of charged species (but not exceeding, say, six) with arbitrary mobilities and bulk concentrations present; general interactions (e.g. generation–recombination) possible between the various positive and negative charged species; arbitrary (i.e., general) blocking–nonblocking, reaction–adsorption conditions for each of the mobile species at the electrode interfaces; and separate treatments of homogeneous and polycrystalline situations. One would like to calculate the direct current I over a wide range of applied potential difference and, at any given applied steady state potential difference, calculate the impedance as a function of frequency.

Unfortunately, this general problem, which usually involves an inhomogeneous distribution of charge within the material, has not been solved. The situation is highly nonlinear and, although the many coupled differential equations and boundary conditions which could be used to specify it mathematically could, in principle, be solved with a large computer, only purely numerical results depending on a very large number of input parameter values (e.g., mobilities, equilibrium concentrations) would be obtained. Of course even such a general, and almost useless, solution would still be approximate since the equations used would themselves still be approximations to the actual physical situation.

Thus far, only much simpler idealizations of the general problem have been solved, and in Section 2.2.3.3 we shall discuss some of their results. When the simpler solutions are thought to be adequate, they may be used to analyze experimental data and obtain estimates of such interesting quantities as electrode charge-transfer reaction rate. Many of the simpler solutions can be represented exactly or approximately by an equivalent circuit, but some yield only a complicated expression for $Z(\omega)$ which cannot always be so represented in a useful manner.

Solutions for unbiased, flat-band situations (i.e., where there are no intrinsic space charge layers at the boundaries) are simplest, and only these will be discussed below except when otherwise noted. We shall present a brief discussion of the supported situation, primarily appropriate for liquids and mixed conductors, and devote more space to results for unsupported materials, since most solid electrolytes involve unsupported ionic conduction under conditions of primary interest. For simplicity, theoretical results for flux, currents, impedances, and other circuit elements will be given in specific form, per unit of electrode area A_c, so this area will not appear directly in the formulas.

The possible behaviors of an electrode–electrolyte interface are variously discussed in the literature in terms of polarizability, blocking or nonblocking character, and reversibility, with usage differing somewhat from one author to another. For clarity and precision we shall use the term *polarizability* to denote the electrical behavior of the electrode–electrolyte interface and the terms *blocking* (or *nonblocking*) and *reversibility* to describe the electrochemical character of the interface. An electrode–electrolyte interface is nonpolarizable if the potential drop across the interface is independent of the current through the interface. It is partially polarizable if the interfacial potential difference is dependent on the current and completely polarizable if it completely prohibits the flow of (faradic) current. An interface is blocking with respect to a given charge-carrying species in the electrolyte if that species cannot cross the interface or exchange charge (in the form of electrons) with the electrode; otherwise it is nonblocking with respect to the given species. A nonblocking interface is generally thermodynamically reversible since, in thermal equilibrium, the electrochemical potential of the species involved in the interfacial charge transfer will obey an equilibrium relation. The interface is kinetically reversible if the rate of the electrode charge-transfer reaction is rapid enough that the equilibrium relation is maintained in the immediate vicinity of the interface as current passes through the system.

The polarization of an electrode–electrolyte interface can result either from the slowness of the electrode reaction, as in the case of nonblocking but kinetically nonreversible electrodes, or from any factor which limits the transport of any of the species participating in the electrode reaction, for example, slow diffusion of the reactant or product species away from the interface or the generation or consumption of one of the species by a slow chemical reaction in the electrolyte.

We shall be concerned primarily with the behavior of ionic charge carriers at the interface, and the electrode–electrolyte combinations to be encountered will fall into two general groups: parent-atom electrodes and redox electrodes. In parent-atom electrodes, charge can cross the interface in ionic form. Electrodes of this type include parent-metal electrodes such as Ag in the solid state cell Ag|AgCl|Ag, in which the electrode serves both as a source of ions and as an electronic conductor, and parent-nonmetal electrodes, as in the cell $Br_2(Pt)|AgBr|Br_2(Pt)$, in which an inert metal phase must be present to serve as the electronic conductor. In redox electrodes, charge crosses the interface in the form of electrons and the reaction may be written in the form

$$\text{Red}^{(z-n_e)+} \underset{k_b}{\overset{k_f}{\rightleftharpoons}} \text{Ox}^{z+} + n_e e^- \qquad (26)$$

where k_f and k_b are forward and reverse reaction rate constants, respectively. Both Ox and Red are usually soluble in the electrolyte, but if $z = n_e$, the Red species is uncharged, and if it is a gas, it may evolve at an electrode and/or diffuse into the electrode, especially if the electrode is somewhat porous. The admittance behavior of more complex electrode reactions (in aqueous electrolytes) than those mentioned above has recently been discussed by Seralathan and de Levie [1987] and is not considered herein.

Finally, while still dealing with interface effects, it is worth stating one of the most important equations of reaction rate electrochemistry, the Butler–Volmer equation (see Vetter [1967], Franceschetti [1982]). Written in terms of flux for a simple redox reaction, it is

$$J \equiv (I/nF) = \left\{ k_f^0 c_{\text{Red}}^H \exp\left[(nF/RT)\alpha\eta_{MH} \right] \right.$$
$$\left. - k_b^0 c_{\text{Ox}}^H \exp\left[-(nF/RT)(1-\alpha)\eta_{MH} \right] \right\} \qquad (27)$$

Here n is the number of moles of electrons involved in the reaction; the rate constants have been assumed to be thermally activated; and k_f^0 and k_b^0 are potential-independent rate constant parameters. The potential-dependent concentrations c_{Red} and c_{Ox} are evaluated at their points of closest approach to the electrode, taken here as the outer Helmholtz plane. Further, α is here a dimensionless symmetry factor often assumed to be 0.5 (the symmetrical barrier case), and η_{MH} is the charge transfer overvoltage effective in driving the reaction away from equilibrium (for $\eta_{MH} = 0$, $J = 0$). The Butler–Volmer equation is usually a good approximation for both biased and unbiased conditions.

2.2.3.2 Supported Situations

Half-Cells. The concept of a supported electrolyte has proven quite valuable in solution electrochemistry by allowing great theoretical simplification at (usually) only a small cost in accuracy. The several (often implicit) assumptions made in treating the electrolyte in a given cell as supported, however, deserve careful attention as they generally do not apply in the case of solid state electrochemical systems. It should also be noted that it is usually possible in solution electrochemistry to use a large, essentially kinetically reversible counterelectrode so that all but a negligible fraction of the applied potential difference falls across the electrode–electrolyte interface of interest. In its simplest form, the supported approach assumes that all the potential difference in the system falls across the compact double layer—approximately one solvent molecule diameter in thickness—at this electrode, and the approach of the electroactive species to the boundary of the compact layer, the outer Helmholtz plane, occurs purely by diffusion. Corrections for the buildup of space charge near the interface (the diffuse double layer) and the ohmic drop across the electrolyte are then made in piecewise fashion as needed. In solid state electrochemistry, one is usually concerned with measurements on a cell with an unsupported electrolyte and identical plane parallel electrodes. The impedance of such a cell will be twice that of its two half-cells. Although most solid electrolytes are essentially unsupported for temperatures of interest, Archer and Armstrong [1980] have suggested that when a solid electrolyte contains immobile anions and two different species of cations of similar mobilities, if one of the ionic species is present in much higher bulk concentration than the other, it can act as support for the other species of mobile ion—a supported case. Also, mixed electronic–ionic conduction can sometimes lead to supported conditions. Thus the supported situation may even be of some direct interest for solid electrolytes.

Although it would be convenient to give supported results for a full cell in order to allow direct comparison with unsupported results, this is generally impractical if one wishes to include properly diffusion effects, which can spread through the entire cell at low frequencies. The usual solutions for the supported case are for a half-cell, apply only at sufficiently high frequencies, and do not involve the electrode separation l except when the bulk resistance is included in the circuit. But for an actual full cell we have seen in Section 2.2.2.2 that at sufficiently low frequencies the diffusion impedance becomes extensive and does involve l. It is thus incorrect to combine two supported half-cell impedances appropriate only at higher frequencies to obtain that for a full cell for all frequencies. Although a solution for a symmetrical full cell has been obtained by Sluyters [1963], it is very complicated, is not given in complex impedance form, and has not been reduced to an equivalent circuit. Further discussion and extension of this work appears in Macdonald [1971b].

Let us now consider only infinitesimal deviations from zero bias equilibrium conditions and assume that the equilibrium distribution of all charges is constant throughout the material, so for equilibrium there is no polarization and electroneutrality applies everywhere. Solution of Fick's laws of diffusion under supported small-signal ac conditions for a simple one-step reaction (single reacting charged species) (Randles [1947], Sluyters-Rehbach and Sluyters [1970], Armstrong et al. [1978], Franceschetti [1982]) leads to the equivalent circuit of Figure 2.2.5. This circuit may be taken to apply to a half-cell of infinite extent to the right of the electrode. It thus does not include any bulk or solution resistance R_∞ which would depend on the finite extent of an actual cell. It does include an infinite-length

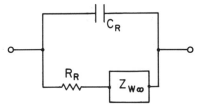

FIGURE 2.2.5. Equivalent circuit for a single electrode and its interface under supported conditions.

Warburg impedance, the charge-transfer reaction resistance R_R, and the capacitance associated with it, C_R, the diffuse double-layer capacitance, sometimes denoted C_{dl}. This circuit, with the addition of a bulk or solution resistance in series with it, is customarily known as the *Randles circuit*. Appreciable discussion of the circuit from an electrochemical viewpoint appears in Macdonald [1971a].

Full-Cell Results. We can generalize the circuit of Figure 2.2.5 to a full-cell situation with identical electrodes as long as Z_W is well approximated by $Z_{W\infty}$, that is, as long as the diffusion length is much smaller than l. One then obtains two contributions to the impedance of the Figure 2.2.5 type, one associated with each electrode. Since the electrodes are taken identical, the two intensive impedances are identical and may be combined to yield a result of twice the individual impedances. When we additionally add a geometric capacitance $C_g \equiv C_\infty$ and the bulk or solution resistance R_∞, we obtain the circuit of Figure 2.2.6.

Although we will always refer to actual circuit elements in an equivalent circuit, we shall, for simplicity, give expressions for these elements per unit area. We shall not, however, usually distinguish between a quantity and its per-unit-area specific form, so, for example, a capacitance-unit area will still be referred to as a *capacitance*. If the separation of the electrodes is l, the geometrical capacitance (per unit area) is given by

$$C_\infty = \epsilon\epsilon_0/l \tag{28}$$

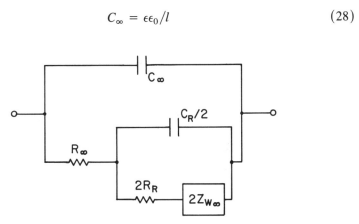

FIGURE 2.2.6. Equivalent circuit for a full cell with two identical electrodes under supported conditions. This circuit only applies when $Z_W \cong Z_{W\infty}$.

where ϵ is the effective dielectric constant of the electrolyte. It is customary to omit C_∞ in supported situations since IS measurements on liquids rarely extend to high enough frequencies for it to affect the overall impedance appreciably. This is not always the case for solid electrolytes where impedance contributions from other elements may be high, especially at low temperatures.

Now ionic conduction in a full cell actually occurs in a region of length $l_{eff} = l - 2l_H$, where l_H is the effective thickness of the inner region, next to the electrode, into which ions cannot fully penetrate. Because of the finite size of ions, the minimum steric but not necessarily electrical value of l_H is an ionic radius. Except for unrealistically thin cells, the distinction between l_{eff} and l is not important for most circuit elements and will usually be neglected hereafter. It should be mentioned, however, that in the study of thin (sometimes monomolecular) membranes in the biological field, using high-molarity liquid electrolyte electrodes, the distinction may be important. We may now write the expression for R_∞ as

$$R_\infty = (l/F) \left(\sum_{i=1}^{m} z_i \mu_i c_i^0 \right)^{-1} \tag{29}$$

where z_i and μ_i are the valence numbers and mobilities of the m charged species of bulk concentration c_i^0 present in the electrolyte. We have followed past work in assigning the ∞ subscripts to C_∞ and R_∞ herein since these elements lead to the semicircle in the complex impedance plane which occurs at higher frequencies than do any other impedance plane structures. This semicircle peaks at $\omega = \tau_D^{-1}$, where $\tau_D = R_\infty C_\infty$ is the dielectric relaxation time of the material, an intensive quantity.

An important quantity in solid and liquid electrolytes is the Debye length L_D, given by

$$L_D = \left[(\epsilon\epsilon_0 RT)^{-1} F^2 \sum_{i=1}^{m'} z_i^2 c_i^0 \right]^{-1/2} \tag{30}$$

where the sum includes mobile charge species only. The Debye length is a measure of the distance in the electrolyte over which a small perturbation in potential or electric field decays. Such a perturbation creates a region of space charge where electroneutrality no longer holds. This region extends only over a few Debye lengths.

The Gouy–Chapman diffuse double-layer differential capacitance C_R which is associated with the charge-transfer reaction resistance R_R appearing in Figure 2.2.6 is given by

$$C_R = \epsilon\epsilon_0/L_D \tag{31}$$

in the absence of bias. An expression taking dc bias into account appears in Macdonald [1954], and a further generalization taking finite-ion-size effects into account as well by means of a lattice gas treatment is presented in Macdonald, Franceschetti, and Lehnen [1980] (see also Franceschetti [1982]).

The supported electrolyte half-cell reaction resistance R_R may be written in the general form

$$R_R = (RT/n^2F^2)(k_i^0 c_i^0)^{-1} \tag{32}$$

where k_i^0 and c_i^0 depend on the specific type of reaction considered. For the parent metal electrode situation, $M | M^{z+}$, where M denotes the metal electrode, $k_i^0 = k_b^0$, a potential-independent Butler–Volmer-type reaction rate parameter, and $c_i^0 = c_{M+}^0$, the bulk equilibrium concentration of the reacting cation. Similarly for a redox situation one finds (Vetter [1967], Gabrielli [1981] $k_i^0 c_i^0 = (I_0/nF) = k_f^0 c_{Red}^0 = k_b^0 c_{Ox}^0$, where I_0 is the exchange current, a measure of the rates at which oxidation and reduction processes occur in equilibrium. It is the common magnitude of the equal and opposite electrical currents associated with oxidation and with reduction (no net current in equilibrium). A more complicated expression for R_R and for infinite-length Warburg impedance under steady state conditions where the dc bias is nonzero has been given by Sluyters-Rehbach and Sluyters [1970].

Diffusion Effects. Next consider small-signal unbiased diffusion effects. We initially discuss the parent-ion equal electrode situation for simplicity. The problem with applying the circuit of Figure 2.2.6 to the supported full-cell situation at low frequencies is that C_R remains an intensive quantity associated with an interface region but $2Z_{W\infty} \to Z_D$, some new diffusion impedance, and becomes extensive at low enough frequencies. It is then not correct to continue to allow the element $C_R/2$ to bridge the series combination of $2R_\infty$ and $2Z_{W\infty}$. In fact, at sufficiently low frequencies where the diffusion effects are extensive, there can only be a single diffusion impedance Z_D for the entire cell. At sufficiently high frequencies Z_D will be well approximated by $Z_{W\infty}$, but the full cell then exhibits an effective diffusion impedance of $2Z_{W\infty}$, not $Z_{W\infty}$, as shown in Figure 2.2.6. There is thus a transition region in frequency as the frequency is decreased in which the effective diffusion impedance goes from $2Z_{W\infty}$ to $Z_D = Z_W$, *not* $2Z_W$, and the $C_R/2$ connection shown in Figure 2.2.6 is also no longer entirely applicable. Let us therefore first consider sufficiently low frequencies that the extensive Z_D is present and the admittance of $C_R/2$ is negligible. We can then consider Z_D alone and its transition to $Z_{W\infty}$. We shall start with the expressions (Franceschetti [1981]) appropriate for Z_D in the parent-metal electrode and redox cases.

In the parent-metal electrode case take the diffusion coefficient of the metal ion M^{z+} as D_M. Then one finds

$$Z_D = Z_W = \left[\frac{RT}{(zF)^2}\right]\left[\frac{l}{D_M c_{M+}^0}\right]\left[\frac{\tanh \sqrt{j\omega l^2/D_M}}{\sqrt{j\omega l^2/D_M}}\right] \tag{33}$$

which should be compared with the result in Eq. (10). Now when $(\omega l^2/D_M) \gg 3$, $Z_W \to Z_{W\infty}$ and one obtains

$$Z_{W\infty} = \frac{RT/(zF)^2}{(c_{M+}^0)(j\omega D_M)^{1/2}} \quad (34)$$

The situation is somewhat more complicated in the redox case because of the presence of the two charged species in the electrolyte [unless $(z - n_e) = 0$]. Then one finds

$$Z_D = Z_{W,Ox} + Z_{W,Red}$$

where

$$Z_{W,Ox} = \left[\frac{RT}{(nF)^2}\right]\left[\frac{l}{D_{Ox} c_{Ox}^0}\right]\left[\frac{\tanh \sqrt{j\omega l^2/D_{Ox}}}{\sqrt{j\omega l^2/D_{Ox}}}\right] \quad (35)$$

$$Z_{W,Red} = \left[\frac{RT}{(nF)^2}\right]\left[\frac{l}{D_{Red} c_{Red}^0}\right]\left[\frac{\tanh \sqrt{j\omega l^2/D_{Red}}}{\sqrt{j\omega l^2/D_{Red}}}\right] \quad (36)$$

and D_{Ox} and D_{Red} are the relevant diffusion coefficients and all concentrations are those in the bulk. When both tanh terms are well approximated by unity, the expression for Z_D reduces to

$$Z_D = Z_{W\infty,Ox} + Z_{W\infty,Red}$$

$$\equiv Z_{W\infty} = \left[\frac{RT}{(nF)^2}\right]\left[(D_{Ox}(c_{Ox}^0)^2)^{-1/2} + (D_{Red}(c_{Red}^0)^2)^{-1/2}\right](j\omega)^{-1/2} \quad (37)$$

the classical result (Armstrong et al. [1978], Gabrielli [1981]). The above results show that we may expect to find two finite-length Warburgs in series (and generally displaced in frequency) at sufficiently low frequencies. It is worth again emphasizing that although it will be the $Z_{W\infty}$ of Eq. (37) which appears in the half-cell circuit, it is $2Z_{W\infty}$ which appears for the full cell under supported conditions at sufficiently high frequencies.

We have attempted to give supported results in a form appropriate for comparison with unsupported ones by considering full-cell conditions. The transition problems discussed above only occur for unstirred (liquid) electrolytes or for solid electrolytes. When a stirred solution or rotating electrode with laminar flow is employed, the l which appears in Z_D and Z_W expressions is replaced by δ_N, where δ_N is the thickness of the Nernst diffusion layer. It decreases as the frequency of rotation of a rotating electrode increases and the experiment is always carried out for conditions where $\delta_N \ll l$.

2.2.3.3 *Unsupported Situations: Theoretical Models*

Introduction. We shall discuss results for unsupported situations under two categories: (1) those that follow directly or indirectly from exact solutions of the small-signal differential equations of charge motion in the material–electrode system; and (2) those which largely arise from empirical analysis of data and often use such ubiquitous distributed elements as CPEs. The first category deals with more idealized situations than the second but generally leads to more detailed results and to more specific relations between macroscopic equivalent circuit elements and microscopic processes occurring in the system. At the present early stage of theoretical analysis of real systems, both approaches have important roles to play.

Most theories of charge transport and interfacial charge transfer in unsupported situations involve a model which assumes a homogeneous material, for example, a single crystal. Here we shall initially discuss the electrical response following from the application of a small-signal ac potential difference to homogeneous materials without applied dc bias or built-in Frenkel space charge layers and with identical plane parallel electrodes—the idealized full-cell situation. Theoretical results are only available so far for conditions where there may be a single species of mobile positive charge and a single species of mobile negative charge present with electroneutrality in the bulk. Results for polycrystalline materials and for homogeneous ones with Frenkel layers or applied dc bias will be discussed later on.

We shall start with a discussion of the exact results obtained from the solution of the most general model yet considered, but one which is still appreciably idealized (Macdonald and Franceschetti [1978]). Then work relaxing some of the idealizations will be discussed. Some of the present results have been included in the solid electrolyte reviews of Archer and Armstrong [1980] and by Franceschetti [1982]. The model of Macdonald and Franceschetti involves mobile positive and negative charges which may arise from three sources: the partial or full dissocation of (a) neutral intrinsic centers, (b) neutral donor centers, and (c) neutral acceptor centers. The model is general enough to include disordered sublattice materials and single crystals with Schottky or Frenkel disorder. Arbitrary amounts of generation–recombination (G/R) are allowed. After dissociation of a neutral center, the resulting positive and negative charges are taken to have arbitrary mobilities; so, for example, a donor center might dissociate to yield an immobile positive charge and a mobile negative charge. We shall denote the mobility ratio for negative and positive charges as $\pi_m \equiv \mu_n / \mu_p$. Although the present model also allows arbitrary valence numbers for the mobile charged species, we shall primarily restrict attention here to the usual uni-univalent case.

Boundary Conditions: Adsorption–Reaction Effects. The Macdonald–Franceschetti model involves relatively general boundary conditions at the electrodes and so includes the possibility of charge transfer reactions and specific adsorption. Because of its generality, however, the model prediction for $Z_t(\omega)$ is very complicated and, in general, cannot be well represented by even a complicated equivalent circuit. The $Z_t(\omega)$ expression, may, however, be used directly in CNLS fitting. Here, for simplicity, we shall consider only those specific situations where an approximate equivalent circuit is applicable. Idealizations involved in the model include the usual assumption of diffusion coefficients independent of field and position, the use of the simplified Chang–Jaffé [1952] boundary conditions, and the omission of all inner layer and finite-ion-size effects. Some rectification of the latter two idealizations will be discussed later.

The Chang–Jaffé boundary conditions involve the physical assumption that the current arising from the reaction of a charge carrier of a given species is proportional to the excess concentration of that species at the interface, that is, for, say, a negatively charged species,

$$I_n = -z_n e k_n (n - n^0)$$ (38)

where z_n is the valence number of the charge carrier, k_n is a reaction rate parameter at the reaction plane, and n^0 is the bulk concentration of the species. The Chang–Jaffé conditions, as compared to the Butler–Volmer equation, are unrealistic in two important respects: there is a complete neglect of the finite size of the charge carriers (i.e., the compact double layer), and it is assumed that the charge transfer rate does not depend at all on the local concentration of the electrode reaction products. These deficiencies are not, however, nearly as limiting as one might at first expect. The neglect of the compact double layer introduces only a small error (which becomes zero at zero frequency) in many solid state situations (Macdonald [1974b], Franceschetti and Macdonald [1977]). The accumulation of the electrode reaction product can be neglected when (a) the product species is a metal atom which is rapidly incorporated into a parent-metal electrode, (b) the product is a gas atom which equilibrates very rapidly with the ambient atmosphere, or (c) the product species is soluble in the electrolyte or electrode and diffuses away from the interface very quickly. If the accumulation of the product species is not eliminated by one of these processes, but the transport of the product is governed by diffusion and therefore is independent of the electric field, it may be incorporated into the Chang–Jaffé boundary condition through the artifice of a complex, frequency-dependent reaction rate constant as described below.

It proves convenient in the theoretical work to use the dimensionless Chang–Jaffé rate parameters

$$\rho_n = (l/2)(k_n/D_n)$$ (39)

and

$$\rho_p = (l/2)(k_p/D_p)$$ (40)

where the k's are effective rate constants and the D's the diffusion coefficients of the negative and positive species. These parameters have in some previous work been given in terms of the alternate equivalent quantities $\rho_n = \rho_2 = r_2/2 = r_n/2$ and $\rho_p = \rho_1 = r_1/2 = r_p/2$. Clearly when $\rho_n = 0$ the electrode is completely blocking for the negative species and for $\rho_n = \infty$ it is completely nonblocking and nonpolarized.

Now since to a good approximation specific adsorption of an ion at an electrode and then a reaction of the adsorbed ion to form a neutral species occur at very nearly the same point in space, one might expect that these sequential interface processes would be largely decoupled from bulk and double-layer effects which occur elsewhere in the system. Some time ago Lànyi [1975] introduced the concept of frequency-dependent complex rate constants, and they have been found very useful in allowing reaction–adsorption effects to be included in a very simple way (Macdonald [1976a], Franceschetti and Macdonald [1977]). In essence, if a $Z_t(\omega)$ solution has been found for a certain situation involving the presence of real, frequency-independent ρ_n and ρ_p boundary parameters, one only needs to change them to complex frequency-dependent quantities to automatically include adsorption effects. No other parts of the solution are affected. As an example, suppose that a negative carrier is adsorbed and the adsorbed species then reacts to form a neutral species whose concentration remains, or is held, constant. One finds that the real ρ_n originally present in the solution need only be replaced by

$$\rho_n = \frac{\rho_{n0} + j(\omega\tau_D)\xi_{na}\rho_{n\infty}}{1 + j(\omega\tau_D)\xi_{na}} \tag{41}$$

where ρ_{n0} and $\rho_{n\infty}$ are the $\omega \to 0$ and $\omega \to \infty$ limits of ρ_n: $\xi_{na} \equiv \tau_{na}/\tau_D$; and τ_{na} is the adsorption relaxation time. In this case $\rho_{n\infty}$ is the rate constant for the first step in the adsorption–reaction sequence in which charge is exchanged between the electrolyte and the adsorbed layer, and ρ_{n0} is a function of both $\rho_{n\infty}$ and the rate constant for the second step in the process, in which charge is exchanged between the adsorbed layer and the electrode. For the case of pure adsorption, ρ_{n0} is zero and ρ_n becomes zero at $\omega \to 0$. In the limit in which the second (reaction) step is much faster than the initial (adsorption) step, the adsorbed layer becomes inconsequential, $\tau_{na} \to 0$, $\rho_{n0} \to \rho_{n\infty}$, and $\rho_{n\infty}$ becomes real and frequency-independent. The quantities ρ_{n0} and $\rho_{n\infty}$ or, equivalently, k_{n0} and $k_{n\infty}$, may be expressed in terms of partial derivatives with respect to various surface concentrations of the small-signal boundary conditions written in terms of current (Franceschetti and Macdonald [1977]). Derivative definitions of this form which depend on Taylor series expansions, appropriate for small-signal conditions, were developed earlier by Armstrong and Henderson [1972] for example. See also Armstrong et al. [1978].

The rate-limiting diffusion of an electrode reaction can also be incorporated into the Chang–Jaffé boundary conditions by a similar approach. In this case, assuming, for example, diffusion through a semiinfinite electrode, the result obtained is

$$\rho_n = \frac{\rho_{n\infty}\sqrt{i\omega D}}{\rho_n' + \sqrt{i\omega D}} \tag{42}$$

where, as before, $\rho_{n\infty}$ is the $\omega \to \infty$ limit of ρ_n, ρ_n' is a rate parameter for the inverse electrode reaction, and D is the diffusion coefficient of the reaction product. For sufficiently large D, diffusion becomes undetectable and $\rho_n = \rho_{n\infty}$. Various adsorption–reaction–diffusion sequences have been considered by Franceschetti and Macdonald ([1979c, 1982]) and Franceschetti ([1982, 1984]).

Dc Response. Before passing to the uni-univalent case which we will consider in detail, let us consider the full dc resistance of the system for arbitrary valence numbers, but only for two (or possibly one) species of mobile charge with equilibrium bulk concentrations n^0 and p^0, valence numbers z_n and z_p, and electrical mobilities μ_n and μ_p. Then electroneutrality in the bulk leads to $z_n n^0 = z_p p^0$. The bulk conductance G_∞ may be expressed as

$$G_\infty = R_\infty^{-1} = G_{\infty n} + G_{\infty p} \tag{43}$$

where

$$G_{\infty n} = (F/l)\,(z_n \mu_n n^0) \tag{44}$$

and

$$G_{\infty p} = (F/l)\,(z_p \mu_p p^0) \tag{45}$$

Let us further define the conductivity fractions (or bulk transport numbers) $\epsilon_n \equiv G_{\infty n}/G_\infty$ and $\epsilon_p \equiv G_{\infty p}/G_\infty$. These quantities may be written in the simple forms

$$\epsilon_n = \left(1 + \pi_m^{-1}\right)^{-1} \tag{46}$$

and

$$\epsilon_p = \left(1 + \pi_m\right)^{-1} \tag{47}$$

under intrinsic conditions, the only case to be considered in detail here. Finally, define the Debye length when only one species of charge is mobile as L_{D1}. The important quantities $M \equiv (l/2)/L_D$ and $M_1 \equiv (l/2)/L_{D1}$ then measure the number of Debye lengths in a half-cell (half a symmetrical cell of full electrode separation l). In the present case where L_D and M refer to a single species of positive and a single species of negative charge mobile, $L_{D1} = \sqrt{2}L_D$.

Let us (apparently arbitrarily) now define the small-signal half-cell adsorption–reaction impedances associated with the positively and negatively charged species as

$$Z_{Rn} \equiv RT/z_n^2 F^2 k_n n^0 \tag{48}$$

and

$$Z_{Rp} \equiv RT/z_p^2 F^2 k_p p^0 \tag{49}$$

where the k's may be complex. Using the Einstein relation $D_j = (RT/F)\,(\mu_j/z_j)$ for the $j = n$ and p species, one readily finds that ρ_j may be written as

$(1/2)(F/RT)(z_j k_j/\mu_j)$, so if ρ_j is complex, so is k_j. Note that the $\omega \to 0$ limits of Z_{Rn} and Z_{Rp} are

$$R_{\theta n} \equiv RT/z_n^2 F^2 k_{n0} n^0 \qquad (50)$$

and

$$R_{\theta p} \equiv RT/z_p^2 F^2 k_{p0} p^0 \qquad (51)$$

where k_{n0} and k_{p0} are related as above to ρ_{n0} and ρ_{p0}.

Now it is often useful to consider normalized quantities in theoretical analysis or even in an equivalent circuit or 3-D plot. We shall, when desirable, normalize impedances with the bulk resistance R_∞, so $Z_N \equiv Z/R_\infty$, and capacitances with C_∞, so $C_N \equiv C/C_\infty$. The normalized expressions for some of the circuit elements defined above simplify considerably and are

$$G_{\infty n N} \equiv G_{\infty n}/G_\infty = \epsilon_n \qquad (52)$$

$$G_{\infty p N} \equiv G_{\infty p}/G_\infty = \epsilon_p \qquad (53)$$

$$G_{\theta n N} \equiv R_\infty/R_{\theta n} = 2\epsilon_n \rho_{n0} \qquad (54)$$

and

$$G_{\theta p N} \equiv R_\infty/R_{\theta p} = 2\epsilon_p \rho_{p0} \qquad (55)$$

We are now finally in a good position to write down the expression for the full-cell complete dc resistance following from the present model, $R_D \equiv Z_t(\omega \to 0)$. We write it here as R_D or R_{DN} rather than R_0 to agree with earlier usage. The exact R_D result applies for arbitrary valences for the mobile charged species, arbitrary mobilities, intrinsic, extrinsic, or intrinsic and extrinsic conduction, and any dissociation–recombination conditions. In unnormalized form it is just

$$R_D \equiv G_D^{-1} = (G_n + G_p)^{-1} \qquad (56)$$

where

$$G_n \equiv (R_{\infty n} + 2R_{\theta n})^{-1} \qquad (57)$$

and

$$G_p \equiv (R_{\infty p} + 2R_{\theta p})^{-1} \qquad (58)$$

These results show that the total dc conductance is made up of a branch G_n, involving negative charge carrier effects only, in parallel with a similar branch involving only positive carrier effects. Each individual branch involves a bulk resistive contribution and two equal adsorption–reaction resistances, one associated with each electrode. The expression for R_D in normalized form, R_{DN} (see, e.g., Franceschetti and Macdonald [1977]), is even simpler, namely,

$$R_{DN} = \left(\frac{\epsilon_n}{1 + \rho_{n0}^{-1}} + \frac{\epsilon_p}{1 + \rho_{p0}^{-1}} \right)^{-1} \tag{59}$$

Note that when $\rho_{n0} = \rho_{p0} = \infty$, one obtains $R_{DN} = (\epsilon_n + \epsilon_p)^{-1} \equiv 1$; so $R_D = R_\infty$, ohmic behavior and thus not very interesting. Of course in the completely blocking $\rho_{n0} = \rho_{p0} = 0$ case, $R_{DN} = \infty$. To set a scale, it is interesting to note that when $\rho_{n0} = \rho_{p0} = 1$, $R_{DN} = 2$ and adsorption–reaction effects have contributed an additional R_∞ resistance to R_D. In general, when $\rho_{p0} = \rho_{n0} \equiv \rho_e$, then $R_{DN} = 1 + \rho_e^{-1}$, a result entirely independent of mobilities and π_m, except indirectly through R_∞ and ρ_e themselves.

Adsorption–Reaction and Reaction–Diffusion Predictions. Next, in order to investigate adsorption and reaction effects more fully, let us consider Z_{Rn} and Z_{Rp} in normalized form. It is straightforward to show that

$$Z_{RnN} = (2\epsilon_n \rho_n)^{-1} = \rho_{n0} R_{\theta nN} / \rho_n \tag{60}$$

and

$$Z_{RpN} = (2\epsilon_p \rho_p)^{-1} = \rho_{p0} R_{\theta pN} / \rho_p \tag{61}$$

If we now substitute the complex ρ_n from Eq. (41) into Eq. (60) and a similar expression for ρ_p into Eq. (61), we readily find that the resulting impedances each lead to a simple ladder network whose hierarchical form is consonant with the sequential processes: adsorption then reaction. But for the full cell there are two identical interface impedances in series. The circuit for a half-cell with total impedance Z_{Rn} is shown in Figure 2.2.7a. The full-cell impedance is just $2Z_{Rn}$. The normalized elements of Figure 2.2.7a are readily found to be given by

$$R_{RnN} = (\epsilon_n \rho_{n\infty})^{-1} \tag{62}$$

$$R_{AnN} = \rho_{nm} / \epsilon_n \rho_{n0} \rho_{n\infty} \tag{63}$$

and

$$C_{AnN} = \xi_n \epsilon_n \rho_{n\infty}^2 / \rho_{nm} \tag{64}$$

where $\rho_{nm} \equiv \rho_{n\infty} - \rho_{n0}$. All the elements in the circuit of Figure 2.2.7 are intensive, as they should be for interface effects. The normalized dc resistance of the circuit is just

$$2Z_{RnN}(\omega \to 0) = R_{RnN} + R_{AnN} = (\epsilon_n \rho_{n0})^{-1} \equiv 2R_{\theta nN} \tag{65}$$

which is as expected. Note that since ρ_{nm} may be either positive or negative, R_{An} and C_{An}, specific adsorption elements, have the same sign and also may be positive or negative, in agreement with earlier discussion. In the absence of the adsorption

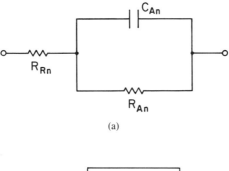

(a)

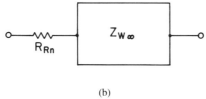

(b)

FIGURE 2.2.7. (a) Equivalent circuit for two identical simple electrode adsorption–reaction processes in series, one at each electrode, with negative charge carriers reacting. Unsupported conditions. (b) Equivalent circuit for two identical reaction–diffusion processes with negative charge carriers reacting. Unsupported conditions.

step, $2Z_{Rn} = R_{Rn}$, the reaction resistance (for two electrodes), since R_{An} is then zero and C_{An} infinite. See Fig. 4.3.25 for some of the complex plane shapes which follow from the present approach.

In like manner, if we substitute the complex ρ_n from Eq. (42) into Eq. (60), we obtain the circuit shown in Figure 2.2.7b, appropriate for a reaction–diffusion sequence without an intermediate adsorption stage. Here R_{RnN} is given by Eq. (62) as before and

$$Z_{WN} = \rho'_n / \epsilon_n \rho_{n\infty} \sqrt{i\omega D} \tag{66}$$

Theoretical Results for Various Cases of Interest. Thus far we have only considered some exact $\omega \rightarrow 0$ results and typical adsorption–reaction interface frequency response for a half-cell or full cell. Let us now turn to further predictions of the complete full-cell model (Macdonald and Franceschetti [1978]), predictions derived from its specific analytical results in several simplified cases and from a large amount of CNLS fitting of various equivalent circuits to the exact model predictions (see, e.g., Franceschetti and Macdonald [1977], Macdonald, Franceschetti, and Meaudre [1977], Macdonald and Franceschetti [1979a], and Macdonald and Hull [1984]). For simplicity, we consider only the uni-univalent case ($z_n = z_p = 1$), intrinsic conduction, and $M \gg 1$. The latter condition excludes the behavior of very thin layers and membranes, but their response has been discussed in the literature. Let us define cases of interest by their [ρ_p, ρ_n, π_m] values. Actual values of ρ_p and ρ_n cited in this way will always be real, but when the symbols are used, they may include complex cases. Because the electrodes are taken identical, ρ_n and ρ_p values apply to both electrodes.

Little has been done on the $[\rho_e, \rho_e, \pi_m]$ case, that where the normalized reaction rates (but not necessarily k_n and k_p) are equal and nonzero. Although this is a situation of small experimental interest except for $\rho_e \cong 0$, a formal expression for its admittance has been given (Macdonald and Franceschetti [1978]) but does not lead to a useful approximate equivalent circuit representation except when $\pi_m = 1$. When $\rho_e = 0$ as well, there is no finite-length Warburg present. Because of its complexity, model predictions for the general $[\rho_p, \rho_n, \pi_m]$ case with both ρ_p and ρ_n values nonzero and noninfinite have been little explored, although it has been found that for a nonzero ρ_n value say, and even for $\pi_m = 1$, as ρ_p increases from zero toward ρ_n a finite-length Warburg arc which appears in the impedance plane rapidly decreases in size (Macdonald [1975]). Note that when a diffusion arc is present, one finds (Macdonald [1974b]) that $C_p(\omega)$, the total frequency-dependent parallel capacitance associated with Y, exhibits $\omega^{-3/2}$ and/or $\omega^{-1/2}$ dependence, quite different from ω^{-2} simple Debye response, yet frequently observed experimentally.

The situation of most experimental interest, especially for solid electrolytes, is defined by $[0, \rho_n, \pi_m]$ or, equivalently, $[\rho_p, 0, \pi_m]$. Thus only one species of charge carrier discharges, but both positive and negative ones may be mobile. The equivalent circuit we believe to be most appropriate in this case is presented in Figure 2.2.8. First, we see the usual elements C_∞, R_∞, R_R, C_A, and R_A already discussed. For this $[0, \rho_n, \pi_m]$ situation, R_R, R_A, and C_A are given by just the unnormalized forms of R_{RnN}, R_{AnN}, and C_{AnN} presented in Eqs. (62)–(64). The additional elements Z_{De}, Z_W, C_R, and $R_{R\infty}$ which appear in the circuit require discussion. But first it should be emphasized that much theoretical analysis and fitting of theory to different equivalent circuits makes it quite clear that the hierarchical ladder network form of this circuit is far more appropriate than either the series Voigt or the parallel Maxwell forms. The ladder network, which leads to a con-

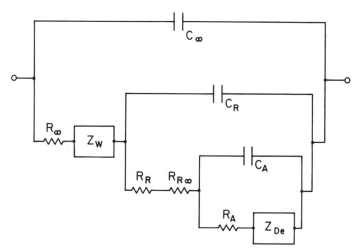

FIGURE 2.2.8. General approximate equivalent circuit (full-cell, unsupported) for the $[0, \rho_n, \pi_m]$ cases applying to a homogeneous liquid or solid material.

tinued fraction expression for the total impedance, ensures that the $R_\infty C_\infty$ arc in the impedance plane will occur at the highest frequencies, followed (with or without overlap) by a $R_R C_R$ arc and then a $R_A C_A$ arc as the frequency decreases. The Voigt circuit, for example, imposes no such requirements. The diffusion arc(s) may actually occur in any frequency range for the present circuit but is usually found at the right of the diagram, the lowest frequency region.

The element Z_{De} is a general diffusion impedance added to account for possible diffusion of uncharged reactants such as oxygen atoms (but not parent-electrode atoms) in the electrodes (for the present full-cell situation it accounts of course for diffusion in both electrodes). When $\rho_n \equiv 0$, it should not appear. In most cases of interest, it will probably be best represented by a finite-length Warburg impedance, so $Z_{De} = Z_{We}$. One might at first be surprised to find Z_{De} in series with the adsorption related resistance R_A rather than in series with C_∞ and R_∞. Although the diffusion process in the electrode does occur after the charge carrier has been transported through the electrolyte, it has only a negligible effect on the potential difference across the electrode, which is essentially zero for a metallic electrode. Rather, it contributes to the interfacial impedance by hindering the discharge of the adsorbed species. If adsorption does not occur to any significant extent then $C_A \to 0$, $R_A \to 0$, and Z_{De} is in series with R_R.

The next new element is a finite-length Warburg impedance Z_W, associated with diffusion of charged particles within the electrolyte. It is primarily present in low-frequency regions where the more mobile or more abundant charged species have time in a half-cycle to rearrange positions so as to screen the less mobile or less abundant charges from the electric field, leaving diffusion as the primary conduction method for such low frequencies. Since this diffusional process occurs in the electrolyte bulk, Z_W is placed in series with R_∞. For the present situation, the diffusion impedance is well approximated by

$$Z_{WN} = Z_W/R_\infty = \pi_m^{-1}\left[\tanh\left(j\Omega H_N^2\right)^{1/2}/\left(j\Omega H_N^2\right)^{1/2}\right] \tag{67}$$

except when $\rho_n = 0$ and $\pi_m = 1$ simultaneously, a situation where $Z_W = 0$. Here $\Omega \equiv \omega\tau_D \equiv \omega R_\infty C_\infty$. The quantity H_N is found to be

$$H_N = (M_e/2)\left(\pi_m^{-1} + 2 + \pi_m\right)^{1/2} \tag{68}$$

We define $M_e \equiv l/2L_{De}$, where L_{De} in the present case is given by

$$L_{De} = \left[\epsilon\epsilon_0 RT/8\pi F^2 q n^0\right]^{1/2} \tag{69}$$

and q takes intrinsic G/R into account and is actually frequency-dependent and complex in the exact theory. Here it will be sufficient to take $q = 1$ when charges of both sign are mobile and usually take $q = 0.5$ when only one species is mobile (see later discussion). Then L_{De} equals either L_D or L_{D1}. Note that when $\pi_m = 1$, $H_N = M_e$. Alternatively, when $\pi_m \to 0$ because $\mu_n \to 0$, there will be no dc path through the circuit and $Z_W \to \infty$.

It should be mentioned that when the product of a reaction at the electrode is neutral, it may possibly diffuse back into the electrolyte as well as into the electrode (Franceschetti [1981]). Then another Z_W appears in series with R_R. How can one distinguish between up to three different Z_W's, all effectively in series with R_∞? By changes in the electrode thickness l_e and the separation of the electrodes l, one should be able to identify a given Z_W arc as arising from diffusion in one or the other region. One can then decide whether a Z_W which depends on l involves charged or uncharged species by changing (if possible) the equilibrium concentration of the neutral species in the electrolyte, either directly or by changing the composition or pressure of the ambient atmosphere. Of course, in most experimental situations only a single Z_W arc appears in the very-low-frequency region (or measurements do not extend to low enough frequencies to show others). As we shall see subsequently, a single Z_W can only arise from neutral-species diffusion if $\mu_p = 0$ and $\mu_n \neq 0$, so $\pi_m = \infty$, the one-mobile case. Similar results appear for $\mu_n = 0$ and $\mu_p \neq 0$.

The next element requiring discussion is C_R, the reaction capacitance, arising from the series combination of equal diffuse double-layer capacitance effects at each electrode. It is usually very well approximated by

$$C_{RN} \equiv C_R/C_\infty = M_e \, \mathrm{ctnh} \, (M_e) - 1 \qquad (70)$$

essentially equal to M_e in the present $M \gg 1$ case, but note that $C_R + C_\infty = C_\infty M_e \, \mathrm{ctnh} \, (M_e)$. When $M_e \gg 1$, the usual situation, this full-cell result is just half of the conventional diffuse double-layer capacitance, an intensive quantity, given in Eq. (31) for the supported half-cell situation.

Now what about the remaining element, $R_{R\infty}$? In the present case, Eq. (59) leads to $R_{DN} = \epsilon_n^{-1}[1 + \rho_{n0}^{-1}] = 1 + \pi_m^{-1} + [\epsilon_n \rho_{n0}]^{-1}$, an exact result. But on omitting Z_{De}, the circuit of Figure 2.2.8 leads to $R_{DN} = R_{\infty N} + Z_{WN0} + (R_{RN} + R_{AN}) + R_{R\infty N} = 1 + \pi_m^{-1} + (\epsilon_n \rho_{n0})^{-1} + R_{R\infty N}$. Thus $R_{R\infty}$ must actually be zero, at least in the $\omega \to 0$ limit, unless expressions for one or more of the other parameters are incorrect in this limit. Macdonald and Hull [1984] found that even when $R_A = 0$ and $R_R = 0$ (taking $\rho_n = \rho_{n\infty} = \infty$) a circuit similar to the present one with the present Z_W could be best fitted to the exact $[0, \infty, \pi_m]$ case with a nonzero $R_{R\infty}$ approximately given by

$$R_{R\infty N} = 2M^{-1}[\pi_m^{-1} - 1] \qquad (71)$$

for $\pi_m \leqslant 1$ and by zero for $\pi_m \geqslant 1$. Thus even in the absence of a normal reaction resistance, CNLS fitting of exact data leads to a nonzero apparent reaction resistance. For large M it is only of importance when π_m is very small (a high-resistance case), since when $R_{R\infty} < 0.01R_\infty$ its effect will be essentially negligible and difficult to resolve even with CNLS fitting. Note also that $R_{R\infty N}$ will always be appreciably smaller than $Z_{WN0} = \pi_m^{-1}$ for $M \gg 1$. Nevertheless, the presence of this element in the circuit and the natural tendency to consider the measured $R_R + R_{R\infty}$ as "the" reaction resistance can lead to incorrect estimates of the rate parameter unless the presence of $R_{R\infty}$ is explicitly recognized (Macdonald and Hull [1984]). The fact that $R_{R\infty}$ should not actually appear at $\omega = 0$, yet is needed in the fitting circuit, is an indication of some inappropriateness in the fitting circuit itself. But for the present it seems the best circuit available.

Let us continue to ignore Z_{De} and investigate two simpler cases. First consider the important completely blocking case $[0, 0, \pi_m]$ where $R_R = \infty$. There is still a Warburg impedance present in general, but it can only contribute to making the normally vertical spur present at low frequencies in the impedance plane and associated with complete blocking show less than vertical behavior over a finite frequency range. But an inadequacy of the present expression for Z_W appears when $\pi_m = 1$; then the exact solution leads to no Z_W but Eq. (67) still yields a nonzero Z_W. The special $[0, 0, 1]$ case must therefore be handled separately until a more complete expression for Z_W is found or unless direct CNLS comparison between data and model predictions is employed.

The Case of Charge of Only a Single-Sign Mobile. The remaining one-mobile case, $[0, \rho_n, \infty]$, is of particular interest for solid electrolytes with only a single-species-of-charge (here negative-charge) mobile. This situation is the most usual one for solid electrolytes, although it should be realized that it is always something of an approximation. At nonzero temperature both positive- and negative-charge species present in a solid material or a fused salt are mobile, although their mobility ratio, $\pi_m \equiv \mu_n/\mu_p$, may be either very large or very small. The relatively immobile species may have so low a mobility at a given temperature that motion of this species is negligible during a half-cycle of the lowest frequency applied. Then the one-mobile approximation will be a good one. Although we have taken ρ_p as zero in the above case designation, its value is immaterial since the positive charges are taken immobile ($\mu_p = 0$) and cannot react at an electrode. In the present case all the ϵ_n's which appear in the defining equations are unity and all π_m^{-1}'s zero. Therefore, as Eq. (67) shows, $Z_W = 0$, and no charged-particle Warburg arc is present, and the only Warburg diffusion response possible must arise from diffusion of neutral particles in the electrodes or the bulk of the material. The exact theoretical results show that in the present case the circuit of Figure 2.2.8 (with $R_{R\infty} = 0$) is completely applicable with all frequency-independent elements given exactly by their values following from the foregoing expressions except that for C_{RN}. When $Z_{De} = 0$ and the time constants are well separated, so that $R_A C_A \gg R_R C_R \gg R_\infty C_\infty$, the circuit of Figure 2.2.8 leads to just three distinct arcs in the Z^* plane.

It is in the partially dissociated one-mobile case that G/R can play a role of some importance (e.g., Macdonald [1953], Macdonald and Franceschetti [1978]). We have already mentioned that the L_{De} which appears in the equation for C_{RN} should usually be taken as L_{D1} ($q = 0.5$) in the one-mobile case. This choice is particularly relevant for fully dissociated charges, such as might arise from the complete ionization of immobile donors. But in the partly dissociated situation, appropriate for intrinsic conduction, G/R can lead to an effective mobility for the immobile charge species (except at dc). Then over some region of frequency as ν

decreases L_{De} changes from L_{D1} to L_D because of the frequency dependence of q. Thus, although the formal expression for C_{RN} given above remains valid for the one-mobile case, one must consider the physical situation to decide whether to use L_{D1} or L_D in M_e. Alternatively, it is more accurate to use the full frequency-dependent expression for q as a part of the definition of C_{RN} in this case. Then CNLS fitting can, in principle, lead to information about the degree of dissociation of intrinsic centers and the associated G/R parameters. Such a procedure would only be justified, however, for excellent data. It was originally thought (Macdonald [1976b]) that G/R might lead to a separate semicircle in the impedance plane, but later work (Macdonald, Franceschetti, and Meaudre [1977]) suggests that it does not for ionic conduction. Thus its effects for the one-mobile case are entirely restricted to C_R only and are relatively small even there. Note that in the two-mobile case, as long as the mobilities are not greatly different, effective mobilization, arising from G/R, of the species with the smaller mobility will still lead to negligible effects.

Some Results for More General and Realistic Situations. Next let us consider the removal of some of the approximations inherent in the foregoing model. For the small-signal flat-band case, it turns out that the half-cell reaction resistance in the supported case, Eq. (32), derived using the Butler–Volmer equation, and the half-cell reaction resistance in the unsupported case, $R_{\theta n}$, Eq. (50), which followed from use of the Chang–Jaffé boundary equations, are essentially identical (Macdonald, [1974a,b]. Furthermore, Franceschetti and Macdonald [1977] and Macdonald and Franceschetti [1979a] later showed that the calculation of the reaction resistance in either the unsupported or supported case gave the same result (because of compensating errors) whether Chang–Jaffé or Butler–Volmer equations were employed, provided, however, that any inner- or compact-layer capacitance C_c present was much larger than the diffuse double-layer capacitance. In addition, a method of transforming an unsupported small-signal impedance solution based on Chang–Jaffé boundary conditions to one employing Butler–Volmer, or even more general boundary equations, was developed. This method obviates the difficult task of solving the small-signal equations ab initio with the new boundary conditions.

There are plausible physical reasons to prefer Butler–Volmer to Chang–Jaffé conditions, especially when a compact layer is present, since the Butler–Volmer equations can account for the p.d. across this layer. Macdonald and Franceschetti [1979a] therefore studied how, for the $[0, \rho_{n0}, \infty]$ case without adsorption, C_R and R_R are changed from the results given here to new values when Butler–Volmer, or even more general boundary conditions, are used instead of Chang–Jaffé conditions and a compact layer of arbitrary constant capacitance was assumed present as well. Theoretical results were given and CNLS fitting of such results to an equivalent circuit were carried out in order to find the simplest adequate modifications needed. This approach not only allowed $\omega \to 0$ modifications to appear but yielded information on changes in the interface impedance over all ω values of interest arising from the presence of C_c and the more general boundary conditions.

Results found from the above approach were surprisingly simple. The present expression for $R_{\theta N}$ or $R_{\theta nN}$ (with $\epsilon_n = 1$) was shown to hold exactly in the Butler–Volmer case, independent of the size of C_{cN}. A simple expression for a new effective full-cell C_R, say C_{Re}, was found in the Butler–Volmer case, namely,

$$C_{ReN} = C_{RN} - (C_{RN} + 1 + \rho_{n0})^2/(C_{RN} + 1 + C_{cN}) \qquad (72)$$

Now in the usual $M \gg 1$ case where $C_{RN} \gg 1 + \rho_{n0}$, this result reduces to just $C_{ReN}^{-1} \simeq C_{RN}^{-1} + C_{cN}^{-1}$, a series combination of the original C_R and the compact-layer capacitance C_c. However, this is just the $\omega \to 0$ result always used in the supported case! Although these results were derived for the $[0, \rho_{n0}, \infty]$ case without adsorption, they should hold quite adequately for the nonadsorption $[0, \rho_{n0}, \pi_m]$ case as well.

When adsorption is present and the effects of a compact layer are included as well, it has been shown (Macdonald et al. [1980]) that in the $[0, 0, \infty]$ case the $\omega \to 0$ expression for the total differential capacitance is more complicated than just C_{Re} and C_A in parallel. One needs first to separate C_c into two series parts so $C_c^{-1} = C_\alpha^{-1} + C_\beta^{-1}$, where C_α is the capacitance between the electrode and the charge centroids of the adsorbed ions (at the inner Helmholtz plane), and C_β is that from this plane to the outer Helmholtz plane, where the diffuse layer of charge begins. Then one obtains the circuit of Figure 2.2.9, which reduces to the above result for C_{Re} when $C_\alpha = \infty$. When $C_\alpha < \infty$, it is not clear how R_R should be added to this circuit since it should bridge C_α and still be in series with C_A. For most solid electrolyte situations, however, it will usually be an adequate approximation to take $C_\alpha = \infty$ in fact and to put R_R in series with C_A, returning to the usual form of the interface part of the circuit (see Fig. 2.2.8). First, there is the probability of the electron wave function spilling out from the surface of a metal electrode and reducing the effective thickness of C_α toward zero (Kornyshev et al. [1982]). Second, for solid materials there will be no inner uncharged layer of solvent material, as in liquid electrolyte situations. This means that the plane marking the beginning of the diffuse layer is nearly as close to the electrode as the plane where adsorption occurs; when $C_{\alpha N} \to \infty$ or it is very large, $C_{\beta N}$ will also be extremely large and may often be neglected compared to C_R, so $C_{Re} \cong C_R$. Inci-

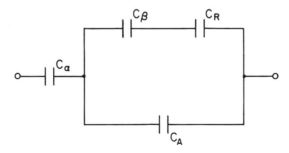

FIGURE 2.2.9. Circuit for the total interface differential capacitance in the $(0, 0, \infty)$ case without adsorption for $\omega \to 0$.

dentally, in the presence of dc bias, C_R (and C_A) are both functions of the effective overpotential and can increase greatly over their flat-band values under some conditions. They are limited in maximum value, however, because of the finite size of ions (Macdonald et al. [1980]). When one finds experimentally that C_{Re} is nearly independent of applied dc bias and temperature, it is likely that it is dominated by C_c rather than C_R.

Now let us briefly consider some results for the non-flat-band case without adsorption (Franceschetti and Macdonald [1979a,b, 1980]). Both transient response and biased small-signal frequency response results have been obtained using computer simulation, that is, numerical solution of coupled sets of partial or ordinary second-order differential equations describing the model. Chang–Jaffé and Butler–Volmer boundary conditions were both employed. Here we shall discuss only the frequency response results. First, the response of cells with $(0, 0, \pi_m)$ conditions at the left electrode and (∞, ∞, π_m) conditions at the right one was considered, leading to essentially half-cell conditions. The dc bias was assumed to arise from either built-in Frenkel space charge regions or an actual applied p.d. No direct current flowed in these completely blocking situations. Second, results were obtained for the full-cell system with $(0, 0, \pi_m)$ conditions at each electrode. Many complex plane Z and Y plots were presented to show how arcs and spurs varied with applied bias. More importantly, it was found that the equivalent circuit of Figure 2.2.8 with $Z_{De} = 0$, $\rho_{n0} = 0$, and no adsorption applied quite well, not only for zero bias (flat-band) but for either polarity of applied bias up to the maximum magnitude applied of about $15(RT/F)$, sufficient bias to make the system very nonlinear. There are no R_A and R_R elements present under these conditions, but Z_W did appear for $\pi_m \neq 1$ and for $\pi_m \neq \infty$ conditions. It is impractical to summarize here all the results found and reported, but the dependences of C_R and the components of Z_W generally varied with bias in reasonable and expected ways. The bulk parameters R_∞ and C_∞ showed negligible variation with bias.

Later, large-bias frequency response results were obtained for one-mobile partially blocking situations with no adsorption and with either $(\rho_{p0}, -, 0)$ for the left electrode and $(\infty, -, 0)$ for the right (half-cell conditions) or $(\rho_{p0}, -, 0)$ for both. Then a direct current can flow, and steady state current–voltage curves for Chang–Jaffé conditions were compared with those for Butler–Volmer ones. Appreciable differences occurred for biases bigger in magnitude than (RT/F). As expected, no diffusion effects were present in the response. The circuit of Figure 2.2.8 was again found adequate to describe the response, here with $Z_{De} = 0$, $R_A = 0$, and $Z_W = 0$. For Chang–Jaffé conditions $R_{\infty N}$ remained very close to its expected unity value and $C_{\infty N}$ was held fixed at unity, but for Butler–Volmer conditions $R_{\infty N}$ and $C_{\infty N}$ were somewhat bias-dependent and differed from unity. The dependences of R_R and C_R on bias were in accord with predictions based on the buildup of charge accumulation or depletion regions near the partly blocking electrodes, and it was found that C_{RN} was smaller in the Butler–Volmer case, because of compact-layer effects, than in the Chang–Jaffé one. For the same reason R_{RN} was less bias-dependent for Butler–Volmer than for Chang–Jaffé.

The foregoing results show the wide scope of the Figure 2.2.8 general circuit. It applies with good approximation for both flat-band small-signal conditions and under equilibrium or nonequilibrium biased conditions provided its elements are properly interpreted to account for the presence or absence of a compact layer and the appropriate type of boundary conditions.

Thus far we have dealt with either half-cells, where the right half-cell boundary involves nonpolarizing, ohmic boundary conditions (∞, ∞, π_m) and the left involves conditions such as (0, ρ_n, π_m) or with full cells with identical boundary conditions at each electrode. Another interesting full-cell case is that of crossed reactions where the left electrode involves (0, ρ_n, π_m) and the right (ρ_p, 0, π_m). Different electrodes are used, so charge of one sign reacts at one electrode and that of the other sign at the other electrode. This double-injection model cannot pass dc and has been analyzed by Glarum and Marshall [1980]. Although their solution is rather complicated, it reduces under conditions of interest ($\pi_m \ll 1$) to just the Z_{DOC} impedance of Section 2.2.2.2, that for an open-circuited transmission line. As already noted, it leads to an ordinary finite-length Warburg arc in the ϵ, not the Z, complex plane. Glarum and Marshall have used this result with some success in analyzing data for iridium oxide thin films (see also Franceschetti and Macdonald [1982]). It is therefore likely that a modification of the general Figure 2.2.8 circuit, useful in some $R_D = 0$ situations, would be to replace Z_W by Z_{DOC}. In fact, the most general modification would be to replace Z_W by Z_D, allowing the possibility of any kind of uniform-transmission-line-like behavior.

2.2.3.4 Unsupported Situations: Empirical and Semiempirical Models

In this section, we shall first discuss some ways in which the theoretical model results and equivalent circuits of the last section may be modified to attempt to account for less ideal conditions than assumed in the theory, conditions often appreciably closer to those found in real material–electrode systems. Then we shall discuss empirical and semiempirical models which may be useful as elements in equivalent circuits used for fitting real IS data.

Possible Circuit and Model Modifications. Further modifications of the Figure 2.2.8 equivalent circuit are often necessary, especially for polycrystalline material. One frequently finds experimentally that one or more of the $R_\infty C_\infty$, $R_R C_R$, or $R_A C_A$ semicircles in the complex plane are depressed so their centers lie below the real axis. It is more probable for the $R_R C_R$ arc to show such depression than the bulk $R_\infty C_\infty$ one for single-crystal material. Such depression may be interpreted in terms of a distribution of relaxation times, possibly arising in the case of the $R_R C_R$ arc from electrode surface roughness and/or porosity (see, e.g., de Levie [1967], Franklin [1975]). Although the exact small-signal solution with identical electrodes actually leads to some arc depression when π_m is very different from unity (Macdonald [1974b]), the amount of depression possible from widely different positive and negative charge mobilities is insufficient to explain most experimental depressions.

In the absence of a fully adequate microscopic theory leading to arc depression in the impedance plane, it has become customary to use the ZARC function defined in Section 2.2.2.2 to describe the depression analytically. This function involves either a resistance and a CPE in parallel or a unified impedance as in Eqs. (21)

and (23). For describing depressed arcs in the ϵ plane, the ϵARC (Cole–Cole) function also defined in Section 2.2.2.2 has long been used. Note that the CPE which appears in both the ZARC and the ϵARC may be associated with a nonuniform transmission line and nonuniform diffusion in a region of infinite extent, but the CPE has not been generalized so far to the finite-length diffusion regime. When an adequate expression is available, it will represent a more complex process than ordinary finite-length (uniform) diffusion, for example, Z_W, a subset of such generalized CPE response. Although this process would be appropriate for distributed (nonuniform) bulk response, since it would change from intensive to extensive behavior with decreasing frequency as Z_W does, the ordinary CPE, which can be taken either intensive or extensive, seems more appropriate for intensive interface processes. Whenever a straight-line spur in the impedance of plane is found with an angle from the real axis different from $\pi/4$, the CPE should replace Z_W.

Although there is no complete derivation of a generalized CPE yet available which arises from nonuniform diffusion (NUD) in a finite-length region, one may heuristically modify the CPE and Warburg diffusion expressions in such a way as to generalize them both. The result is

$$Z_{\mathrm{NUD}} = R_0 \tanh \left[A_0 R_0 (j\omega)^\psi \right] / \left[A_0 R_0 (j\omega)^\psi \right] \tag{73}$$

where R_0, which might be R_∞ for bulk behavior, is taken extensive and we require $0 < \psi \le 1$. For $\psi = 0.5$, this expression reduces to just Z_W when one takes $A_0 = l_e / R_0 \sqrt{D}$. For $A_0 R_0 \omega^\psi \gg 1$, $Z_{\mathrm{NUD}} \simeq Z_{\mathrm{CPE}}$. Further, for any ψ, $Z_{\mathrm{NUD}} \rightarrow R_0$, an extensive quantity, when $A_0 R_0 \omega^\psi \ll 1$, and, for $\psi < 1$, to $[A_0 (j\omega)^\psi]^{-1}$ for $A_0 R_0 \omega^\psi \gg 1$. This result is intensive, as it should be if A_0 is taken intensive, as it is in the Warburg limit. Note that while both the ZARC impedance and Z_{NUD} involve $R_0 A_0$, they will have somewhat different shapes at the lowest frequencies where they approach the real axis. Futher, the present expression for Z_{NUD} is just an empirical stopgap result and is probably most useful for $0 < \psi \le 0.6$. We shall denote this heuristic generalization as the generalized finite-length Warburg model (GFW).

The fact that the current ungeneralized CPE has no dc path (while a generalized one would) can lead to problems in using the CPE in hierarchical circuits such as that of Figure 2.2.8. Whenever it seems appropriate to replace an ordinary capacitor in the circuit by a CPE, no problem arises. But one cannot replace a resistor needed as part of a dc path by a CPE and still maintain the dc path; all one can do is put a CPE in parallel with the resistor, producing a ZARC function, or perhaps to use Z_{NUD}. But there is a problem in using such elements in hierarchical circuits. Although the real and imaginary parts of, say, a ZARC function could be separated, and the imaginary part used in place of an ideal capacitor in an hierarchical circuit and the real part in place of an ideal resistor, there is no physical justification for such separation.

In the polycrystalline case, one must consider the processes which occur within an individual single-crystal grain and what happens at the grain boundaries, taking into account that there is almost certainly a distribution of grain sizes and orientations present. Since the response is a three-dimensional average of the response of a great many interacting grains, one expects that the bulk response for composite materials will both be more complicated than the $R_\infty C_\infty$ semicircle expected for a perfect homogeneous material and may often be described by a distribution of relaxation times, either discrete or continuous. Since one usually finds that it is experimentally impossible to distinguish results arising from a continuous distribution and its approximation by, say, 10 or more discrete relaxations, it is often easiest to use the continuous distributions, since many less parameters need be specified. For the distorted and displaced arcs which are usually seen in the impedance plane for composite materials such as ceramics, it of course makes no difference in which order the elements representing this overall bulk response appear in the equivalent circuit (Voigt, not hierarchical connection), and it has also been customary to try to represent the response by one or more ZARC functions in series. No charged-particle Warburg response will be present if charge of only a single sign is mobile.

Although is does not seem reasonable to build a hierarchical ladder network circuit using the separated real and imaginary parts of a distributed element such as the ZARC, we can still achieve considerable generality and flexibility if we form a circuit using *only* unified distributed elements as in the three-level circuit of Figure 2.2.10. Here DE represents a general distributed element, one like the ZARC which can well approximate either an ideal resistor or an ideal capacitor in limiting cases of its fractional frequency dependence exponent ψ. Thus, in Figure 2.2.10, the odd-numbered DEs could, in the limit, be taken as capacitors, and the even-numbered ones as resistors. In practical cases, however, one would often find it necessary to choose some of the DEs as nonideal distributed elements. Note that if the electrodes were nonblocking, one would need to ensure a dc path through the circuit by, for example, taking the even-numbered DEs as resistors, ZARCs, or some other unitary or composite nonblocking distributed element.

Further Empirical and Semiempirical Models. Although various empirical distributed-element models have already been discussed, particularly in Section 2.2.2.2, the subject is by no means exhausted. Here we briefly mention and discuss various old and new elements which may sometimes be of use in a fitting circuit such as that of Figure 2.2.10. Complex plane plots of IS data by no means always yield perfect or depressed semicircular arcs; often the arc is unsymmetric and cannot be well approximated by the ZC. An unsymmetrical impedance plane arc usually exhibits a peak at low frequencies and CPE-like response at sufficiently high frequencies. The reverse behavior is not, however, unknown (Badwal [1984]). An expression originally proposed in the dielectric field by Davidson and Cole [1951] yields ordinary asymmetric behavior. Its I_k generalization is

$$I_k = [1 + js]^{-\psi_k} \tag{74}$$

with $s \equiv \omega\tau$ and $0 \leqslant \psi_k \leqslant 1$, and it reduces to symmetric Debye response for $\psi_k = 1$. This model will be denoted by DC. A further empirical dielectric-area ap-

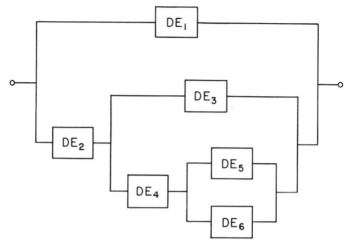

FIGURE 2.2.10. General equivalent circuit showing hierarchical structure and involving general distributed elements.

proach, due to Kohlrausch [1854] and Williams and Watts [1970], yields transient response of the fractional exponential form $\exp{(-t/\tau_0)^\psi}$, with $0 < \psi \leqslant 1$. Only complicated series expressions and tables exist for its small-signal response, but it yields frequency domain response generally rather similar to that of Eq. (74). It will be denoted by WW. See Macdonald and Hurt [1986] for an accurate WW fitting method.

Jonscher [1974, 1975a,b, 1980, 1983], in an extensive series of papers, has also worked primarily in the dielectric area, independently emphasized the importance and ubiquity of constant phase response, and proposed and demonstrated the utility of three different empirical frequency response functions in IS data fitting. These three equations, termed *universal dielectric response* by Jonscher, were originally expressed in terms of the imaginary part of the complex dielectric susceptibility, χ''. They may alternatively be expressed, of course, in terms of ϵ'' or Y'. Further, they may all be generalized to the I_k'' ($k = \epsilon$ or Z) representation. Finally, it has been found (Macdonald [1985d]) that two of them may be written in full complex form, not just as I_k''.

The three generalized Jonscher equations may be expressed as

$$I_k = B_{k0}(j\omega)^{-\psi_k} \tag{75}$$

$$I_k = B_{k1}\left[(j\omega\tau_{k1})^{-\psi_{k1}} + (j\omega\tau_{k2})^{-\psi_{k2}}\right] \tag{76}$$

and

$$I_k'' = -B_{k2}\left[(\omega/\omega_{kp})^{-\psi_{k3}} + (\omega/\omega_{kp})^{-\psi_{k4}}\right]^{-1} \tag{77}$$

It is clear that Eq. (75) is just the CPE and (76) is a combination of two CPEs (in parallel for $k = \epsilon$ and in series of $k = Z$). Of course, the ω_{kp} of Eq. (77), which denotes a peak frequency, could be replaced by $\tau_{kp} \equiv \omega_{kp}^{-1}$. The possible range of all the exponents is $(0, 1)$. In Jonscher's $k = \epsilon$ "universal dielectric response" case, one has $\psi_\epsilon = 1 - n$, $\psi_{\epsilon 1} = 1 - n_1$, $\psi_{\epsilon 2} = 1 - n_2$, $\psi_{\epsilon 3} = m$, and $\psi_{\epsilon 4} = 1 - n$. If we further choose $\psi_Z = n$, $\psi_{Z1} = n_1$, and $\psi_{Z2} = n_2$, Eqs. (75) and (76) yield the same frequency dependence exponents at the admittance level when a single term dominates. For example, Eq. (75) yields $Y_\epsilon \propto (j\omega)^n$ and $Y_Z \propto (j\omega)^n$ for the above choices. We shall term Eq. (76) the *generalized second Jonscher equation* (GJ2) and Eq. (77) the *generalized third Jonscher equation* (GJ3).

The minus sign in Eq. (77) arises because we have defined I_k with a plus sign as $I_k = I_k' + iI_k''$. When $\omega = \omega_{kp}$ in Eq. (77), $|I_k''|$ reaches a maximum. Further, when $\psi_{k3} = \psi_{k4}$, this equation reduces to the long-known Fuoss–Kirkwood [1941] form, yielding a symmetrical curve for $-I''$ vs. log (ω/ω_{kp}). We shall denote this special form of the GJ3 as the GFKJ equation. Although no fully complex general expression consistent with (77) is available, when $\psi_k \equiv \psi_{k3} = \psi_{k4}$ complex forms have been given for various fractional values of ψ_k. Jonscher and his collaborators have shown that the χ'' forms of Eqs. (75)–(77) can fit a great deal of dielectric and conductive system data. Unfortunately, the fits never used CNLS, and no ordinary nonlinear least squares fits of χ'' giving fitted parameter estimates and standard deviation estimates have been presented. Further recent discussion of "universal dielectric response" appears in Macdonald [1985d].

Some time ago Almond, West, and Grant [1982], Bruce, West, and Almond [1982], and Almond and West [1983b] specialized the χ'' form of Jonscher's Eq. (76) for hopping conduction situations to obtain

$$\sigma(\omega) = k(\omega_p + \omega_p^{1-n}\omega^n) \tag{78}$$

where $\sigma(\omega)$ is the ac conductivity, K is a temperature-dependent constant, and ω_p was identified as the thermally activated ionic hopping frequency ν_H. Now Eq. (78) may be rewritten at the Y level as

$$Y'(\omega) = G_0[1 + (\omega/\omega_p)^n] \tag{79}$$

Next, generalization of Eq. (79) to the complex plane yields

$$Y(\omega) = G_0[1 + (j\omega\tau_0)^n] \tag{80}$$

which is fully consistent with (79) when

$$\omega_p = \left\{\tau_0[\cos(n\pi/2)]^{1/n}\right\}^{-1} \tag{81}$$

or, equivalently,

$$\tau_0 \equiv \omega_0^{-1} = [\cos(n\pi/2)]^{-1/n}/\omega_p \tag{82}$$

Finally, Eq. (80) yields

$$Z = R_0 / [1 + (j\omega\tau_0)^n]$$ (83)

where $R_0 = G_0^{-1}$. This expression is just the long-known ZARC [compare Eq. (20)]. Thus, it appears that the principal new element in the Almond–West work is the identification of ω_p as the hopping frequency. This interesting suggestion has been examined at some length recently (Macdonald and Cook [1985]), with the conclusion that the case is not proven so far. If, in fact, the hopping frequency ν_H is directly involved in the ZARC when it is applied to hopping conduction situations, it seems most plausible that $\nu_H = \omega_0 = \tau_0^{-1}$, or perhaps $\omega_0/2\pi$, rather than ω_p. Equation (81) shows that $\omega_p \to \infty$ as $n \to 1$, an unlikely result and one avoided by the choice of ω_0 instead.

Several of the empirical model responses discussed above have been given a more theoretical basis (see Macdonald [1985c,d] for references), but they still suffer from two important weaknesses. They do not generally lead to physically realistic response at both high- and low-frequency extremes, and they do not lead to any predictions for possible temperature dependence of the fractional exponent(s) ψ. A semiempirical theory whose frequency response results are briefly discussed below does, however, avoid these weaknesses. Since any real material will have a largest (τ_∞) and a smallest (τ_0) response time, response at longer (shorter) times then these will be determined by these limiting responses (for a single type of physical process). But such single-time-constant behavior leads to frequency response proportional to ω for $\omega \ll \tau_\infty^{-1}$ and to ω^{-1} for $\omega \gg \tau_0^{-1}$. Although simple Debye behavior with $\tau_0 = \tau_\infty = \tau$ also leads to limiting $\omega^{\pm 1}$ response, here τ_0 and τ_∞ may differ greatly, and for the range $\tau_\infty^{-1} < \omega < \tau_0^{-1}$, non-Debye fractional exponent response may appear and usually does so.

Note that CPE response fails the above test of physical realism at both frequency extremes; so does ZC response for $\psi < 1$. On the other hand, Davidson–Cole low-frequency-limiting response is realistic, but not its high-frequency-limiting response. One might reasonably ask, if all these models are not entirely physically realistic, why are they discussed and used for fitting? The reason is that it is rare for a single-response process, say that associated with an electrode reaction (R_R and C_R in the ideal nondistributed case), to be so isolated in its frequency range that one can follow its response alone to very high or low (relative) frequencies. Because of the usual presence of other processes yielding response near or even overlapping in frequency that of the process of immediate concern, one cannot usually follow the response of the process in question very far into its wings where $\omega^{\pm 1}$ limiting response finally must appear. Further, one usually finds that experimental limitations preclude measuring far into the high-frequency wing of the highest-frequency process present or far into the low-frequency wing of the lowest-frequency process present. In essence, what we can't measure doesn't matter—at least until we can measure it! Nevertheless, a theoretical model which does incorporate proper limiting behavior is clearly superior in that respect to one that doesn't.

Although dielectric response data often leads to temperature-independent ψ's (so that the time—temperature or frequency–temperature superposition law holds), this is by no means always the case (Jonscher [1983]). Further, conductive-system response, as in ionic hopping conductors, often leads to appreciable temperature dependence of ψ. Suprisingly, ψ_ϵ and ψ_Z temperature responses, when apparent, are usually found to be quite different, with ψ_ϵ increasing with increasing temperature and ψ_Z decreasing.

Fitting Ambiguity and a New Semiempirical Model. Although all the models we have discussed in this section and in Section 2.2.2.2 are distinct and separate, and although they may be associated with different physical processes, it turns out that there is a high degree of practical fitting ambiguity between most of them. Response differences between several unsymmetric models with the same ψ value are demonstrated in Figure 2.2.11. Here Debye response is included for comparison and DAE$_1$ (which involves the parameter ϕ rather than ψ) refers to the semiempirical distribution-of-activation-energies model discussed below. But the situation is different when "data" derived from one model involving a given ψ, say ψ_a, are fitted by CNLS to another model, yielding a ψ estimate for this model, say ψ_b. It turns out, as we shall demonstrate below (see also Macdonald [1985d]), that when ψ_a and ψ_b are allowed to be different, one model can often fit another within 1% or so (usually better than most experimental data are known) over quite wide frequency and magnitude ranges. When such ambiguity is present, as it usually is for practical less-than-perfect data, it will often be easiest to fit with the simplest model, whether or not it is physically reasonable for the material–electrode system considered, and then relate the fitting results to a more appropriate, but more complex, model.

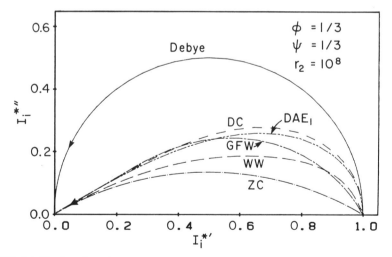

FIGURE 2.2.11. Complex plane response of the normalized I_i^* response of various distributed models.

Such a more complex model is the DAE, involving an exponential distribution of activation energies. Its rationale and results are described in detail in Macdonald [1963, 1985c,d]. Let us distinguish three forms of it. First is the DAE_1, which involves a single exponential density distribution and leads to unsymmetrical response (Macdonald [1985c]) Second is the DAE_2, which involves two joined, complementary exponential distributions and leads to symmetric behavior in the complex plane. Finally, the general DAE involves two joined, noncomplementary exponential distributions and spans the range of shapes from the DAE_1 to the DAE_2 (Macdonald [1985d]). Of course, the DAE is most generally given in normalized I_k form.

Although the frequency response of the DAE model can only be expressed in integral form (associated with a hypergeometric function) for arbitrary ϕ, relatively simple closed-form response has been given for many values of ϕ. Such closed-form, discrete-ϕ response is useless, however, for accurate CNLS fitting using this model. Therefore, the full-integral DAE model as well as nearly all of the other distributed element models discussed so far, have been built into the general CNLS fitting program available from Dr. J. R. Macdonald. Thus, any of the models can be used to fit experimental frequency response data or "data" derived from another model.

Some of the model-fitting ambiguity mentioned above is demonstrated in the next figures. Further discussion of DAE–Jonscher ambiguity appears in Macdonald [1985d]. First, it is worthwhile to categorize the models discussed by their complex plane symmetry as in Table 2.2.3. The symmetric and asymmetric curves give closed arcs in the complex I_k plane, but the CPE and GJ2 yield only open spurs in this plane (and, as mentioned earlier, cannot be normalized in the usual I_k way). The fitting ambiguity with which we are concerned here applies only within a given column of Table 2.2.3. We cannot expect to get a good fit of asymmetric WW data, for example, with a symmetric model such as the ZC. Note that all models will generally show some region of frequency where CPE-like response appears. In this region, the CPE model is clearly sufficient. It is not this ambiguity with which we are concerned but rather with the holistic response, that which includes regions beyond and below that where CPE response alone dominates.

We have found that any symmetric model of Table 2.2.3 can be very well fitted by any other symmetric model. Figures 2.2.12 and 2.2.13 show two-dimensional

TABLE 2.2.3. Summary of Main Models Discussed, Showing Their Symmetry Characteristics in the Complex Plane

Symmetric	Asymmetric	General
ZC	DC	CPE
GFKJ	GFW	GJ2
DAE_2	WW	DAE
	DAE_1	

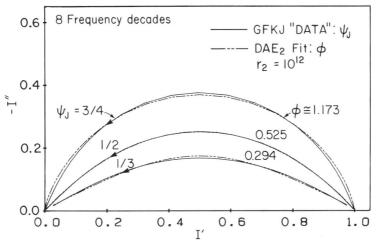

FIGURE 2.2.12. Complex plane comparisons of the response of the GFKJ and DAE$_2$ models when the DAE$_2$ is fitted to GFKJ response with CNLS.

plots obtained from full CNLS fitting of the DAE$_2$ and ZC models to GFKJ model ''data'' for several different ψ values. The fits are so good that only with extremely accurate experimental data (better than those usually available) could one decide unambiguously between any of the three symmetric models on the basis of CNLS fitting alone. Further, the DAE has been found to yield a very good fit of the GJ2 model as well (Macdonald [1985d]).

Figures 2.2.12 and 2.2.13 of course do not show frequency response explicitly. It is found to agree exceptionally well also at each point, at least until one moves

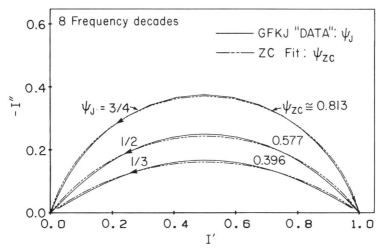

FIGURE 2.2.13. Complex plane comparisons of the response of the GFKJ and ZC models when the ZC is fitted to GFKJ response with CNLS.

far away from the peak frequency. Some results for DAE_2 fitting to GFKJ data for $\psi_J = \frac{1}{3}$ are shown in Figure 2.2.14. The unity weighting used in Figures 2.2.12 and 2.2.13 yields a better fit near the peak, and proportional weighting (see Section 3.2.2) leads to better agreement in the skirts of the curve. Here $s \equiv \omega\tau_0$ is a normalized frequency variable (Macdonald [1985c,d]).

The situation is somewhat less ambiguous for the special asymmetric models of Table 2.2.3. Although the DC and DAE_1 models can well fit each other, neither one can fit WW model response adequately over a wide ψ_i range. For $0.7 \leqslant \psi_{WW} \leqslant 1$, the DAE_1 can fit WW results quite well, but better fits are obtained for this region and below using the general DAE model to fit WW "data." Figure 2.2.15 shows the results of such CNLS fitting plotted in the complex plane, and Figure 2.2.16 shows them with 3-D plotting (see Section 3.2.1). Again the fits are so close that only with superb data could one unambiguously discriminate between the two models. One reason to prefer the various DAE models to the others, however, is that the former yield explicit temperature dependences for the ϕ_i's which enter the model, while no such ψ_i temperature dependence is a part of the other models. When the actual fractional frequency response exponents observed in a set of experimental data are found to vary with temperature, as they often do for

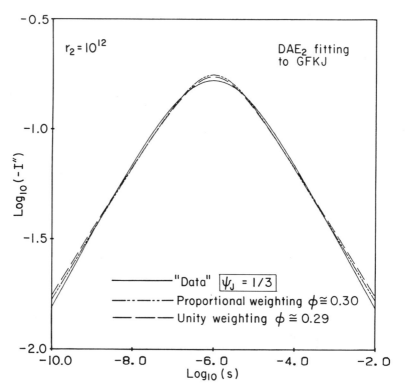

FIGURE 2.2.14. Frequency response curves comparing the results of fitting the DAE_2 model to GFKJ "data." The logarithm of $-I''$ is plotted vs. the logarithm of a normalized frequency variable s.

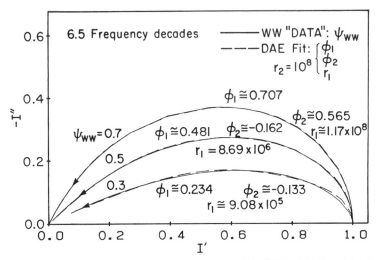

FIGURE 2.2.15. Complex plane comparisons of the response of the WW and DAE models when the DAE is fitted to WW response with CNLS.

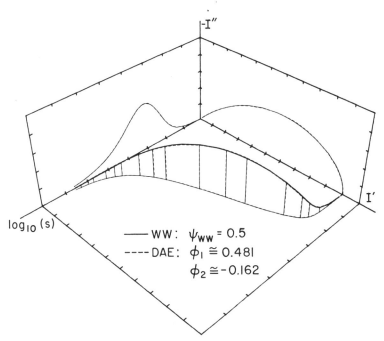

FIGURE 2.2.16. Three-dimensional plot with perspective showing the excellent agreement between WW "data" and the results of fitting these data with the DAE model.

both conductive and dielectric systems, it is thus natural to try DAE fitting and see if the ϕ_i estimates found depend on temperature in one of the ways predicted by the theory (Macdonald [1985c,d]). If such agreement is established, much can be learned about the detailed response of the system. Finally, it should be mentioned that the exponential DAE model has recently been simplified and its predictions compared to those of a symmetric or asymmetric Gaussian DAE model (Macdonald [1987]). Again fitting ambiguity is sometimes found.

This section was written by J. Ross Macdonald and Donald R. Franceschetti.

CHAPTER THREE

MEASURING TECHNIQUES AND DATA ANALYSIS

Michael C. H. McKubre
Digby D. Macdonald
J. Ross Macdonald

3.1 IMPEDANCE MEASUREMENT TECHNIQUES

3.1.1 Introduction

Until the advent of digital computers, all electrochemical studies involved the processing and analysis of analog signals in either the time domain or the frequency domain. Typical examples of analog signal analysis include the use of ac coupled bridges and of Lissajous figures for determining interfacial impedance. In both instances, the desired information (e.g., balance of a bridge) is obtained in purely analog format, and no need exists for converting signals into digital form.

When describing analog instrumental methods, it is convenient to classify techniques according to the type of excitation functions employed, particularly with respect to the independent variable. For example, frequency domain impedance measurements are carried out using a small-amplitude sinusoidal excitation with frequency as the independent variable. Alternatively, the perturbation and response may be recorded in the time domain with time as the independent variable, and the impedance as a function of frequency can then be extracted by time-to-frequency conversion techniques such as Laplace or Fourier transformation. Time domain methods characteristically use digital-processing techniques; frequency domain methods have traditionally used analog techniques, although digital processing is becoming common in synthesis and analysis of sinusoidal signals.

The application of a sine wave excitation to a system under test often is the easiest method of determining the system transfer function. Here we are concerned with measuring or inferring a transfer function for an electrochemical cell as a first step in determining reaction mechanistic and kinetic parameters (Macdonald [1977], Gabrielli [1981], Macdonald and McKubre [1981], Macdonald and McKubre [1982])

By way of review, the transfer function of a system can be determined as the output divided by the input

$$G(j\omega) = X_{out}(j\omega)/X_{in}(j\omega) \tag{1}$$

For the special case where the output signal is the system voltage and the input (or excitation function) is the current, the transfer function is the system impedance

$$G(j\omega) = E(j\omega)/I(j\omega) = Z(j\omega) \tag{2}$$

Since the output may be changed in both amplitude and phase with respect to the input, we must express the impedance as a complex number:

$$Z(j\omega) = Z' + jZ'' \tag{3}$$

where primed and double-primed variables refer to in-phase and quadrature components, respectively.

It is important to note that we are using the formalism of linear systems analysis; that is, Eq. (2) is considered to hold independently of the magnitude of the input perturbation. Electrochemical systems do not, in general, have linear current–voltage characteristics. However, since any continuous, differentiable function can be considered linear for limitingly small input perturbation amplitudes (Taylor expansion), this presents more of a practical problem than a theoretical one.

In the following section we present a number of standard methods of measuring a system impedance or a frequency domain transfer function. In applying any of the methods described, the perturbation must be of a sufficiently small magnitude that the response is linear. Although the condition of linearity may be decided from theoretical considerations, (Bertocci [1979], McKubre [1981], McKubre [1983], McKubre and Syrett [1986]), the most practical method is to increase the input signal to the maximum value at which the response is independent of the excitation function amplitude.

3.1.2   Frequency Domain Methods

3.1.2.1   *Audio Frequency Bridges*

In the past, impedance measurements using reactively substituted Wheatstone bridges at audio frequencies have been the easiest to accomplish. Consequently, great emphasis has been placed historically on electrochemical processes having characteristic impedance spectra in the audio frequency range 20–20,000 Hz, namely, double-layer capacitive and moderately fast reaction kinetic effects at plane parallel electrodes.

The mathematics and methodology of such measurements are well understood (Hague [1957], Armstrong et al. [1968]). However, considerable use still may be made of passive audio frequency bridge measurements in this age of active circuitry, principally in high-precision applications. Following a brief review of bridge circuits, we will restrict our discussion to the limitations imposed by the use of each type of bridge, since these will influence the point at which an experimentalist will select a more complex measuring device.

Figure 3.1.1 shows schematically the familiar representation of an audio frequency bridge adapted for use with an imposed dc potential. The condition of balance for the bridge shown is

$$Z_x = (R_1/R_2)Z_s \tag{4}$$

where subscripts x and s refer to unknown and standard impedances, respectively. A variety of RCL combinations are possible for Z_s; in the commonly used Wien bridge (Hague [1957]) Z_s takes the form of series variable resistance and capacitance standards, which are adjusted alternately until the real and imaginary components of the voltage at the null detector simultaneously are zero. For this null condition the real and imaginary components of the unknown impedance may be calculated as

$$Z_x' = (R_1/R_2)R_s \tag{5}$$
$$Z_x'' = (R_1/R_2)/\omega C_s \tag{6}$$

The form of Eqs. (5) and (6) has led to the widespread and unfortunate practice of tabulating and plotting measured impedance data in terms of the complex pair

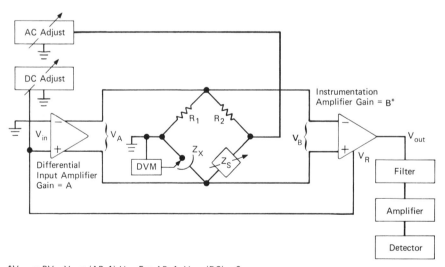

$^*V_{out} = BV_B - V_R = (AB-1) V_{in}$. For AB=1, V_{out} (DC) = 0.

FIGURE 3.1.1. Audio frequency bridge modified to include working electrode dc potential control.

$(R_s, j/\omega C_s)$ even when a Wien bridge has not been used. The impedance notation (Z', jZ'') is significantly less ambiguous and will be used here.

High-Frequency Limitations. The upper operating limit is imposed primarily by reactivity and nonlinearity of available resistive standards (chiefly inductive effects) and the effects of stray capacitive shunts. By using a Wagner earth (Hague [1957], Armstrong et al. [1968]), the latter effect can often be reduced sufficiently to allow sensible measurements at frequencies up to 10^5 Hz. However, the importance of Wagner earthing varies greatly with the magnitude of the impedance being measured (Hague [1957]). In general, elimination of stray capacitance is most important at high frequencies when measuring small capacitances or large resistances (i.e., for small-area electrodes).

Low-Frequency Limitations. The null detection system traditionally used with an audio frequency bridge consists of an amplifier, filter, and ac voltmeter. This combination imposes three limitations at low frequencies:

- Null detection with a magnitude voltmeter or oscilloscope is most sensitive when the resistive and reactive components of the unknown impedance are of the same magnitude, since the total bridge out-of-balance signal contains terms proportional to each. For an impedance bridge used to measure the electrical properties of electrochemical cells, this fact imposes a limit on accuracy at low frequencies since the reactive terms, which are primarily capacitive, dominate the cell admittance with decreasing frequency. Increasing the gain to observe the resistive component more precisely results in saturation of the detection system with the reactive out-of-balance signal.

- A significant source of noise at the detector may result from harmonic distortion originating in the oscillator or caused by nonlinearity in the system under test or in subsequent amplifiers. In such cases, the signal at balance consists mainly of the second harmonic. At high frequencies, this signal can be removed effectively by appropriate signal conditioning with bandpass, low-pass, or notch filters (McKubre and Macdonald [1984]). At low frequencies, however, analog filters of bandwidth less than 10 Hz are less easy to construct and control.

- Another major source of noise at low frequencies is mains pickup. This may amount to hundreds of millivolts superimposed on the test signal unless major efforts are made at shielding and ground loop suppression. Usually, unless an adequate notch filter is used in addition, the experimentalist must be satisfied with reduced precision at frequencies below about 100 Hz.

These three effects can be reduced to a large extent by using a phase-sensitive detector (PSD) to measure separately the real and imaginary components of the bridge out-of-balance signal. By separate amplification of the in-phase and quadrature components, differential sensitivities in excess of 100:1 can be attained.

The advantages and limitations conferred by the use of PSDs are described in Section 3.1.2.6.

In normal operation, a PSD is completely insensitive to the second harmonic, but most commercial instruments have the additional facility of being able to select a reference signal at twice the fundamental frequency. By this means the extent of second-harmonic distortion can be measured. This distortion often reflects not an error signal (i.e., noise) but an expected response induced by nonlinearity of the system under test (McKubre [1981], McKubre [1983], McKubre and Syrett [1984]).

In addition, and unlike traditional bandpass filters, a PSD has a bandpass characteristic with bandwidth that decreases with decreasing frequency and frequently can be used within ±5 Hz of 50- or 60-Hz mains pickup.

When phase-sensitive null detection is used, the practical low-frequency limit becomes a function of the particular form of bridge chosen. For the Wien bridge, this limit is imposed by the selection of suitably large adjustable capacitance standards at frequencies below about 20 Hz.

Limitations of Imposed Potential. A considerable limitation on the use of this form of bridge is that it necessitates the use of a two-terminal cell. Although it is often possible to construct a cell in which the working-electrode impedance greatly exceeds that of the counter electrode, potentiostatic conditions cannot be established adequately with this type of bridge. Closely associated with this limitation is the fact that in normal use, the cell current and voltage vary with the settings of the resistive and reactive standards.

In electrochemical applications, these combined limitations may be severe. Figure 3.1.1 shows one of a variety of possible methods by which an imposed working-electrode dc potential can be adjusted to the desired value without influencing the detector circuit. The method shown can be used at frequencies less than the normal operating frequency limit of ac coupled amplifiers.

3.1.2.2 Transformer Ratio Arm Bridges

The high-frequency limitation imposed on the operation of reactively substituted Wheatstone bridges by unavoidable stray capacitances prompted the development of the transformer ratio arm bridge (Calvert [1948]). By substituting a transformer for orthodox ratio arms, a bridge was produced for which the impedance ratio is proportional to the square of the number of turns and which was capable of accepting heavy capacitive loads with virtually no effect on the voltage ratio.

The operation of a transformer ratio arm bridge is shown schematically in Figure 3.1.2. Briefly, voltages 180° out of phase are fed from the secondary winding of the input "voltage" transformer to the cell or unknown impedance and to resistance and capacitance standards. The "arms" of the bridge consist of a series of ratio taps of the primary windings of an output "current" transformer. The standard and unknown impedances are connected to the output transformer in such a way that a detector null is achieved when the sum of the flux induced by the unknown and standard currents in the output transformer is zero. In this condition

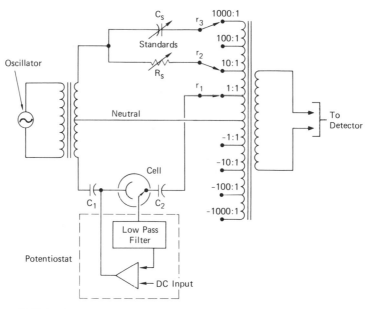

FIGURE 3.1.2. Transformer ratio arm bridge with dc potentiostatic control.

$$\frac{r_1}{Z_x} = \frac{r_2}{R} + j\omega Cr_3 \tag{7}$$

for all V_{in}, where r_1, r_2, and r_3 are ratios (usually decade), separately selected. The advantages of using this type of bridge are as follows:

- Error resulting from the impurity of standard variables can be virtually eliminated. Because ratios are selectable over a wide range (usually 1000:1), standards can be small. Also, with decade-spaced transformer ratios, standards need be variable only over a range of about 11:1. Consequently, standards can be used that closely approximate ideality (e.g., air-gap capacitors and nonreactively wound metal resistance), and one standard can be used to measure a wide impedance range.

- By the use of precision transformers as ratio arms, one can obtain highly accurate ratio values that are essentially independent of frequency well into the megahertz range.

- The bridge is highly insensitive to the presence of stray capacitance. Figure 3.1.3 shows the reason: C_1, C_2, and C_3 can cause no measurement error—C_1 because it merely produces a reactive potential drop that is common to the unknown and standard circuits, and C_2 and C_3 because at balance no potential drop appears across them. Now C_u represents the capacitance across the unknown terminals and its effect is canceled by trimming capacitor C_t on the standard side. Here C_t is adjusted at each measurement frequency by discon-

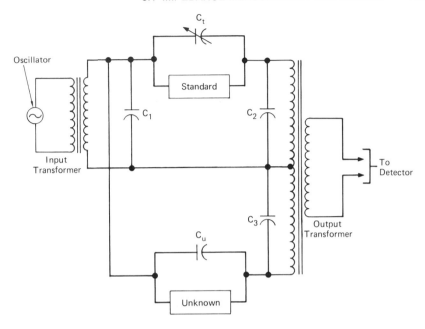

FIGURE 3.1.3. The effect of stray capacitances in the transformer ratio arm bridge.

necting the standard and balancing the bridge. Similarly, effects of the stray capacitances to earth virtually disappear if the neutral terminal is grounded (Calvert [1948]; see also Figure 3.1.2).

• Impedances may be measured in all four quadrants by selecting positive or negative ratios. Of particular importance is the use of pure capacitive standards to measure unknowns with a positive (inductive) reactance.

High-Frequency Limitations. In normal use for electrochemical cells, the effective upper operating limit is imposed by effects external to the bridge. These effects, which have been described in detail by Armstrong et al. [1968], consist primarily of transmission line effects in connecting cables, the effect of residual series inductance in leads and the cell, and (normally desired) impedance dispersion effects of solid electrodes. In the latter group, edge effects (Sluyters [1970]) and transmission line effects due to surface roughness (deLevie [1963], [1965*b*], [1967]) become dominant with increasing frequency. In electrochemical systems for which the interfacial impedance is the desired parameter, measurement precision becomes limited by the dominance of the uncompensated electrolyte resistance in the total measured impedance. This effect has prompted the use of very small electrodes for which the ratio of uncompensated resistance to interfacial impedance is reduced (Zeuthen [1978]).

Series leakage inductances in the transformers within the bridge result in an impedance measurement error that is proportional to frequency. This effect has been examined by Calvert [1948], but is seldom likely to impose high-frequency limitations in electrochemical applications.

Low-Frequency Limitations. The use of input and output transformers results in cell current and voltage, and thus detector signals, that decrease with decreasing frequency. This effect becomes apparent only at low audio frequencies and imposes a practical lower limit of the order of 100–200 Hz with commercial bridges.

Limitations of Potential Control. The limitations of potential control for a transformer ratio arm bridge are similar to those imposed in classical bridge measurement. That is, it is not possible to apply the ac potential via a reference electrode and potentiostat circuit only to the interface of interest. The measured impedance necessarily includes series terms associated with the lead and electrolyte resistances and the counter electrode impedance.

Dc potentials can be applied to the interface of interest by using a circuit of the form shown within the dashed lines in Figure 3.1.2, since at moderate frequencies the low-pass filter will not observe the ac component. However, direct current must be excluded from the bridge windings by the use of blocking capacitors C_1 and C_2. The impedance of these also will be included in the measured "cell" impedance.

3.1.2.3 *Berberian–Cole Bridge*

An active null admittance measuring instrument that incorporates many of the advantages of the transformer ratio arm technique, while obviating many of the disadvantages of passive bridges, has been reported by Berberian and Cole [1969]. Figure 3.1.4 shows a form of this bridge modified to measure impedance and to remove some of the limitations of the earlier instrument (McKubre [1976]).

The basic operation is as follows. The external variable decade standards are R_1 and C, while R' and R'' are internal and fixed. With reference to Figure 3.1.4, at all times,

$$i_1 + i_2 + i_3 = 0 \tag{8}$$

$$i_1 = AV_A/R_1 \qquad (V_A = IZ) \tag{9}$$

$$i_2 = AV_A(j\omega C) \qquad (V_A = IZ) \tag{10}$$

$$i_3 = BV_B/R' \qquad (V_B = -IR') \tag{11}$$

where Z is the impedance between the working electrode and the reference electrode and I is the current flowing through the cells. Therefore, for the condition of balance at the summing point

$$BIR'/R'' = AIZ/R_1 + AIZ(j\omega C) \tag{12}$$

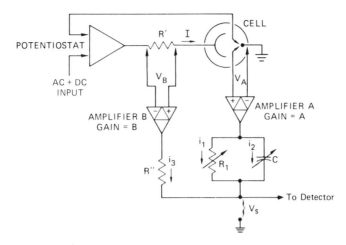

FIGURE 3.1.4. Modified Berberian–Cole bridge shown as a three-terminal interfacial-impedance-measuring system, with potentiostatic control of the working electrode.

Removing I and solving for the unknown impedance yields

$$Z = \frac{BR'R_1}{AR''} \frac{(1 - j\omega R_1 C)}{1 + \omega^2 R_1^2 C^2} \tag{13}$$

The advantages of this method apply principally at low (audio and subaudio) frequencies. It is important to note that the device shown schematically in Figure 3.1.4 is a bridge only in the sense that external variables are adjusted to produce an output null.

The principal advantages of the Cole–Berberian bridge are as follows:

- Because of the use of buffer amplifiers, null adjustment does not vary the potential across (or current through) the unknown impedance, as is the case for classical and transformer ratio bridge measurement.

- Measurements can be made on two, three, or four terminal cells, allowing the isolation of the impedance component of interest from the total cell impedance. This is not possible with a passive bridge, and it is frequently infeasible to construct a cell for which the impedance of interest is much greater than all series terms. This is particularly difficult when measuring the impedance of an electrode of large area, when measuring impedance in a highly resistive electrolyte, or when the impedance of interest is that of a highly conductive electrolyte.

- Measurements can be made effectively down to 0 Hz. Because the bridge shown in Figure 3.1.4 is direct coupled, the low-frequency limits are those of the null detection system and the patience of the experimenter.

- Measurements can be made in the presence of a dc bias under potentiostatic control, without the use of blocking capacitors.

- Impedance can be measured over an extremely wide range, from below 10^{-3} Ω to over 10^9 Ω.
- Error resulting from the impurity of standards can be virtually eliminated because standards can be selected according to ideality, not magnitude of the components.
- By using differential gain for the real and reactive standards, a suitable range of measurement can be selected for each impedance component separately. This feature is incorporated in Figure 3.1.5.
- Impedances may be measured in all four quadrants (RC, $-RC$, RL, $-RL$) using resistance and capacitance standards alone.

Because the gains of amplifiers A and B perform the same function as the ratios in a transformer ratio arm bridge, the two techniques have many features in common.

High-Frequency Limitations. Inaccuracies at high frequencies can occur because of errors in the gain functions A and B with decreasing amplifier open-loop gain

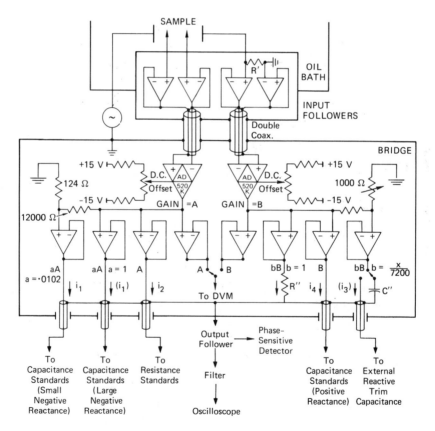

FIGURE 3.1.5. Schematic diagram of a working (modified) Berberian–Cole bridge shown as a four-terminal impedance-measuring system.

(McKubre and Macdonald [1984]). Figure 3.1.5 shows, schematically, a practical bridge of the Berberian–Cole type. Gain errors in the voltage followers are negligible, and, since amplifiers A and B are identical devices and their gains appear as a ratio in Eq. (13), inaccuracies in this term are partially compensated. Nevertheless, the upper operating frequency limit for the bridge shown in Figure 3.1.5 is about 10 kHz, depending somewhat on the magnitude of the unknown impedance. This device is capable of 0.01% measurement accuracy for both impedance components between 1.0 Hz and 1 kHz, and 0.1% accuracy in the peripheral decades (0.1–1.0 Hz, 1–10 kHz).

Low-Frequency Limitations. As stated previously, the low-frequency operating limit is imposed by the detection system. At frequencies down to 0.5 Hz, a two-component PSD performs an ideal null detection function (McKubre [1976]). At frequencies below 0.1 Hz, a low-pass filter and oscilloscope or picoammeter can be used (Berberian and Cole [1969]).

Potential Control. Although it is possible to impose ac potentiostatic control at the interface of interest, the presence of a dc bias will result in a signal in the active bridge circuits. Dc offset must be adjusted to near zero to prevent overloading in subsequent gain stages. For a cell under dc potentiostatic control, this requirement may necessitate frequent offset adjustment of the current amplifier B.

3.1.2.4 *Considerations of Potentiostatic Control*

An essential element of electrode kinetics is the characteristic dependence of electrode reaction rate on the electrode potential. Thus, for many electrode studies, the use of the potentiostatic control is the most convenient method of obtaining relevant kinetic and mechanistic parameters. A limitation of passive bridge methods in general is their inflexibility with regard to potential control, so that in many cases, the experimenter must forgo the advantages of simplicity and sensitivity associated with bridge measurement to impose ac and/or dc potentiostatic control at a single interface. The "direct" methods permit effective potential control while retaining the relative simplicity of operation of many of the bridge techniques.

If the cell current and voltage are measured with regard to their magnitude and phase relations, the impedance can be determined directly from Eq. (2). Figure 3.1.6 shows, in simplified form, a circuit that allows the direct measurement of impedance under potentiostatic control.

It is necessary at the outset to separate phase shifts associated with the cell impedance from those attributable to the potentiostat control loop. Commercial

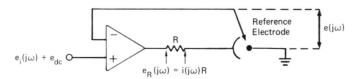

FIGURE 3.1.6. Direct measurement of interfacial impedance under ac and dc potentiostatic control.

potentiostats normally are optimized for fast step response, and the potentiostatting function becomes substantially in error for sinusoidal inputs, with increasing frequency. Analyses have been performed of the frequency-dependent errors introduced by the potentiostatting function for a variety of potentiostats with varying loads (Brown, et al. [1968], McKubre and Macdonald [1984]). However, the fidelity of the potentiostatting function with respect to an ac test signal superimposed on a dc control level is seldom of significance *provided that* the system under test is linear (the fundamental assumption of the use of ac methods in electrochemical kinetic studies) and that the ac voltage is measured directly as the potential difference between the working and a suitably placed reference electrode, and *not* at the input to the potentiostat.

In fact, so-called high-speed potentiostats are often undesirable for use in high-frequency impedance measurements at an electrode–aqueous-electrolyte interface. The reactive impedance at such an interface reduces to that of the double-layer capacitance at limiting high frequencies. Thus, one may have 3 A$/$cm^2 of out-of-phase current flowing at 10^6 Hz to an electrode with 50 μF$/$cm^2 of double-layer capacitance if the potentiostat is able to maintain a 10 mV ac perturbation at that frequency. Such high current densities may result in severe nonlinearities, and one often will prefer the reduced amplitude and phase shift of a narrow-bandwidth potentiostat when the voltage is measured at the point of interest (e, not e_i in Figure 3.1.6).

3.1.2.5 Oscilloscopic Methods for Direct Measurement

By recording $e(j\omega)$ and $i(j\omega)$ (as the voltage drop across a series resistance R_s; see Fig. 3.1.6) with a twin-beam oscilloscope, the magnitude of the impedance can be calculated from the ratio of the two peak-to-peak voltages and the directly observed phase angle. Figure 3.1.7 shows the oscilloscope traces for $e(j\omega)$ and $e_R(j\omega)$ that result from the imposition of a sine wave between the working and reference electrodes.

The real and imaginary components can be calculated (with reference to Fig. 3.1.7) as

$$|Z| = \frac{R_s\left|e(j\omega)\right|}{\left|e_R(j\omega)\right|} \tag{14}$$

$$Z' = |Z|\cos(\phi) \tag{15}$$

$$Z'' = |Z|\sin(\phi) \tag{16}$$

The time base of available storage oscilloscopes limits low-frequency measurements to about 10^{-2} Hz. High-frequency limitations are imposed by effects external to the oscilloscope, principally stray capacitance and transmission line effects in the leads and cell. Measurements can often be made at frequencies above 10^5 Hz.

The primary limitation of this technique is precision. Oscilloscope linearity is seldom better than 1%, and it is difficult to measure phase angles directly with a

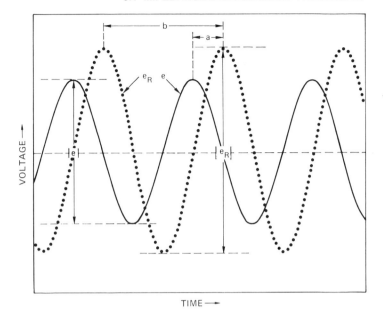

TIME ⟶

FIGURE 3.1.7. The direct measurement of impedance using a twin-beam oscilloscope.

precision of better than 2°. Measurements usually can be accomplished with an uncertainty in Z' and Z'' of $\pm3\%$ of $|Z|$.

A single-beam oscilloscope or an "$X-Y$" recorder also can be used to measure impedance parameters directly by the method of Lissajous figures.

Elimination of t between expressions for e and i of the form

$$e = |e| \sin (\omega t)$$
$$i = |i| \sin (\omega t + \phi)$$

leads to an equation of the form of an ellipse when e and i are plotted orthogonally (e applied to the "X" plates and i applied to the "Y" plates), and the components of the impedance can be calculated from the dimensions of the ellipse. With reference to Figure 3.1.8,

$$|Z| = \Delta e / \Delta i \tag{17}$$

$$\sin (\phi) = \Delta i' / \Delta i = \alpha\beta / (\Delta i \Delta e) \tag{18}$$

where Z' and Z'' can be calculated from Eqs. (15) and (16).

Limitations of oscilloscopic recording are essentially those of precision as described above for two-channel measurement. However, since time is not an explicit variable, time base limitations do not apply when recording Lissajous figures. Low-frequency limitations are imposed by electrochemical instabilities in the system under test and electrical instabilities (particularly dc offset drift) in the

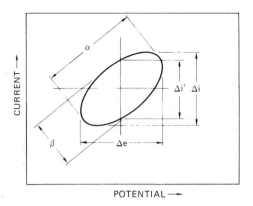

FIGURE 3.1.8. Lissajous figure for the evaluation of impedance.

attendant circuitry. Electromechanical "X–Y" recording can be used to achieve a precision better than 1% of $|Z|$ at frequencies from 1 Hz to below 10^{-3} Hz.

Considerable caution is necessary when applying this last method. Electrochemical systems are susceptible to external noise pickup. The use of high gain, without appropriate electrical filtering, to amplify low-level sine-wave voltage and current perturbations may result in severe errors in the dimensions of the ellipse traced on an electromechanical "X–Y" plotter because the mechanical damping of the plotter may disguise the fact that the input amplifiers are overloaded by the "high"-frequency (> 10 Hz) noise envelope. This effect is shown schematically in Figure 3.1.9 for 50- or 60-Hz mains pickup in the "Y" amplifier. Errors may, of course, occur in both channels. This phenomenon often is reflected as skewing or tracking of the recorded ellipse, but may result in a stable erroneous trace. To prevent errors in the calculated impedance values, appropriate electronic low-pass

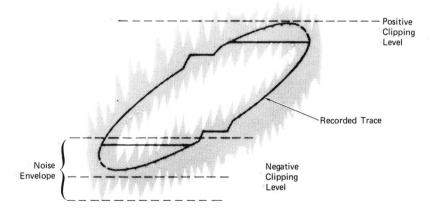

FIGURE 3.1.9. The errors in electromechanical Lissajous figure recording due to the presence of input noise.

or notch (50- or 60-Hz mains pickup) filtering must be used at an early stage of amplification.

3.1.2.6 Phase-Sensitive Detection for Direct Measurement

The real and imaginary components of a voltage can be measured directly with respect to a reference signal with a phase-sensitive detector (PSD). Because of the requirements for linearity, small input signals must be used to measure electrochemical impedances, and noise problems often make it impractical to use either e or e_R (Figure 3.1.6) as a reference signal. Accordingly, e and e_R must be measured alternately in terms of a coherent reference signal of arbitrary phase and the impedance determined from the complex quotient

$$Z = \frac{e}{e_R} R = \frac{(e' + je'')}{(e'_R + je''_R)} R \tag{19}$$

To understand the advantages inherent in this method, it is appropriate to discuss briefly the detection technique.

Phase-sensitive detection may be accomplished by the sequential operation of multiplexing and time-averaging circuits. The multiplexer serves effectively to multiply the input sine wave e_i with a reference square wave e_{ref}. We can represent e_{ref} in terms of its Fourier components,

$$e_{\text{ref}} = \frac{4}{\pi}\left[\sin{(\omega_r t)} + \frac{1}{3}\sin{(3\omega_r t)} + \frac{1}{5}\sin{(5\omega_r t)} + \cdots\right] \tag{20}$$

and the input sine wave can be written as $e_i = |A^0|\sin{(\omega_i t + \phi)}$, where $|A^0|$ is the input signal amplitude, ω is the angular frequency, and subscripts r and i refer to the reference and input signals.

The multiplexer output will be

$$e_{\text{mpx}} = e_{\text{ref}}\, e_i = \frac{2|A^0|}{\pi}\left\{\cos{\left[(\omega_i - \omega_r)t + \phi\right]}\right.$$

$$\left. + \frac{1}{3}\cos{\left[(\omega_i - 3\omega_r)t + \phi\right]} - \frac{1}{3}\cos{\left[(\omega_i + 3\,\omega_r)t + \phi\right]} + \cdots\right\}$$

$$\tag{21}$$

In normal practice, ω_r and ω_i are derived from a common source (i.e., $\omega_r = \omega_i$), and the multiplexer output is

$$e_{\text{mpx}} = \frac{2|A^0|}{\pi}\left[\cos{(\phi)} - \cos{(2\omega_r t + \phi)} + \frac{1}{3}\cos{(-2\omega_r t + \phi)}\right.$$

$$\left. - \frac{1}{3}\cos{(4\omega_r t + \phi)} + \cdots\right] \tag{22}$$

Only the first term in Eq. (22) is time-independent and, when applied to the time-average circuit, will result in a nonzero output,

$$e_{out} = \frac{2}{\pi} \left| A^0 \right| \cos (\phi) \qquad (23)$$

This is obviously a phase-sensitive dc output voltage, which is a maximum at $\phi = 0$.

The PSD output is frequency-selective since the time average of Eq. (21) for $\omega_r \neq \omega_i$ is zero. The important exception to this statement is for $\omega_i = 3\omega_r$, $5\omega_r$, $7\omega_r$, and so forth. That is, a PSD responds to odd-order harmonics of the input signal. This contribution diminishes with the order of the harmonic.

Time-averaging may be accomplished by analog or digital means. In the vast majority of commercial instruments, an analog low-pass smoothing circuit is used with a front-panel-adjustable time constant. This arrangement offers the advantage of simplicity and flexibility in high-frequency operation. The upper frequency limit is commonly 10^5 Hz. The low-frequency limit of analog time-averaging devices is imposed by the practical details of low-pass filter design (Sallen and Key [1955]) smoothing capacitor ideality, current leakage in buffer amplifiers, and external asynchronous (nonrandom) noise effects. The low-frequency limit of commercial instruments is commonly in the range 0.5–10 Hz. Impedance usually can be measured with 0.1% precision in both components over the specified frequency range.

By using digital integration methods, the low-frequency response can be extended to below 10^{-3} Hz. In this method, the average is taken digitally over an integral number of cycles (McKubre and Hills [1979]). At very low frequencies, information relating to e and i taken over a single cycle can be used to calculate the real and imaginary impedance components with a precision of 0.1%.

3.1.2.7 Automated Frequency Response Analysis

In general, direct methods can be used to acquire impedance data significantly more rapidly than bridge methods. This is particularly true for digitally demodulated, phase-sensitive detectors, for which only a single cycle is required. Nevertheless, in unstable systems, such as rapidly corroding specimens, acquisition rate is an important consideration, and a major criticism of PSD methods is that these must be performed frequency by frequency. Fortunately, this often is not a serious hindrance when such equipment is automated. In the past decade, a number of experimenters have used automated "frequency response analyzers" as digitally demodulated, stepped-frequency impedance meters. Typical of this class are the Solartron 1170 and 1250 series frequency response analyzers (FRAs).

FRAs determine the impedance by correlating the cell response $S(t)$ with two synchronous reference signals, one of which is in phase with the sine-wave perturbation and the other shifted $90°$ in phase (Gabrielli, [1981], Gabrielli and Keddam [1974], Armstrong et al. [1968], Armstrong et al. [1977]). A typical FRA is shown schematically in Figure 3.1.10. The sine-wave perturbation function $P(t)$

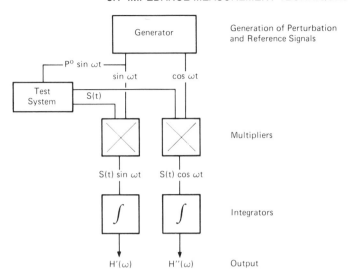

FIGURE 3.1.10. Schematic of transfer function analyzer.

applied to the cell may be represented as

$$P(t) = P^0 \sin{(\omega t)} \tag{24}$$

where P^0 is the amplitude and ω is the frequency. Likewise, the cell response may be written as

$$S(t) = P^0 |Z(\omega)| \sin{[\omega t + \phi(\omega)]} + \sum_m A_m \sin{(m\omega t - \phi_m)} + N(t) \tag{25}$$

where $|Z(\omega)|e^{j\phi(\omega)}$ is the transfer function of the cell and the first term on the right side of Eq. (25) is the fundamental component. However, because of the nonlinear nature of electrochemical systems, the response will also contain harmonics. Also, electrochemical studies are normally carried out in environments electronically "cluttered" by signals due principally to pickup from main power sources. The harmonic and noise contents of the cell response are represented by the second and third terms, respectively, on the right side of Eq. (25).

The real and imaginary components of the impedance are given by the integrals

$$H'(\omega) = \frac{1}{T} \int_0^T S(t) \sin{(\omega t)} \, dt \tag{26}$$

$$H''(\omega) = \frac{1}{T} \int_0^T S(t) \cos{(\omega t)} \, dt \tag{27}$$

Substituting Eqs. (26) and (27) into Eqs. (24) and (25), we obtain

$$H'(\omega) = P^0|Z(\omega)| \int_0^T \sin\left[\omega t + \phi(\omega)\right] \sin(\omega t)\, dt \tag{28}$$

$$+ \frac{1}{T} \int_0^T \sum_m A_m \sin(m\omega t - \phi_m) \sin(\tau t)\, dt$$

$$+ \frac{1}{T} \int_0^T N(t) \sin(\omega t)\, dt$$

$$H''(\omega) = P|Z(\omega)| \int_0^T \sin\left[\omega t + \phi(\omega)\right] \cos(\omega t)\, dt \tag{29}$$

$$+ \frac{1}{T} \int_0^T \sum_m A_m \sin(m\omega t - \phi_m) \cos(\omega t)\, dt$$

$$+ \frac{1}{T} \int_0^T N(t) \cos(\omega t)\, dt$$

If the noise is completely random (i.e., asynchronous), then the last integrals in Eqs. (28) and (29) are equal to zero provided that they are carried out over infinite time. If the integration is carried out over N_f periods of the sinusoidal perturbation, the equivalent filter selectively is given by (Gabrielli [1981])

$$\Delta f/f_1 = 1/N_f \tag{30}$$

where f_1 is the center frequency in hertz and Δf is the bandwidth. For example, if the integration is carried out over 10 periods, then at $f_1 = 1000$ Hz and 1 Hz, Δf is 100 Hz and 0.1 Hz, respectively. On the other hand, if the integration is carried out over 100 periods, the bandwidths are reduced to 10 Hz and 0.01 Hz. Clearly, the ability of a transfer function analyzer to reject asynchronous noise improves greatly as the number of periods over which the integration is performed is increased. However, the price is an excessively long data acquisition time, during which the stability condition may be violated (see Section 3.1.2.9). Figure 3.1.11 shows the transfer function of an FRA as a function of the number of integration cycles performed.

As far as the harmonics are concerned, the integrals in Eqs. (28) and (29) may be expanded to read

$$\int_0^T \sin(m\omega t - \phi_m) \sin(\omega t)\, dt = \cos(\phi_m) \int_0^T \sin(\omega t) \sin(m\omega t)\, dt \tag{31}$$

$$- \sin(\phi_m) \int_0^T \sin(\omega t) \cos(m\omega t)\, dt$$

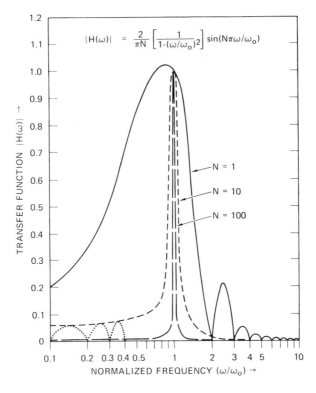

FIGURE 3.1.11. Frequency response analyzer transfer function vs. normalized frequency, as a function of number of integration cycles.

$$\int_0^T \sin\left(m\omega t - \phi_m\right) \cos\left(\omega t\right) dt = \cos\left(\phi_m\right) \int_0^T \cos\left(\omega t\right) \sin\left(m\omega t\right) dt \qquad (32)$$

$$- \sin\left(\phi_m\right) \int_0^T \cos\left(\omega t\right) \cos\left(m\omega t\right) dt$$

Also noting that

$$\int_0^{k\pi T} \sin\left(nx\right) \sin\left(mx\right) dx = \begin{cases} 0 \text{ if } m, n \text{ integers, } m \neq n \\ k\pi/2 \text{ is } m, n \text{ integers, } m = n \end{cases}$$

$$\int_0^{k\pi T} \sin\left(nx\right) \cos\left(mx\right) dx = \begin{cases} 0 \text{ if } m, n \text{ integers, } m + n \text{ even} \\ 2kn/(m^2 - m^2) \text{ if } m, n \text{ integers, } m + n \text{ odd} \end{cases}$$

$$(34)$$

Then the integrals involving the harmonics in Eqs. (31) and (32) are identically equal to zero provided that the integrals are carried out over multiples of 2π. Accordingly, FRAs effectively reject the harmonics. Application of the above identities to the fundamental components in Eqs. (28) and (29) therefore yields the real and imaginary outputs from the integrators as

$$H'(\omega) = P|Z(\omega)| \cos [\phi(\omega)] \tag{35}$$

$$H''(\omega) = P|Z(\omega)| \sin [\phi(\omega)] \tag{36}$$

which may be scaled to give directly the real and imaginary components of the cell impedance.

FRAs are also readily used to determine the harmonics contained within the output from the cell. This is done by multiplying the reference signal to the multipliers (but not to the cell) by the harmonic coefficient (2 for the second harmonic, 3 for the third, and so forth). The ability of FRAs to characterize the harmonics provides a powerful tool for investigating nonlinear systems; a topic that is now being actively developed (McKubre [1983]).

FRAs provide a very convenient, high-precision, wide-bandwidth method of measuring impedances in electrochemical systems. Commercial instruments are available which provide up to $4\frac{1}{2}$ digits of precision in the real and imaginary components, in frequency ranges covering 10^{-4} to 10^6 Hz. These are direct-measuring devices and therefore are not susceptible to limitations on imposed potentiostat control.

The primary limitation is one of cost. The basic FRA may cost on the order of $20,000 (1986), and an additional investment for microcomputer and data storage facility is necessary to accommodate the high rates of data collection made possible by the use of a FRA. A more subtle difficulty often occurs, as these devices are capable of operating with $4\frac{1}{2}$ digits of precision (the data dutifully recorded by the microcomputer) whether or not the instrument is connected correctly, or at all, to the electrochemical cell. Considerable familiarity with electrical systems is necessary in order to get accurate impedance data, particularly at higher frequencies. It is highly desirable that an oscilloscope be used in parallel with the two input channels of an FRA, in order to monitor continuously the form of the input and output signals.

3.1.2.8 Automated Impedance Analyzers

There are a number of automated and semiautomated "impedance analyzers" on the market. Although these are intended primarily for network and network component analysis, they have a limited applicability for measurements in electrochemical systems.

Generally, this class of ac analyzer operates with a so-called autobalance bridge. The desired signal (comprising both ac and dc components) is applied to the unknown impedance, as is shown in Figure 3.1.12a. The current follower effectively constrains all the current flowing through the unknown impedance i_r to flow through the range resistor R_r, presenting a virtual ground at the terminal marked "low" (for a further description of the use of current followers, see McKubre and Mac-

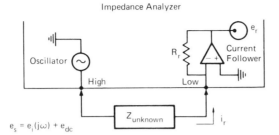

(a) Electrical Connections

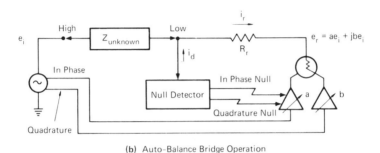

(b) Auto-Balance Bridge Operation

FIGURE 3.1.12. Direct measurement of impedance using an impedance analyzer.

donald [1984]). For this condition the impedance can be measured as

$$Z_{\text{unknown}} = R_r \frac{e_i}{e_r} \tag{37}$$

The complex ratio e_i/e_r is measured in a manner very similar to that described in Sections 2.2 and 2.3 for transformer ratio and Berberian–Cole bridge circuits, in which in-phase and quadrature fractions of the input signal are summed with the unknown output signal (current) until the result is zero. One method of accomplishing this is shown in Figure 3.1.12b. The oscillator that produces the input perturbation signal e_i also outputs in-phase and quadrature (90° out-of-phase) reference signals that are proportional in amplitude to e_i. These are fed to a summing circuit and summed with the unknown current until the current to the detector, i_d, is zero. At this condition the low-potential terminal is at ground voltage, and

$$Z_i = \frac{R_r}{a + jb} \tag{38}$$

Thus, the unknown impedance can be determined directly from the value of the range resistor, R_r, and the attenuation factors a and b imposed by the null detector to achieve the null condition.

The advantages of this method are that relatively high speed and high precision are attainable. Being a null method, the effects of stray capacitances are somewhat reduced, although, unlike in a "true" bridge, currents do flow through the unknown impedance at the null condition. This method is usable up to very high frequencies (tens or hundreds of megahertz), well beyond the range of interest in aqueous electrochemistry.

The intrinsic disadvantage of this method is its two-terminal nature: the facts that a dc potential cannot be applied to the electrode of interest with respect to a suitable reference electrode and that the potential e_i across the specimen varies during the balance procedure. Since the in-phase and quadrature null signals usually are derived from a PSD, instruments of this type are limited at low frequencies to approximately 1 Hz due to the instability of analog filters with longer time constants.

3.1.2.9 The Use of Kramers–Kronig Transforms

The use of a frequency domain transformation first described by Kramers [1929] and Kronig [1926] offers a relatively simple method of obtaining complex impedance spectra using one or two ac multimeters. More important, retrospective use of Kramers–Kronig (KK) transforms allows a check to be made on the validity of an impedance data set obtained for a linear system over a wide range of frequencies. Macdonald and Urquidi-Macdonald [1985] recently have applied this technique to electrochemical and corrosion impedance systems.

The KK transforms of interest in analyzing corrosion and electrochemical systems are

$$Z'(\omega) - Z'(\infty) = \left(\frac{2}{\pi}\right) \int_0^\infty \frac{xZ''(x) - \omega Z''(\omega)}{x^2 - \omega^2} \, dx \tag{39}$$

$$Z'(\omega) - Z'(0) = \left(\frac{2\omega}{\pi}\right) \int_0^\infty \left[\left(\frac{\omega}{x}\right) Z''(x) - Z''(\omega)\right] \frac{1}{x^2 - \omega^2} \, dx \tag{40}$$

$$Z''(\omega) = -\left(\frac{2\omega}{\pi}\right) \int_0^\infty \frac{Z'(x) - Z'(\omega)}{x^2 - \omega^2} \, dx \tag{41}$$

$$\phi(\omega) = \left(\frac{2\omega}{\pi}\right) \int_0^\infty \frac{\log|Z(x)|}{x^2 - \omega^2} \, dx \tag{42}$$

$$R_p = \left(\frac{2}{\pi}\right) \int_0^\infty \frac{Z''(x)}{x} \, dx \tag{43}$$

These equations show that the real component of the impedance can be calculated from the imaginary component and vice versa, the phase angle $\phi(\omega)$ can be computed from the magnitude of the impedance, and the polarization resistance (R_p) can be extracted from the imaginary component [Eq. (43)] in addition to being derived directly from the real component of the impedance:

$$R_p = Z'(0) - Z'(\infty) \tag{44}$$

The use of these expressions to validate impedance data will not be discussed in detail here (see Section 3.4 and Macdonald and Urquidi-Macdonald [1985], Urquidi-Macdonald et al, [1985]). Instead, we note that if the magnitude of the impedance is measured over an effectively infinite bandwidth, then the real and imaginary components can be calculated. These data are then used to compute the polarization resistance directly from Eq. (44) and indirectly from the imaginary component according to Eq. (43). The application of this latter method to calculating R_p in concentrated potassium hydroxide solution at 25°C is shown in Figures 3.1.13 and 3.1.14. In the first figure, the complex-plane diagram for this system is shown, illustrating the inductive behavior at low frequencies and the extrapolation to the real axis to determine $Z'(0)$ and $Z'(\infty)$. In the second figure, the imaginary component is plotted as a function of log x, together with a fitted polynomial

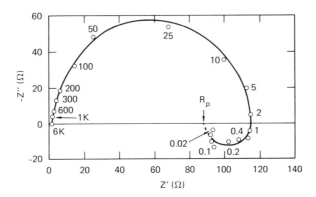

FIGURE 3.1.13. Impedance diagram for Al–0.1 P–0.1 In–0.2 Ga–0.01 Tl alloy in 4M KOH at 25°C and at the open-circuit potential (−1.760 V vs. Hg/HgO). The parameter is frequency in Hz.

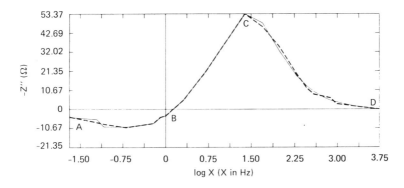

FIGURE 3.1.14. Plot of Z'' vs. log X (X in Hz) for Al–0.1 P–0.1 In–0.2 Ga–0.1 Tl alloy in 4 M KOH at 25°C under open-circuit conditions ($E = −1.76$ V vs. Hg/HgO). Dashed line = polynomial fit [Eq. (45)], solid line = experimental data.

$$Z''(x) = \sum_{k=0}^{n} a_k x^k \tag{45}$$

which is then used to evaluate the integral in Eq. (40). This procedure yields a value for R_p of 82.4 Ω, compared with 90 ± 5 Ω determined from the real components. This difference is insignificant from a corrosion-monitoring viewpoint; it probably arises from changes in the interface during the period of data acquisition.

As indicated earlier, KK transforms not only can be used to check a data set for internal consistency, but also provide a simple method of obtaining impedance data. Briefly, the magnitude of an unknown impedance is often very easily measured as the scalar ratio of the magnitudes of the voltage across, and current passing through, the unknown element. It is considerably more difficult to measure the phase information, but this can be calculated from the impedance magnitude spectrum using the appropriate KK transformation.

A circuit to allow the measurement of an unknown impedance magnitude spectrum is shown schematically in Figure 3.1.15; although a two-terminal configuration is shown, there is no reason that this method cannot be applied with potentiostatic control. To obtain an impedance magnitude spectrum, the frequency of the ac oscillator is simply stepped or swept through the desired frequency range, the ac voltage and current recorded, and the ratio taken as a function of frequency. Since a machine transformation is necessary to obtain the phase spectrum (which can be used in conjunction with the magnitude information to yield the real and imaginary components in Cartesian coordinates), the most practical implementation of this method is to use a computer interfaced with, for example, an IEEE-488 controlled oscillator and multimeters to output frequencies and record the current and voltage information directly (see Section 3.1.3.3 for a description of the IEEE-488 interface).

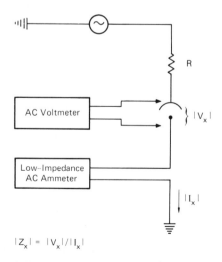

$$|Z_x| = |V_x|/|I_x|$$

FIGURE 3.1.15. Circuit to obtain an impedance magnitude spectrum.

The major advantages of this method are simplicity and low cost; a practical system can be configured for under $5000 if a microcomputer is used to synthesize the sine wave. Since ac amp/voltmeters operate basically as dc devices (following a rectifying front end), they can operate to very high frequencies, and reliable measurements can be made well into the megahertz frequency range, the high-frequency limitation usually depending only on stray capacitances and transmission line effects external to the measurement circuit.

Practical limitations are imposed at low frequencies, however, where the rectification–smoothing function necessary to transduce the ac voltage magnitude to a dc level becomes inaccurate. Ac voltmeters typically become seriously in error at frequencies below 20 Hz. To obtain an accurate KK transform, it is necessary to extend the measurement frequency range significantly beyond the limits of frequency needed to elucidate the equivalent circuit under test. Thus, the method described here is not appropriate for aqueous electrochemical systems for which the diffusional impedance is prominent. This method can be useful for systems in which the lowest frequency of interest is greater than 50 Hz or so, as is usually the case for solid ionic conductors, oxide films, and semiconductor surfaces.

A more subtle limitation is imposed by the use of the method described here in that all of the four assumptions implied in the use of KK transforms are subsumed when the magnitude-to-phase transformation is made. That is, the unknown impedance is given the properties of linearity, invariance, and causality whether or not they apply, and there is no independent check of this assumption. In current practice, this limitation is not very severe since experimenters frequently report and draw conclusions from impedance data sets, normally derived, that have not been subjected to the scrutiny of the KK rules or other simple tests of experimental validity.

3.1.2.10 *Spectrum Analyzers*

Spectrum analyzers are instruments that are optimized to characterize signals in the frequency domain; the requirements of signal analysis are subtly different from those of linear network analysis, the former requiring low noise and low distortion over a wide range of frequencies (bandwidth), the latter being optimized to give accurate amplitude and phase measurements over a wide range of input–output voltages (dynamic range). Nevertheless, spectrum analyzers can be used to measure impedances rapidly at audio and higher frequencies, using a variety of input excitation functions. In this section we will describe the functioning of the three major classes of spectrum analyzer: parallel filter, swept filter, and dynamic.

The classical function of a spectrum analyzer is to measure the power (or amplitude) of a signal at a number of discrete points, or in discrete frequency bands, within a defined frequency range. Normally, the frequency bands are linearly or, more commonly, logarithmically spaced within the spectrum of interest. A very simple method to accomplish this goal is to apply the unknown signal to a parallel array of filters, each tuned to pass a defined (and narrow) frequency band. If these bandpass filters are arranged to be uniformly spaced with minimal overlap, as shown in Figure 3.1.16a, then the output of one or more voltmeters applied se-

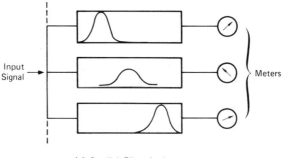

(a) Parallel-Filter Analyzer

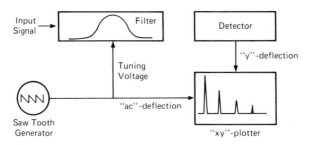

(b) Swept-Filter Analyzer

FIGURE 3.1.16. Spectrum analyzers.

quentially or simultaneously to the parallel array of "N" filters will be an "N"-point analog of the input frequency spectrum. The advantages of this method are simplicity and speed. If, however, a highly accurate analog of the spectrum is needed, then a large number of closely spaced, narrow-bandwidth filters are required. As the bandwidth is reduced, such filters become expensive and unstable, and the cost of such an analyzer becomes greater as the resolution is increased.

One way to avoid the need for a large number of expensive filters is to use only one filter and to sweep it slowly through the frequency range of interest. If, as in figure 3.1.16b, the output of the filter is plotted against the frequency to which it is tuned, then one obtains a record of the spectrum of the input signal. This swept-analysis technique is commonly used in radio frequency and microwave spectrum analysis. However, the filter has a finite reponse time, and the narrower the bandwidth of the filter, the longer it takes to respond. To avoid amplitude errors, one must sweep the filter slowly through the frequency range of interest, and the advantage of speed afforded by spectrum analyzers is compromised.

There is a basic trade-off between parallel- and swept-filter spectrum analyzers. The parallel-filter analyzer is fast but has limited resolution and is expensive. The swept-filter analyzer can be cheaper and have higher resolution, but the measurement takes longer (especially at high resolution). Furthermore, since the swept-filter analyzer does not observe all frequencies simultaneously, it cannot be used to analyze transient events.

A disadvantage common to both classes of spectrum analyzer discussed so far is that they do not measure absolute amplitudes accurately, and they do not measure phase at all. Although this last limitation can be circumvented by the use of KK transformations (see Section 2.9), these instruments generally are poor choices for linear circuit (ac impedance) analysis.

In recent years, another kind of analyzer has been developed which offers the best features of parallel- and swept-filter spectrum analyzers. So-called dynamic signal analyzers use analog-to-digital conversion followed by frequency-to-time-domain transformation, usually using hard-wired computational machines, to mimic the function of a parallel-filter analyzer with hundreds of filters, and yet are cost-competitive with swept-filter analyzers. In addition, dynamic spectrum analyzers are capable of measuring amplitude *and phase* accurately; these are basically time domain instruments, and their function will be discussed in Section 3.1.4.

3.1.3 Time Domain Methods

3.1.3.1 *Introduction*

With the advent of high-speed digital computers, a clear trend toward digital signal processing has become apparent. The advantage of digital over analog data processing is purely mathematical; a far wider range of mathematical computations can be performed in the digital mode than on analog signals. Digital signal processing using hard-wired devices has also expanded rapidly over the past decade and is likely to find even more application in the years to come.

Since the world of electrochemistry is an analog one, the use of digital computation methods must be preceded by analog-to-digital conversion. One of the most important experimental aspects of this process is the method by which the computer interacts with the analog experiment: the computer interface. Having achieved the digital state, the range of computational algorithms used to extract ac impedance information is very diverse. Although a thorough discussion of these topics is beyond the scope of this chapter, in this section we discuss briefly the techniques that are now in common use for analog-to-digital conversion, computer interfacing, and digital signal processing, with reference to the measurement of the ac impedance parameters.

From the definitions given in Section 4.3.2, it is apparent that the interfacial impedance can be calculated from the perturbation and response in the time domain, in which the excitation can be any arbitrary function of time. In principle, any one of several linear integral transforms can be used (Macdonald and McKubre [1981]) to convert from the time domain into the frequency domain, but the two most commonly used are the Laplace and Fourier transforms:

$$F(s) = \int_0^\infty F(t)\, e^{-st}\, dt \qquad (46)$$

$$F(j\omega) = \frac{1}{\sqrt{2\pi}} \int_{-\infty}^\infty F(t)\, e^{-j\omega t}\, dt \qquad (47)$$

where s is the Laplace frequency. Noting that $s = \sigma + j\omega$, Eq. (46) leads to

$$F(j\omega) = \int_0^\infty F(t)\, e^{-j\omega t}\, dt \qquad (48)$$

which is referred to as a single-sided Fourier transform. By transforming the time domain voltage $[E(t)]$ and current $[I(t)]$ to yield the frequency domain quantities $[E(j\omega)$ and $I(j\omega)]$, the impedance may be calculated as

$$Z(\omega) = \frac{E(j\omega)}{I(j\omega)} \cdot \frac{I^*(j\omega)}{I^*(j\omega)} \qquad (49)$$

where $I^*(j\omega)$ is the complex conjugate of $I(j\omega)$.

As noted above, any arbitrary time domain excitation can be used to measure the system impedance provided that the excitation is applied and the response recorded over a sufficiently long time to complete the transforms over the desired frequency band. Thus, potential and current steps and various noise excitations have been extensively used (Sierra-Alcazar [1976], Pilla [1970, 1972, 1975], Doblhofer and Pilla [1971], Smith [1966, 1971, 1976], Creason and Smith [1972, 1973]), particularly in the field of ac polarography. More recently, these same methods have been applied in corrosion science (Pound and Macdonald [1985]; Smyrl ([1985a,b]; Smyrl and Stephenson [1985]) to obtain impedance spectra, but more importantly to estimate polarization resistance for rapidly corroding systems. In the work of Pound and Macdonald [1985], various time-to-frequency transformation techniques were evaluated, including the discrete Fourier transform, the fast Fourier transform (FFT), the Laplace transform, and an algorithm that duplicates the mathematical operation of an FRA (see Section 3.1.2.7). All these techniques involve the recording of the perturbation and response in digital form in the time domain before signal processing in either software or hardware. Regardless of the mode of processing, the accuracy of transformation depends critically on acquiring data records having the desired characteristics of length and sampling frequency.

3.1.3.2 Analog-to-Digital (A/D) conversion

The conversion of analog signals into digital form, and ultimately into binary-word representation, is now a common practice in electrochemistry, particularly for interfacing analog instruments, such as potentiostats, with digital recording and processing equipment (e.g., computers). The essential operation desired is to convert the value of an analog signal into a binary word whose magnitude is proportional to the signal being sampled. This process involves two operations: sampling and quantization. The first involves momentarily "freezing" the analog signal in time to produce a discrete value. This value is then converted to its binary representation during the "quantization" step, after which the cycle is repeated.

Sampling is normally achieved by using "sample and hold" amplifiers of the type shown schematically in Figure 3.1.17. In this circuit, a signal to the analog switch (e.g., 4066 CMOS) connects the analog input to the amplifier. Provided that the capacitance to ground is sufficiently small, the capacitor will charge to the analog input voltage with good fidelity. Removal of the control signal effectively disconnects the input from the capacitor so that the analog output assumes the value of the input at the instant the switch was opened. The above cycle is then repeated, with the sample rate being determined by the control signal from the clock; it is necessary in the case of A/D conversion, however, that the hold time be sufficiently long for the quantization step to generate an accurate digital representation of the analog input.

A number of quantization techniques are available, and the selection of the optimum analog-to-digital converter (ADC) for a particular application is properly based on considerations of resolution (precision), accuracy (initial and drift with time and temperature), ease of interfacing, cost, and convenience (availability, size and power requirements). To select an ADC, it is useful to understand exactly what types are available and how they work. The listing in Table 3.1.1, although far from complete, does include the most popular ADCs, especially those currently produced in an integrated circuit form.

We will discuss here only successive-approximation and integration ADCs in any detail. The tracking A/D and voltage-to-frequency converter can be looked on as variations of the successive-approximation and the integration design techniques; in these types, the digital data is available on a virtually continuous basis.

The principle of the successive-approximation technique is shown in Figure 3.1.18. In this particular example we wish to convert the analog input voltage (10.3 V) into its floating-point, 8-bit binary form to the nearest 0.0625 V. The process involves eight successive steps, in which the field is divided into halves and a bit of 1 or 0 is assigned to each step, depending in which half of each field the analog value lies. For example, in the first step 10.3 lies between 8 and 16, so that 1 is assigned as the most significant bit. However, in the second conversion

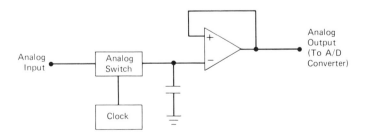

FIGURE 3.1.17. Schematic diagram of a sample-and-hold amplifier.

TABLE 3.1.1. Property of Common Analog-to-Digital Converters

Type	Advantages	Disadvantages	Typical Uses
Successive approximation	High speed Flexibility	Precision expensive Susceptible to noise	Multiplexing 100 Hz – 1 MHz/ channel
Integrating	High accuracy Low cost Low sensitivity to noise	Low speed	DC – 100 Hz Digital voltmeters
Tracking (counter– comparator)	High tracking speed	Susceptible to noise	DC – 100 Hz
Multicomparator (flash)	High speed High resolution	Expensive	1MHz and up
Voltage–frequency converter	Fast response Continuous output	Moderate precision	Telemetry

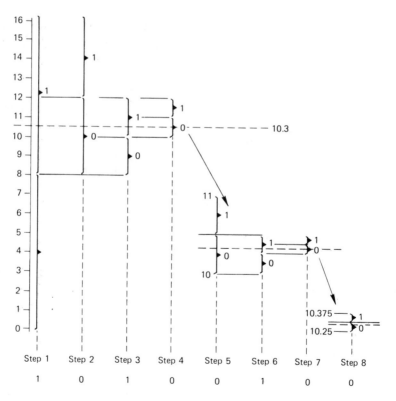

FIGURE 3.1.18. Successive-approximation conversion of an analog signal to its 8-bit binary representation. Analog input = 10.3, equivalent binary output = 1010.0100 ≡ 10.25.

step the analog input lies in the 8–12 field rather than in the upper half (12–16), so that 0 is assigned to the second most significant bit. This process is repeated until the desired precision is achieved. Clearly, only n steps are required to quantize an unknown voltage into its $\{A\}_n$ binary form.

A block diagram of a successive-approximation ADC is shown in Figure 3.1.19. The circuit converts each successive approximation into an analog signal Y, which is equivalent to the center of each division shown in Figure 3.1.18. The analog input (A) is then compared with Y in the following cycle; if A is greater than Y, the comparator swings to positive saturation and a "1" is loaded into the register.

The principal advantage of the successive-approximation technique is high speed, and conversion rates in the megahertz range are possible. The principal disadvantages are limited accuracy and precision. Accuracy is limited because, as with all wide-bandwidth (high-speed) devices, the technique is very susceptible to external noise sources, and a noise spike coinciding with any of the more significant bit conversions can result in large errors; to some extent this difficulty can be removed by averaging multiple conversions, at the expense of speed. The precision is limited by the number of bits converted. Thus a 16-bit ADC (8, 10, 12, 14, and 16 bits are commonly used) will have a minimum uncertainty of 1 part in 2^{16} or 0.0015% and can achieve that level of precision only if the voltage being measured is close to the ADC maximum. However, even when using a 2-V device, the sensitivity of a 16-bit ADC is 0.03 mV, which is adequate for most electrochemical measurements.

A second A/D conversion technique offers greater immunity to noise and almost unlimited precision, as well as reducing the need for sample-and-hold circuitry at the signal input. The concept of the "dual-slope" or up/down integrating ADC is simple. A current proportional to the input signal charges a capacitor for a fixed length of time; the capacitor is then discharged by a current proportional to a reference voltage until the starting point is crossed. Figure 3.1.20 shows a schematic representation of the implementation of this technique. The input voltage e_i is applied to the input resistor R_i. With the switches in position 1, op amp

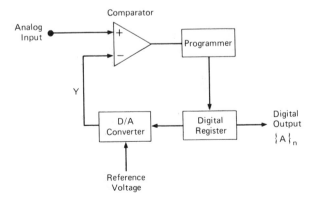

FIGURE 3.1.19. Schematic diagram of a successive-approximation A/D converter.

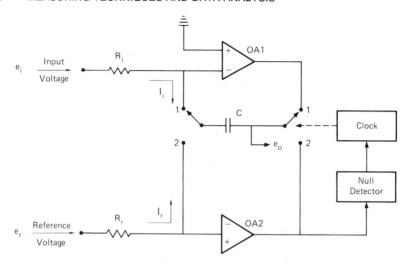

FIGURE 3.1.20. Schematic diagram of an up/down integrating D/D converter.

OA1 forces a current $I_i = e_i/R_i$ to charge the plates of capacitor C. Thus, with the switches in position 1, and after time t_i, the output voltage e_o will be given by

$$e_o = e_{initial} + \frac{1}{RC} \int_0^{T_i} e_i \, dt \tag{50}$$

where $e_{initial}$ is the output at $t = 0$. For simplicity, if the initial voltage is zero and the integral is replaced by the average in the input time window, $\bar{e}_i$,

$$e_o = \frac{\bar{e}_i \, t_i}{R_i C} \tag{51}$$

After an accurately clocked interval t_i, the clock sets the switches to position 2, and the negative reference current is applied to discharge the capacitor. Thus,

$$e_o = \frac{\bar{e}_i t_i}{R_i C} - \frac{e_r t}{R_r C} \tag{52}$$

and at the condition of null, when $e_o = 0$,

$$e_i = \frac{R_i t}{R_r t_i} e_s \tag{53}$$

and the averaged input voltage can be calculated very precisely from the accurately known values of resistances and time.

Although good-quality components (especially the capacitor) must be used for reasonable accuracy, only the reference need be an expensive, high-quality component. Speed is an obvious limitation because of the long count time required. (For example, one must count to 2000 and effectively do 1000 successive comparison tests at the null detector to achieve 3-digit or 10-bit resolution.)

Dual-slope integration has many advantages. Conversion accuracy is independent of both the capacitor value and the clock frequency, because they affect both the upslope and the downramp in the same ratio. The averaging mode and the fixed averaging period also grant excellent immunity to noise, and an integrating ADC has "infinite" normal-mode rejection at frequencies that are integral multiples of $1/t_i$. In practical terms, if the ADC is set to integrate over exactly n cycles of some extraneous and periodic noise source (e.g., main frequency) then the integral will be zero, as if the spurious signal had been completely filtered out at the input.

Throughput rate of dual-slope converters is limited to somewhat less than $\frac{1}{2}t_i$ conversions per second; the sample time t_i is determined by the fundamental frequency to be rejected. For example, if one wishes to reject 60 Hz and its harmonics, the minimum integration time is 10.167 ms, and the maximum conversion rate is somewhat less than 30 Hz.

3.1.3.3 *Computer Interfacing*

The details of computer interfacing are so intimately connected to the details of programming itself that a discussion of arbitrary, low-level interfacing is best suited to a treatise on software than to one on hardware. Interfaces that operate at a high level, with their details of operation obscured to the user by a "driver" program, however, are of significant importance to the experimenter interested in the implementation of ac methods. Most common among the high-level interfacing systems is the general-purpose interface bus (GPIB), also known as the IEEE-488 (or IEC-625 or HPIB*). This interface standard is becoming capable of almost universally connecting computers with digital multimeters, transient recorders, Fourier analyzers—in short, with all those tools needed to implement the ac impedance method in the time domain.

IEEE standard 488–1978 interface represents a highly flexible, moderate-speed system that is well suited to general laboratory use. The IEEE-488 interface bus (IB) consists of 16 signal lines. Eight lines are used for data, five for bus management, and three lines are used to establish a temporary communication link, or "handshake," between two devices that are properly attached to the bus. Because there are eight data lines, an 8-bit byte can be communicated in each handshake cycle. Thus 16-bit, 24-bit, etc. words (either instructions or data) can be communicated with sequential handshake cycles. This method is often referred to as "bit parallel, byte serial" transmission (Colloms [1983]).

A very large number of devices can be connected simultaneously to the interface bus. Each device is given a unique address, which is used in establishing a handshake. Handshake is used to ensure that data is transferred from a source to one or more designated acceptors. Figure 3.1.21 shows a portion of the bus structure, and Figure 3.1.22 shows the handshake sequence in detail. The three signals used

---

*Hewlett-Packard Interface Bus.

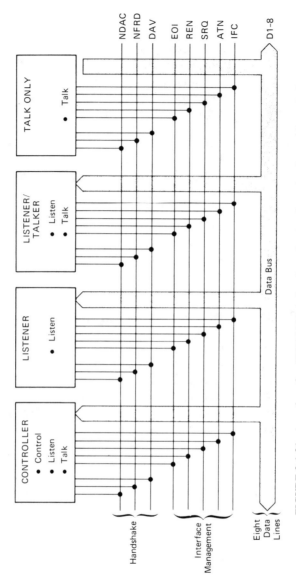

FIGURE 3.1.21. A section of an IEEE-488 interface bus, showing the major classes of user devices.

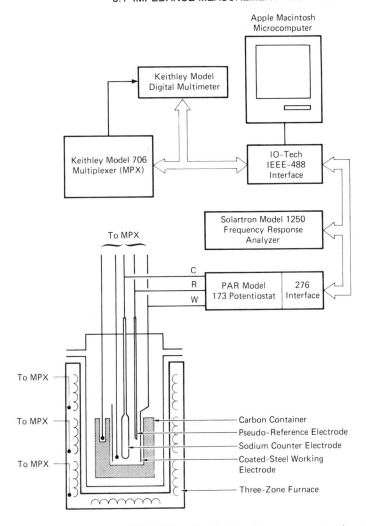

FIGURE 3.1.22. Use of the IEEE-488 interface for dc and ac measurement and control.

for handshake are: data valid (DAV), not ready for data (NRFD), and no data accepted (NDAC). The DAV line is driven by the sender, while the NFRD and NDAC lines are driven by the receiver. The handshake procedure ensures that all listeners are ready to receive data, that the data on the eight data lines is valid, and that the data has been accepted by all listeners. Data will be sent only as fast as it can be accepted by the slowest receiver.

The IEEE-488 IB is designed to interface with the four major types of devices shown in Figure 3.1.21. A master or controller sends commands over the bus, using the bus control (uniline) and data lines (multiline). Normally, one controller is present (e.g., a computer or microprocessor), but if more are present, only one may exercise control at any time. A controller issues a system initiation command, interface clear (IFC), and designates which devices are talkers and which are listeners. The controller has complete control of the attention line (ATN). When ATN is "true," the controller is issuing messages or commands.

A listener receives data over the IB, following an acceptor handshake. Addressed listeners respond to controller commands. Listen-only devices are intended for use in a circuit with no controller. An example of a listener might be a digitally controlled analog potentiostat.

A talker is a device capable of sending data over the IB to a controller or listener. An unaddressed talk-only device, such as a digital voltmeter, may represent a problem in a circuit with a controller, since this device may continue talking when the controller requires attention (ATN "true"), resulting in a garbled message.

Most commercial devices intended for use with an IEEE-488 IB are combined talker–listeners, capable of receiving instructions, setting the data collection mode and experimental conditions, and returning data to the controller.

IEEE-488 systems of considerable complexity have been developed for electrochemical data acquisition and experimental control. One such system, shown in Figure 3.1.22, uses a microcomputer to monitor temperature and dc signals with an IEEE-488 multiplexer and multimeter, to control an IEEE-488 potentiostat, and to output and input data for an IEEE-488 frequency response analyzer, in order to measure impedances in a sodium/sodium-polysulfide cell at elevated temperatures (McKubre and Sierra [1985]).

3.1.3.4 Digital Signal Processing

In principle, any one of several integral transforms can be used to convert data collected in the time domain into the frequency domain for subsequent analysis. Because of the similarity of the various transformation techniques, it is convenient first to discuss briefly the interrelationships of the various transform functions.

The general linear integral transformation of a function $F(t)$ with respect to a kernel $K(t, q)$ is given as (Bohn [1963], Crain [1970])

$$\overline{F}(q) = \alpha \int_a^b K(t, q) F(t) \, dt \qquad (54)$$

where a and b define the transform interval and q is the transformation variable. The kernels and the limits of integration frequently adopted for the transforms of interest are summarized in Table 3.1.2. It is clear that the methods are very closely related. In particular, the reader will note that the imaginary-axis Laplace transformation

$$\tilde{F}(j\omega) = \int_0^\infty F(t) e^{-j\omega t} \, dt \qquad (55)$$

is in fact a single-sided Fourier transform. Also, since the form of the Laplace variable ($s = \sigma + j\omega$) dictates that both frequency domain and transient responses

TABLE 3.1.2. Linear Integral Transforms

Transform	Kernel $K(t, q)$	α	a	b
Laplace	e^{-st}	1	0	∞
Fourier				
Infinite	$e^{-j\omega t}$	1	$-\infty$	$+\infty$
Infinite	$e^{-j\omega t}$	$\dfrac{1}{2\pi}$	$-\infty$	$+\infty$
Single-sided	$e^{-j\omega t}$	1	0	$+\infty$
Segment	$e^{-j\omega t}$	1	0	$+\infty$

can be obtained, it is clear that the Fourier transform is a special case of the more general Laplace transform. The one remaining linear transform of interest, the "Z" transform, to our knowledge has not been used for the analysis of interfacial impedance. Accordingly, this transform will not be considered in this discussion.

In this chapter, we are concerned with machine implementation of transform techniques, either software or hardware, to obtain data in a form convenient for ac impedance analysis. In recent years, the advent of hard-wired (dedicated) Fourier transform units (Reticon [1977]) and the fast Fourier transform algorithm, or FFT (Cooley and Tukey [1965], Hartwell [1971]), have concentrated practical interest almost exclusively on the Fourier transform.

Again, in the formalism of linear systems analysis, the transfer function is the mathematical description of the relationship between any two signals. In the special case where the signals of interest are the input (current excitation) and output (voltage response) of a linear electrical system, the transfer function is equivalent to the system impedance.

Mathematically,

$$G(j\omega) = \frac{E(j\omega)}{I(j\omega)} = \frac{\tilde{F}[E(t)]}{\tilde{F}[I(t)]} = Z(j\omega) \tag{56}$$

where $G(j\omega)$ is the system transfer function, $\tilde{F}$ denotes the Fourier transform, E is the system voltage, and I is the current. The variable $j\omega$ indicates that this is a complex frequency domain parameter, and t indicates a time domain parameter. Equation (56) indicates that the ratio of the Fourier transforms of the measured time domain voltage and current is equal to the impedance.

Obviously, two stages of data manipulation are required to obtain $Z(j\omega)$ as a function of frequency from the response of the system to an arbitrary time domain perturbation: first, the input and response functions must be sampled and recorded in the time window of interest, then the transform of each must be computed and the complex ratio calculated. In hard-wired Fourier transform units, the acquisition subsystem is an integral part of the unit, and this function normally can be ignored. If the experimenter is using a computer or microcomputer to perform these functions, the concepts of analog-to-digital conversion and computer interfacing (described briefly in Sections 3.1.3.2 and 3.1.3.3) must be used.

With either hard-wired or software-programmed logic, the most common method of obtaining $\tilde{F}[E(t)]$ and $\tilde{F}[I(t)]$ uses the FFT algorithm first devised by Cooley and Tukey [1965] as a method for obtaining a discrete digital approximation of the infinite Fourier transformation from a finite data record. The digital nature of the transformation, however, and the finite length of the time record give rise to a number of properties of the FFT that must be recognized in order to minimize distortion of the derived impedance data (Smith [1976], Creason and Smith [1972, 1973]).

The FFT algorithms demand that the time record contain 2^n words, where n is an integer. This requirement is easily satisfied by simply adjusting the digitizing sampling rate and/or the length of the record. However, the sampling theorem states that the highest-frequency component that can be completely characterized in terms of amplitude and phase must have a frequency of less than half the sampling rate. On the other hand, the lowest frequency that is accessible is the reciprocal of the total sampling period. These limitations are readily illustrated by considering a standard FFT array of 1024 words. If this array is collected over 0.7 s, the lowest frequency is 1.43 Hz, whereas the highest frequency is $0.5 \times (1024/0.7) = 731.4$ Hz. Clearly, somewhat less than three decades of frequency are accessible from a single 1024-point FFT. This may be construed as a serious limitation of the FFT algorithm, but it is possible to apply the transformation to successive segments, thereby extending the total frequency range to many orders of magnitude.

The finite length of the data record may cause broadening of the Fourier spectrum relative to the actual spectrum. The phenomenon, which is frequently referred to as "leakage," may be minimized (but not eliminated) by increasing the length of the time record as much as possible or by modifying the way in which the time record is truncated. Also, the leakage error can be reduced to zero if the waveform is periodic within the time record since the components whose frequencies match those computed are not subjected to leakage error.

A third source of error is due to a phenomenon known as "aliasing," which arises because of the discrete nature of the data record. In this case, the error is induced by components whose frequencies are greater than the $\frac{1}{2}x$ sampling rate maximum imposed by the sampling theorem. These higher-frequency components are incorrectly included as lower-frequency components when executing the FFT. Aliasing is easily avoided by simply ensuring that the data-sampling frequency is greater than twice the highest frequency in the exciting waveform. This can be achieved by using a low-pass filter to remove the unwanted high-frequency components, thereby giving rise to the use of bandwidth-limited excitation.

A number of other operational problems exist when using the FFT algorithm. The most important of these, as far as electrochemistry is concerned, is due to the inherently nonlinear nature of the system. When Eq. (56) is used to measure the impedance with an arbitrary time domain input function (i.e., not a single-frequency sinusoidal perturbation), then the Fourier analysis will incorrectly ascribe the harmonic responses due to system nonlinearity, to input signal components which may or may not be present at higher frequencies. As a consequence, the "measured" impedance spectrum may be seriously in error.

Up to this point we have described methods in which impedance is measured in terms of a transfer function of the form given by Eq. (56). For frequency domain methods, the transfer function is determined as the ratio of frequency domain voltage and current, and for time domain methods as the ratio of the Fourier or Laplace transforms of the time-dependent variables. We will now describe methods by which the transfer function can be determined from the power spectra of the excitation and response.

In addition to Eq. (56), the transfer function $G(j\omega)$ can be calculated for the cross-power spectra of the input and the output, which in turn can be calculated from the linear spectra of the input and output. Thus,

$$G(j\omega) = \frac{P_{yx}(j\omega)}{P_{xx}(j\omega)} = \frac{S_y(j\omega)\,S_x^*(j\omega)}{S_x(j\omega)\,S_x^*(j\omega)} \tag{57}$$

where $P_{yx}(j\omega)$ is the average cross-power spectrum of the input and output, $P_{xx}(j\omega)$ is the average power spectrum of the input, $S_x(j\omega)$ and $S_y(j\omega)$ are the linear spectra of the input and output, respectively, and * denotes the complex conjugate.

By invoking the equivalence of Eqs. (56) and (57), it is apparent that the information required to calculate the operational impedance is contained in the input and output linear magnitude spectra S_x and S_y. In practice these are cumbersome to compute. The power and cross-power spectra give the same basic information, are faster to compute, and can be applied to measurements to which linear magnitude spectra cannot (Roth [1970]).

Calculations of power spectra are most conveniently performed via the correlation functions. The auto- and cross-correlation functions for time domain input $[x(t)]$ and output $[y(t)]$ functions are

$$R_{xx} = \frac{1}{T}\int_0^T x(t)\,x(t+\tau)\,dt \tag{58}$$

$$R_{xy} = \frac{1}{T}\int_0^T x(t)\,y(t+\tau)\,dt \tag{59}$$

where T is the time interval over which the correlation is required and τ is a time displacement or delay. In essence, the correlation function yields a time-averaged quantity having greatly improved signal-to-noise characteristics. The value of self- and autocorrelation before transformation is therefore clear.

The significance of the correlation functions in transfer function analysis becomes apparent from the following equations:

$$P_{xx}(\omega) = \tilde{F}\left[R_{xx}(t)\right] \tag{60}$$

$$P_{yx}(\omega) = \tilde{F}\left[R_{yx}(t)\right] \tag{61}$$

Thus, it is possible to calculate the frequency domain power spectra [and hence

$Z(\omega)$] from the Fourier-transformed auto- and cross-correlation functions. The application of correlation techniques for the determination of electrochemical impedance data has been used by Blanc et al. [1975], Barker [1969], and Bindra et al. [1973], using both random noise input functions and internally generated noise.

A number of significant advantages are inherent in this method:

- The correlation technique is an averaging method and thus affords the same type of insensitivity to asynchronous system noise as phase-sensitive detection.

- In common with other transform methods, $G(j\omega)$ is determined for all frequencies simultaneously and in the time required for the lowest frequency alone by conventional methods. Thus, impedance can be measured down to relatively low frequencies in time- varying systems, and impedance parameters can be measured as a function of time in, for example, a rapidly corroding environment.

- Correlation analysis can be performed on internally generated noise in the complete absence of an external excitation function. Because the ionic events that produce this noise are not synchronized to an external trigger, the correlation function in this case contains no phase information, but may be considered analogous to the magnitude of the impedance. This technique is potentially an extremely powerful one, allowing equilibrium and steady state conditions to be approached very closely.

- The coherence function provides an internal check on the validity of the measurement. In this regard, it is important to note that methods which determine impedance as the ratio of the imposed input to the observed output do so without regard to the degree of causality between the two signals. Thus, for example, in a system exposed to mains noise or containing electrolyte pumped in an oscillatory or peristaltic fashion, a component of the output signal power results from frequencies characteristic of the environment or system but not of the applied input. Another frequent cause of error in a measured electrochemical impedance is nonlinearity of the interfacial reaction impedance at large perturbations. Thus excitation at frequency ω_0 results in harmonic distortion and a component of output power at frequencies $2\omega_0$, $3\omega_0$, and so on, which may invalidate the "impedance" measured at these frequencies.

The coherence function $\gamma_{xy}^2(\omega)$ can be calculated to determine the validity of a transfer function measurement if the extent of extraneous input and nonlinearity is not known. This function is defined as

$$\gamma_{xy}^2(\omega) = \frac{\overline{P_{xy}(\omega)}^2}{P_{xx}(\omega)\,P_{yy}(\omega)} \tag{62}$$

where bars denote average quantities and P_{yy} is the auto power spectrum of the output signal $y(t)$. Coherence function values range between 0 and 1. A coherence value of 1 means there is only one input and the system is linear.

The primary limitation of this method of impedance measurement is cost.

3.1.4 Conclusions

In an age of computerized instrumentation, ac impedance and other measurement results are often presented to the user with four or more digits of precision, with little reminder of the intrinsic limitations of the measurement or computational techniques used. Even when considerable care is given to the electrical connections of the system under test and to analysis of subsequently produced data, the operation of the instrument is often transparent to the user. Since the choice of analyzer may determine acquisition precision, time, and other important parameters of data collection, it is of some value to the careful experimenter to understand as fully as possible the method of operation of his impedance analyzer. It is hoped that this chapter is useful in contributing to this understanding.

As a final note of caution, even carefully performed experiments may be subject to systematic error; the system under test may be intrinsically nonlinear, or it may be subject to periodic oscillations, to drift with time, or to other extraneous effects. The results of such perturbations may not be obvious to the experimenter, even when the input and output waveforms are closely monitored with an oscilloscope. It is therefore desirable that impedance data be screened routinely for systematic error. Two screening methods, the Kramers–Kronig integrals and the coherence function, have been described in this chapter, and their use is recommended.

This section was written by Michael C. H. McKubre and Digby D. Macdonald.

3.2 DATA ANALYSIS

3.2.1 Data Presentation and Adjustment

3.2.1.1 *Previous Approaches*
In this section we shall first summarize a number of previous methods of data presentation and then illustrate preferred methods. A common method of showing data has been to plot the imaginary parts (or sometimes their logarithms when they show considerable variation) of such quantities as Z, Y, M, or ϵ vs. ν or log (ν). More rarely, real parts have been plotted vs ν. Such plotting of the individual parts of Z or M data has itself been termed *impedance* or *modulus spectroscopy* (e.g., see Hodge et al. [1976], Almond and West [1983*b*]). As mentioned earlier, however, we believe that this approach represents only a part of the umbrella term *impedance spectroscopy* and that complex plane and 3-D plots can much better show full-function frequency dependence and interrelationships of real and imaginary parts.

Let us consider some of the above plots qualitatively for the simplest possible cases: a resistor R in parallel or in series with a capacitor C. Let the single time constant $\tau \equiv RC$ and define the Debye function $D(\omega\tau) = D' - jD'' = [1 + j\omega\tau]^{-1}$, which leads to a semicircle in the complex plane. Then $D'' = (\omega\tau)/[1 + (\omega\tau)^2]$. A plot of this function vs. ω yields a peak at $\omega_m = \tau^{-1}$ and an eventual dropoff proportional to ω at $\omega \ll \omega_m$ and to ω^{-1} at $\omega \gg \omega_m$. Thus the final slopes of log (D'') vs. log $(\omega\tau)$ are $+1$ and -1. It is easy to show that for the parallel connection; $Z = RD(\omega\tau)$, $-Z'' = RD''(\omega\tau)$, $M = (C_c/C)(j\omega\tau)D(\omega\tau)$, and $M'' = (C_c/C)D''(\omega\tau)$. For the series connection, one finds $Y = (j\omega C)D(\omega\tau)$, $Y'' = GD''(\omega\tau)$, $\epsilon = (C/C_c)D(\omega\tau)$, and $\epsilon'' = (C/C_c)D''(\omega\tau)$. These results demonstrate that under different conditions $-Z''$, M'', Y'', and ϵ'' all exhibit $D''(\omega\tau)$ response. Further, in some sense the pairs Z and $M = (j\omega C_c)Z$, and ϵ and $Y = (j\omega C_c)\epsilon$ are closely related. Real materials often do not lead to IS results of the simple $D(\omega\tau)$ semicircle form, however, but frequently involve a distorted or depressed semicircle in the complex plane which may arise from not one but several relaxation times or from a continuous distribution of relaxation times. Under such conditions, curves of $-Z''$ vs. ν, for example, are often appreciably broader than that following from $D(\omega\tau)$.

It has been customary in much past work to plot $-Z''$ or M'' vs. ν and either not give estimates of parameters, such as R and C, leading to the response or to estimate them roughly by graphical means. It is often found that the frequency at the peak, ω_m, is at least approximately thermally activated. But this frequency involves both R and C, quantities which may be separately and differently thermally activated. Thus instead of using the composite quantity ω_m, which may be hard to interpret properly, we believe it to be far preferable to find estimates of all the parameters entering into an equivalent circuit for the situation. As we shall see, such estimates can best be obtained from CNLS fitting.

One other type of plotting has been very common in the dielectric field in earlier years, namely plotting of tan (δ) vs. ν, where $\delta = -\theta = \tan^{-1}(\epsilon''/\epsilon')$. Thus tan $(\delta) = \epsilon''/\epsilon'$. Since energy loss is proportional to ϵ'' and energy storage to ϵ', δ is often termed the *loss angle*. When ϵ is given by Eq. (18) in Section 2.2 with $\psi_\epsilon = 1$, so there is only a single time constant and $D(\omega\tau)$ response is involved, the simplest case of interest, one readily finds that

$$\tan(\delta) = \frac{(\epsilon_s - \epsilon_\infty)(\omega\tau)}{\epsilon_s + \epsilon_\infty(\omega\tau)^2} \tag{1}$$

which again gives a Debye-type peaked curve, with maximum value $(\epsilon_s - \epsilon_\infty)/[2\sqrt{\epsilon_s\epsilon_\infty}]$ occurring at $\omega\tau = \sqrt{\epsilon_s/\epsilon_\infty}$. But here again there is no longer any good reason to plot tan (δ) vs. ν when much more about the total response can be learned from a 3-D plot of ϵ or Y. Incidentally, the use of tan(δ) plots is entirely absent from a recent compendium of dielectric theory and behavior (Böttcher and Bordewijk [1978]).

3.2.1.2 Three-Dimensional Perspective Plotting
It should be clear from the above discussion that we strongly believe that all IS data should be examined and presented using three-dimensional perspective plot-

ting. One then automatically obtains the plots of real and imaginary parts as projections in the coordinate planes, a normal complex plane plot in the real–imaginary plane and an overall 3-D curve showing the response in proper 3-D perspective. Sometimes when the variation of the quantity plotted is extreme, it is desirable to replace the real and imaginary axes by log(real) and log(imaginary) axes.

A typical simple circuit is shown at the top of Figure 3.2.1, and 3-D plots for its impedance response appear in the middle and bottom of the figure. The two 3-D graphs are for different viewing angles. Note that a log(frequency) axis has been added at right angles to the ordinary $-\text{Im}(Z)$, $\text{Re}(Z)$ complex plane plot, allowing frequency response to appear explicitly. We shall sometimes use f and sometimes ν to indicate frequency in the succeeding graphs. The heavy curve in Figure 3.2.1 is the 3-D response line, its vertical projections to the log (f)–$\text{Re}(Z)$ plane are shown dotted, and the curve in the complex plane plot is actually a semicircle here.

There is an alternative to the usual 3-D plotting of Z in rectangular form which we have been discussing. Instead, we may express a quantity such as Z or M in polar form, involving, say, $|M|$ and θ. If one defines the three axes in a 3-D plot as $(X, Y, Z) = (\theta, |M|, \log(\nu))$, one will obtain a different looking 3-D plot and different projection curves. If the $|M|$ axis is replaced by log $|M|$, one even has a 3-D Bode plot. Note that $\ln(M) = \ln[|M| \exp(j\theta)] = \ln|M| + j\theta$, a complex number, so the $\ln|M|$, θ plane is a consistent complex plane, whereas that involving M and θ is not since M and θ do not form the parts of a complex number. Thus, there is some reason to prefer the second of these approaches to the first. The use of the projection curves $\ln|Z|$ vs. ν and θ vs. ν has been proposed and encouraged recently by Cahan and Chen [1982]. Although there may be instances where both the standard rectangular and polar 3-D representations are worthwhile plotting and examining, we believe that the standard rectangular one with either real and imaginary axes or log axes is usually quite sufficient and shows the entire frequency response of the function considered in a way that yields projection curves of the kind which have been widely used in the past and about which there exists a larger body of interpretative knowledge.

There still remains the problem of which ones of the four functions Z, M, Y, and ϵ to plot in 3-D. Certainly, one should always plot that one of these which was directly measured. If, in addition, all four are plotted and examined, a good general rule, emphasis should be put on the one (or ones) which show the most detail and cover the entire frequency range with best resolution.

We shall illustrate the above by showing results for IS impedance data for a single-crystal Na β-alumina with blocking gold electrodes, measured at T = 83K. These data (Almond and West [1981], Bruce et al. [1982]) were kindly provided by Dr. West [1983]. This set, and data for eight higher temperatures, have recently been reanalyzed with CNLS fitting (Macdonald and Cook [1984, 1985]).

Figure 3.2.2 shows a 3-D impedance plot of the data, using straight lines between data points. Solid dots are employed for the data points themselves and open circles for their projections in the three planes. The log (ν) scale starts at 1 here and in the succeeding plots, and its scale interval (between tickmarks) is always

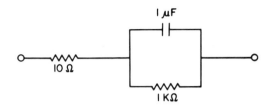

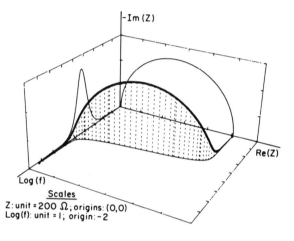

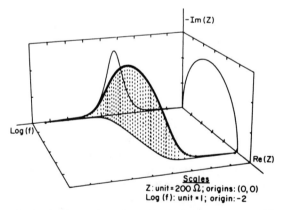

FIGURE 3.2.1. A simple circuit and 3-D plots of its impedance response. The 3-D plots are for different viewpoints. (Reprinted from J. R. Macdonald, J. Schoonman, and A. P. Lehnen, *Solid State Ionics* **5,** 137–140, 1981.)

also 1. The zero points of $-Z''$ and Z' occur at the origin. Three important conclusions may be drawn from these results. First, it is clear that much of the higher-frequency data are too small to be adequately resolved in this plot. Second, it seems likely that the lowest-frequency point is in error. Third, although this anomaly shows up clearly in the 3-D curve and in the complex-plane curve, it is entirely

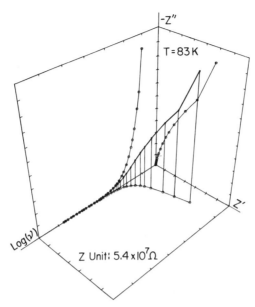

FIGURE 3.2.2. A 3-D plot of the impedance–log(ν) response of single-crystal Na β-alumina at $T =$ 83K. (Reprinted from J. R. Macdonald and G. B. Cook, *J. Electronanal. Chem.* **193**, 57–74, 1985.)

absent from the $-Z''$ vs. log (ν) and the Z' vs. log (ν) projection curves. Were these the only curves plotted, as would often be the case in the absence of 3-D plotting, the anomaly would not be discovered. More 3-D plots are presented in Section 3.2.2.

3.2.1.3 *Treatment of Anomalies*
What should one do about anomalies of this kind, points which do not seem to lie close to a smooth curve? First, if the experiment can be repeated, that should be done, and averaged or best data used. In the present instance, measurement at a few more frequencies between the present lowest and next lowest point would yield intermediate points which would help clarify whether the last point is badly off or not. If the experiment cannot be repeated, then outliers of appreciable magnitude, such as the lowest-frequency point in the present plot, should be omitted (or weighted very low) in subsequent CNLS fitting.

Figure 3.2.3 presents a log-impedance 3-D plot of the same data. It has the virtue of allowing all the data to appear with the same relative resolution, but it clearly reduces anomalies. Although we could show the higher-frequency data alone on a separate regular 3-D plot to achieve better resolution, it turns out for the present data that a 3-D M plot (Fig. 3.2.4) covers the full data range with adequate resolution. Here we see two more important anomalous regions not apparent on the earlier curves or in the papers of Almond and West. The one in the middle arises (West [1983]) from overlapping data taken with two different measuring devices which evidently gave inconsistent results in their regions of overlap. This anomaly only shows up clearly in the 3-D curve and in the M' vs. log (ν)

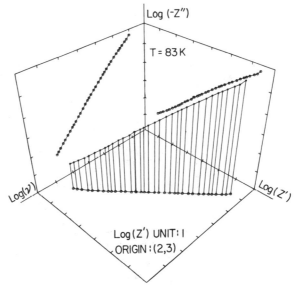

FIGURE 3.2.3. A 3-D plot of the same data as that of Figure 3.2.2, but with logarithmic transformation of the impedance. (Reprinted from J. R. Macdonald and G. B. Cook, *J. Electronanal. Chem.* **168**, 335–354, 1984.)

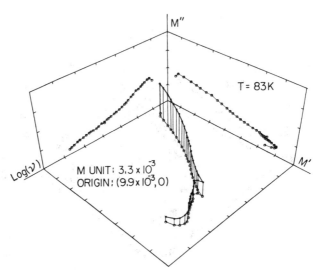

FIGURE 3.2.4. A 3-D modulus function plot of the Na β-alumina, T = 83K data. Here $M = j\omega C_c Z$. (Reprinted from J. R. Macdonald and G. B. Cook, *J. Electroanal. Chem.* **193**, 57–74, 1985.)

curve. It was apparently not recognized until the present 3-D M plotting was carried out (Macdonald and Cook [1984]). In this instance, the plot shows that it would certainly have been desirable to recalibrate the measuring instruments and repeat the experiment. Second, there is a probable anomaly for the highest-frequency three points. Although M plots emphasize high frequencies as compared to Z plots, there is no physical reason to believe that M' should begin to decrease at high frequencies (Macdonald and Cook [1984]), and it would be difficult to justify putting in elements in a fitting equivalent circuit which would cause it to do so. Therefore, in subsequent CNLS fitting of these data, at least the highest-frequency three points should be omitted.

Thus far we have recommended that when appreciable outliers or other anomalies appear in IS data, the experiment be repeated one or more times or, when this is impractical, particularly anomalous points be omitted. But there is one further procedure which has often been found useful in other areas and could be applied here as well. This procedure is smoothing; it should, however, only be carried out after the averaging and pruning mentioned above. All IS frequency data curves should, ideally, follow smooth curves with no abrupt changes in slope, except at possibly sharp cusps where two arcs join each other in the complex plane (see, e.g., Fig. 1.3.1e). Even here no sharp change is present unless the processes represented by the two arcs occur in widely separated frequency regions, and even then it is not present in the 3-D curve itself.

When large anomalies have been eliminated, it is still often valuable to reduce the remaining smaller deviations (noise, experimental error) from a smooth curve by smoothing. The resulting smoothed data will then generally lead to more accurate estimates of the parameters involved in a CNLS fitting. There are many methods available for smoothing. We suggest that the holistic approach of smoothing using B-splines calculated by a least squares approximation would be particularly appropriate for the type of complex data obtained in IS (see de Boor [1978]). Unfortunately it does not yet seem practical to apply this procedure to both real and imaginary parts simultaneously, as would certainly be most appropriate; instead, one would smooth, say, Z' vs. ν [or possibly log (ν)] and Z'' vs. ν results separately and then recombine the results to obtain $Z(\omega)$. This procedure would have another virture. Besides producing an analytic approximation to the Z data, it would yield $\partial Z / \partial \nu$ results with no further calculation. These results, now expressed in analytic form, in turn would simplify CNLS fitting of the smoothed $Z(\omega)$ data to a specific model or equivalent circuit. Much work still remains to be done to develop this approach, which could, in fact, be incorporated as an optional initial part of a CNLS fitting program. It may sometimes prove useful, when major anomalies (which cannot be directly remedied) appear more evident in M or Y 3-D plots than in the Z plot, to carry out the actual smoothing on the transformed data, then convert the smoothed results back to smoothed Z for use in CNLS fitting.

3.2.2 Data Analysis Methods

3.2.2.1 Simple Methods

When nonoverlapping semicircular arcs appear in, say, the impedance plane, one can directly estimate the associated R and C values from the left and right intercepts of the arc with the real axis and the value of ω at the peak of the arc, $\omega_m =$

$(RC)^{-1}$. This procedure is quite adequate for initial estimates, but it yields no uncertainty measures for the parameters and does not check that the frequency response along the arc is consonant with that for an R and C in parallel. Further, experimental arcs rarely approximate exact semicircles well. There have been many graphical methods proposed for the analysis of impedance and dielectric data (e.g., Cole and Cole [1941], Vetter [1967], Macdonald [1974a,b], and Böttcher and Bordewijk [1978]). These methods often consist of plotting some function of the data vs. some function of frequency with the expectation of obtaining a straight line whose parameters may be related to the desired parameters of the equivalent circuit. Although these parameters may be estimated from the straight line by ordinary least squares fitting, this is not usually necessary if the estimates are to be used as initial values in subsequent CNLS fitting. Often subtraction of the effects of some estimated parameters is used to help in the estimation of further parameters. But subtractive methods are notoriously inaccurate. Again, all quantities are usually estimated without using all the available data simultaneously and without yielding uncertainties.

An improved geometrical, iterated-subtractive curve fitting method for resolving two or more overlapping arcs (which may be depressed) has been described by Kleitz and Kennedy [1979] and has been further developed and automated by Tsai and Whitmore [1982]. An algebraic method of estimating parameters for systems which exhibit pseudoinductance (i.e., negative capacitance and resistance) and lead to an arc in the first quadrant of the impedance plane followed by an arc in the fourth quadrant has been proposed by D. D. Macdonald [1978c] (see also Macdonald and Franceschetti [1979b]). Finally, a sophisticated least squares method for finding the best circle through a set of data points in the plane appears in Macdonald, Schoonman, and Lehnen [1982]. For obtaining initial parameter estimates, the simple methods described in Section 1.3 are usually sufficient.

3.2.2.2 Complex Nonlinear Least Squares

None of the above methods uses all the data simultaneously, and they are often restricted to the analysis of limited situations (e.g., two possibly overlapping arcs). Nevertheless, when applicable, these methods are useful for initial exploration of the data and for initial parameter estimates for use in CNLS fitting. The much more general and powerful method of complex nonlinear least squares was first applied to IS in the work of Macdonald and Garber [1977], and further discussion of the method and demonstrations of its high resolving power and accuracy appears in Macdonald, Schoonman, and Lehnen [1982]. The very flexible and general computer program for weighted CNLS fitting described herein is available at nominal cost from J. R. Macdonald. Later, a somewhat similar CNLS program was developed by Tsai and Whitmore [1982], but it does not include the very important feature of allowing arbitrary or analytical weighting of the data and has no built-in procedure for avoiding or recognizing local minima in the sum of squares to be minimized.

Why should one be interested in using complex nonlinear least squares fitting? After all, as already mentioned, the Kronig–Kramers (KK) relations (Macdonald and Brachman [1956]) ensure that if one knows the real or imaginary part of a minimum-phase function over all frequencies, one can calculate the value of the other part at any frequency. This may suggest to the reader that ordinary nonlinear least squares fitting of data for either the real or imaginary part vs. ν should be sufficient. But we never have data over all frequencies, and all experimental data are contaminated with error. Thus, application of the KK relations to the real or imaginary part of actual experimental data often does not allow the measured values of the other part (which contain errors) to be calculated very closely. It therefore appears reasonable to attempt to fit all of the data simultaneously to a given model, a model which involves a set **P** of unknown parameters (e.g., circuit elements) which may enter nonlinearly in the formula for the measured function of frequency, impedance, admittance, and so on. Although the real and imaginary parts of this physically realizable function are connected in a holistic, averaged way with each other through the KK relations, it is usually a good approximation to assume that the random (nonsystematic) errors in each of these quantities are uncorrelated with each other. Since practical fitting models nearly always are minimum-phase and thus satisfy the KK relations, the achievement of a good CNLS fit of data to the model automatically ensures that the KK relations hold, and it is thus unnecessary to carry out the lengthy integrations necessary to check KK satisfaction directly. When no appropriate mathematical model is available, however, CNLS fitting cannot be used. In such cases, KK analysis turns out to be particularly useful, as demonstrated in the discussion of Section 4.3.5.

Complex nonlinear least squares avoids most of the weaknesses of earlier methods since it fits all the data simultaneously and thus yields parameter estimates associated with *all*, rather than half, the data. In addition, it provides uncertainty estimates for all estimated parameters, showing which ones are important and which unimportant in the model or equivalent circuit used for fitting; and finally, it allows one to fit a very complex model, one having 5, 10, or even more unknown (free) parameters. None of the other methods can do this adequately, especially when several of the time constants of the model are close together.

Here, we shall briefly describe a slightly more general fitting method than ordinary weighted CNLS, but for simplicity we shall still refer to it as CNLS. Consider a theoretical model expression $f_t(\omega; \mathbf{P})$ which is a function of both angular frequency ω and a set of model parameters **P**. Consider $i = 1, 2, \ldots, k$ data points associated with ω_i. We need not specify the number of parameters here, but we do assume that they enter f_t nonlinearly in general. Now suppose that f_t is separated into two parts, $f_t^a(\omega; \mathbf{P})$ and $f_t^b(\omega; \mathbf{P})$, which both depend on the same set of parameters. For $f_t(\omega; \mathbf{P}) = Z_t(\omega; \mathbf{P})$, for example, the two parts might be $f_t^a = Z_t'$ and $f_t^b = Z_t''$, or $f_t^a = |Z_t|$ and $f_t^b = \theta$, where θ is the phase angle of Z_t. The least squares procedure involves minimizing the sum of squares function

$$S = \sum_{i=1}^{k} \left\{ w_i^a \left[f_{ei}^a - f_t^a(\omega_i; \mathbf{P}) \right]^2 \right.$$
$$\left. + w_i^b \left[f_{ei}^b - f_t^b(\omega_i; \mathbf{P}) \right]^2 \right\} \tag{2}$$

where w_i^a and w_i^b are the weights associated with the ith data point and f_{ei}^a and f_{ei}^b are experimental data values. When, say, $f_t^b(\omega; \mathbf{P}) \equiv 0$, the procedure reduces to ordinary nonlinear least squares.

Since the above generalization of the ordinary nonlinear least squares method is so minor, it is a simple matter to modify a standard nonlinear least squares program to implement Eq. (2). In any *nonlinear* least squares procedure, however, there is a problem of ensuring that the minimum found in parameter space is the absolute minimum. This problem, which may become serious when the number of free parameters is large, has been attacked in the work of Macdonald, Schoonman, and Lehnen [1982] by using two programs in series. The first does not require inversion of the system matrix and thus nearly always converges. The second program, which uses the results of the first as input, does not usually converge, however, unless it can find a set of parameter values which lead to an absolute minimum in S or at least a good local minimum. The first program often helps it to do so. In spite of the power of this serial method, the achievement of an absolute minimum, giving the least squares estimates of the parameters as well as their uncertainty estimates (standard deviation estimates), is always simplified and facilitated by using the best available estimates of the parameter values as input to the first program. Some of the simpler analysis methods mentioned above are often useful in providing such initial estimates.

The procedure described above is not really a CNLS approach unless f^a and f^b are the real and imaginary parts of a complex variable. But as we have seen, $|Z|$ and θ are not, although $\ln |Z|$ and θ and Z' and Z'' are. Since we sometimes are interested in fitting data in the $|Z|$, θ form rather than $\ln |Z|$, θ, this distinction is worth making, although it makes no formal difference in the minimization of Eq. (2).

3.2.2.3 *Weighting*
The problem of what weights to use is not always an entirely well-defined one. The simplest choice, termed *unweighted* or *unity-weighted*, is to set all w_i^a and w_i^b values equal to unity. But if values of f_e^a and f_e^b vary over several orders of magnitude, as is often the case, only the larger values will contribute appreciably to the sum S, resulting in poor parameter estimates. A reasonable procedure, when it can be done, is to replicate the experiment 5 or 10 times and determine the w_i's from the standard deviations from the mean for each point. The general relations to use are $w_i^a = (\sigma_i^a)^{-2}$ and $w_i^b = (\sigma_i^b)^{-2}$, where the σ's are the experimental standard deviations.

When replication is impractical and there is no direct information on the best individual w_i's to employ, it has usually proved most satisfactory to assume that the relative errors of the measured quantities are constant. This approach has been termed proportional or P weighting. It is equivalent to setting

$$\sigma_i^a = gf_{exi}^a \quad \text{and} \quad \sigma_i^b = gf_{exi}^b \tag{3}$$

where g is a proportionality constant which is usually taken to be unity and whose value makes no difference in the parameter estimates. Here P weighting is particularly needed when the data exhibit large variation. In the Z case, S would become

$$S = S_Z = \sum_{i=1}^{k} \left\{ \left[\frac{Z'_{ei} - Z'_t(\omega_i; \mathbf{P})}{Z'_{ei}} \right]^2 + \left[\frac{Z''_{ei} - Z''_t(\omega_i; \mathbf{P})}{Z''_{ei}} \right]^2 \right\} \tag{4}$$

An alternative choice, which yields nearly the same parameter estimates if the relative errors are small (i.e., the squared terms above), is to take $\sigma^a = gf^a_t(\omega_i; \mathbf{P})$ and $\sigma^b_i = gf^b_t(\omega_i; \mathbf{P})$.

3.2.2.4 *Which Impedance-Related Function to Fit?*
The next problem is which function Z, Y, M, or ϵ to fit. The answer is that it is most sensible from a statistical point of view to fit the data in measured rather than transformed form. Suppose that Z is measured in rectangular form. When both Z and the associated $Y = Z^{-1}$ data are separately fitted with P weighting, it is found that there are often significant differences between the parameter estimates obtained from the two fits. This is not unexpected; the operation of taking an inverse (complex or not) on data with errors generally introduces a bias in the fitted results; it is for this reason that the directly measured results should be fitted directly.

In most automated measurements the rectangular components are measured directly, but sometimes the modulus and phase angle are directly obtained. These are the two quantities, $f^a_e = |f_e|$ and $f^b_e = \theta_e = \measuredangle f_e$, which would then appear in Eq. (2) for S. Again in the absence of measured uncertainties, P weighting would usually be most appropriate. In this case $\sigma^a_i = g|f_{ei}|$ and $\sigma^b_i = g\theta_{ei}$. It is worth mentioning that it has sometimes been suggested that with data in rectangular form a modified P weighting be used in which $\sigma^a_i = \sigma^b_i = g|f_{ei}|$. Such weighting leads to parameter estimates from Z and Y fitting which are generally much closer together than those obtained with ordinary P weighting. Thus with this weighting it makes no significant difference whether the data are fitted to the model in Z or Y form. Since, however, this weighting seems physically unrealistic and blurs a distinction which we in fact expect, it seems to have little to recommend it. Further, when ordinary P weighting is used for fitting of data in either rectangular or polar form, it is easy to show that the ω factors occurring in M or ϵ cancel out of the S function, and thus fitting of Z and M then yield exactly the same set of parameter estimates and relative fitting residuals, and fitting of Y and ϵ also yield the same set of parameter estimates and relative residuals. Of course, the Z and Y sets will be different unless the data are exact for the model considered. Thus when P weighting is employed, one should fit to Z or Y depending on which was measured directly and should fit in rectangular or polar form, again depending on which form is directly measured.

3.2.2.5 *The Question of "What to Fit" Revisited*
The remaining problem is what equivalent circuit or equation to fit by CNLS (see Sections 2.2.2.3 and 2.2.3.4). If it is expected that the data arise from an experiment described by a known analytic model, then of course fitting to the $Z(\omega)$ or $Y(\omega)$ predictions of this model would be appropriate. In the more usual case where a complete, appropriate model is unknown, the first step is to examine 3-D plots of the data and attempt to identify the effects of specific processes appearing in

different frequency regions. An equivalent circuit may then be put together to try to describe these processes and their interactions (see Sections 2.2.3.3 and 2.2.3.4). When CNLS fitting with this circuit is then carried out, one would hope to find little evidence of systematic error (leading to large, serially correlated residuals), small relative standard errors for all free parameters, and small relative residuals for the data points, so the overall standard deviation of the fit, s_f, is small. If relative standard deviations of the parameters are of the order of 30% or more, the associated parameters are not well determined by the data and should be removed from the equivalent circuit. Generally, one would keep modifying the equivalent circuit and doing CNLS fitting until the above criteria are as well satisfied as possible, under the general criterion of using as simple an equivalent circuit with as few individual elements as practical.

3.2.2.6 Deconvolution Approaches

There is an alternative sometimes worth trying to some of the above trial-and-error procedure. Suppose that the impedance-plane plot shows a wide arc which is not exactly a displaced semicircle. It might possibly be best described by several discrete time constants not too far apart (e.g., Armstrong et al. [1974], Badwal and de Bruin [1978]) or by a continuous distribution of time constant. In the first case, the equivalent circuit would involve several individual parallel RC's in series, and in the second it might involve one or more ZARC functions (CPE and R in parallel) in series. Although the best of these choices could be discovered by carrying out several CNLS fits, a more direct method would first be to use deconvolution of the $Z''(\omega)$ data to find an estimate of the distribution function of time constants implicit in the data (see Franklin and de Bruin [1983], Colonomos and Gordon [1979]). Such a distribution, if sufficiently accurate, will separate out the various time constants present, even if they are completely invisible in 3-D plots, and by the width of the individual relaxations apparent in the distribution suggest whether they may be best described by discrete circuit elements or by continuous distributions in the frequency domain. From the values of relaxation time τ where relaxation peaks occur, τ_p, one may also calculate the approximate frequency region $\omega_p = \tau_p^{-1}$ where the relaxation produces its maximum effect. These results may then be used to construct an appropriate equivalent circuit and estimate initial values of the parameters for subsequent CNLS fitting.

The basic equations for obtaining the distribution of relaxation times, $g_Z(\tau)$, at, say, the impedance level, start with the defining relation (Macdonald and Brachman [1956])

$$Z(\omega) = R_0 \int_0^\infty \frac{g_Z(\tau)\, dt}{1 + j\omega\tau} \tag{5}$$

where R_0 is the $\omega \rightarrow 0$ value of $Z(\omega)$. This relation can be put in convolution form by several transformations. Let us use normalized quantities and pick some frequency ω_0 which is approximately the central value of all frequencies measured.

Let $\omega_0 = 2\pi\nu_0$, $\tau_0 \equiv \omega_0^{-1}$, $\omega\tau_0 \equiv \exp(-z)$, $\tau \equiv \tau_0 \exp(s)$, and $G_7(s) \equiv \tau g_Z(\tau)$. We have here introduced the new logarithmic variables s and z. Then Eq. (5) becomes

$$Z(z) = R_0 \int_{-\infty}^{\infty} \frac{G_Z(s)\, ds}{1 + j \exp\left[-(z - s)\right]} \qquad (6)$$

This equation may now be separated into real and imaginary parts, each giving an independent expression involving $G_Z(s)$, the desired quantity. Although both may be used, the imaginary part of Z generally shows more structure than the real part, and it is customary to calculate $G_Z(s)$ from Z'' data rather than from Z' data. The expression for Z'' following from Eq. (6) is

$$Z''(z) = -(R_0/2) \int_{-\infty}^{\infty} G_Z(s)\, \text{sech}\,(z - s)\, ds \qquad (7)$$

now in standard convolution form. The process of deconvolution to find $G_Z(s)$ and thus $g_Z(\tau)$ is generally a complicated one but can be carried out by computer when needed. Two different methods are described by Franklin and de Bruin [1983] and Colonomos and Gordon [1979].

3.2.2.7 Examples of CNLS Fitting
As a first example of CNLS fitting, the circuit shown in Figure 3.2.5 was constructed with lumped elements whose values were measured on an impedance bridge (top figures) (see Macdonald, Schoonman, and Lehnen [1982]). This circuit

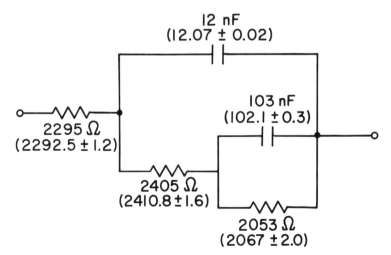

FIGURE 3.2.5. Test circuit involving lumped circuit elements. Nominal values are the numbers on top, while those in parentheses are CNLS estimates. (Reprinted from J. R. Macdonald, J. Schoonman, and A. P. Lehnen, *Solid State Ionics* **5**, 137–140, 1981.)

leads to very little structure in either the Z or Y 3-D plots shown in Figures 3.2.6 and 3.2.7. The bottom numbers shown for each element in Figure 3.2.5 are the CNLS Y-fitting values estimated from all the data; they are probably more accurate than the nominal values. Clearly the resolution and accuracy are very good here

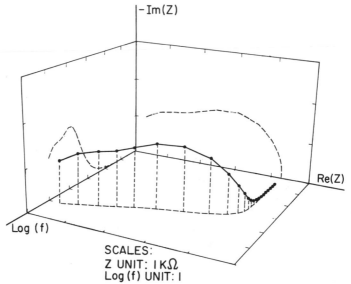

FIGURE 3.2.6. Perspective 3-D plot of the Z response of the circuit of Figure 3.2.5. (Reprinted from J. R. Macdonald, J. Schoonman, and A. P. Lehnen, *Solid State Ionics* **5,** 137–140, 1981.)

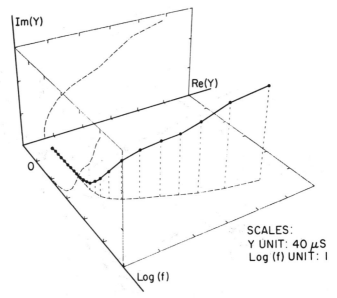

FIGURE 3.2.7. Perspective 3-D plot of the Y response of the circuit of Figure 3.2.5. (Reprinted from J. R. Macdonald, J. Schoonman, and A. P. Lehnen, *Solid State Ionics* **5,** 137–140, 1981.)

in spite of the appearance of little structure in the 3-D plots. But this was a situation where the proper circuit was initially known.

Figure 3.2.8 shows the results of fitting impedance data for β-PbF$_2$ at 474K (Macdonald, Schoonman, and Lehnen [1982]) with CNLS to the circuit at the top of the figure. Initially the form of the circuit which would best fit the data was unknown, so several different ones were tried. The use of a CPE in the circuit, as shown, allowed quite a good fit to the data to be obtained and led to well-determined parameter estimates. The deviations between the dotted and dashed projection lines at the lowest frequencies show that the fit is not perfect in this region, however.

Finally, Figure 3.2.9 shows a 3-D log impedance plot for the same data shown in Figure 3.2.3. Here the fit (dotted curves) appears to agree very well indeed with

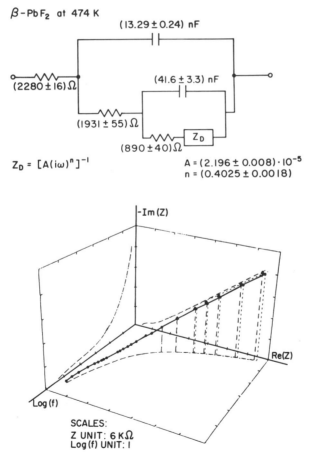

FIGURE 3.2.8. At the top is a circuit used to fit β-PbF$_2$ data at 474K. Parameter values and their standard deviations estimated from CNLS fitting are shown. The bottom part shows a 3-D perspective plot of the Z data (solid line and short dashes) and predicted values and curves (long dashes). (Reprinted from J. R. Macdonald, J. Schoonman, and A. P. Lehnen, *Solid State Ionics* **5,** 137–140, 1981.)

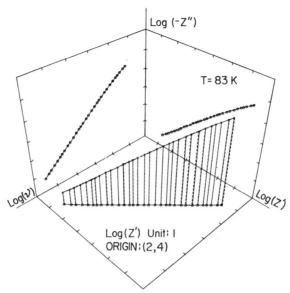

FIGURE 3.2.9. Perspective 3-D Z plot showing both data (solid lines) and CNLS fit (dotted lines) for Na β-alumina data at 83K. Compare Figure 3.2.3 for the data alone. (Reprinted from J. R. Macdonald and G. B. Cook, *J. Electronanal. Chem.* **168**, 335–354, 1984.)

the data, but an M plot (as in Figure 4) showing both data and fit predictions shows regions of appreciable disagreement (Macdonald and Cook [1984, 1985]). Since it has become customary to plot curves such as log ($-Z''$) vs. ν or log (ν) and log (Z') vs. ν or log (ν) when data variation is large and to compare data and fitted results in this form, we wish to warn that this is often insufficient; discrepancies of some importance may be obscured by the logarithmic transformation. Even though several 3-D plots with different scales may be required to show data of wide variability plotted linearly, it is generally a good idea to examine them all when data and predicted values are to be compared. Regions of discrepancy then yield immediate information about needed modifications in the fitting circuit. Finally, although the idea hasn't been much exploited as yet, it is a simple matter to plot relative residuals themselves in 3-D. For Z data one would plot ($-Z''_e + Z''_t)/(Z''_e$) on the vertical axis and $(Z'_e - Z'_t)/Z'_e$ on the horizontal real axis and employ the usual log (ν) axis. For P weighting, one would expect the resulting 3-D curve to be well bounded (magnitudes of most residuals comparable), but the plot should clearly show any regions of correlated residuals, indicating the presence of systematic errors and the need for improving the fitting circuit.

3.2.2.8 Summary and Simple Characterization Example
In the IS field, workers ordinarily spend much time and effort in preparing materials and measuring them under closely controlled conditions. But frequently their subsequent attention to data analysis is quite inadequate and does not do justice to the work done. Although some kind of a data presentation is usually included, it

is rare when it incorporates sufficient plots to resolve all the data well and show its shapes for different kinds of plots. We have tried herein to illustrate some of the virtues of 3-D perspective plotting and strongly urge its increased use in the IS field.

Even in the rare cases where the data are presented adequately, one often finds either no discussion of an appropriate model or equivalent circuit or just the statement that the data fit a given equivalent circuit without either a comparison of the original data and the circuit predictions or a listing of estimates of the values of the parameters in the equivalent circuit. In most cases where such a listing is included, the estimates have been obtained by approximate methods and the reader is given no measure of their accuracy and applicability. Finally, even when an equivalent circuit is presented and used, one rarely finds any discussion of why it is the most appropriate circuit to use or any comparison with other possible fitting circuits.

In summary, we first advocate 3-D plotting of data in various forms, followed by data adjustment and smoothing when warranted. Then crude approximate analysis methods may be employed to estimate initial values of the parameters which enter into an equivalent circuit thought appropriate. Next, CNLS fitting with weighting should be carried out using this and any other possibly likely allied circuits in order to find the simplest and best-fitting one. Then the data and the fitting results should be compared in 3-D and the final parameter estimates and their estimated standard deviations presented.

Even when all the above procedures have been completely carried out, there is a final stage of analysis which should always be included when possible. This stage is the essence of characterization: the passage from good equivalent-circuit macroscopic element estimates to estimates of microscopic parameter values. It is, of course, unnecessary if one is fitting data directly to an impedance function involving microscopic parameters. But in the more usual case of fitting to an equivalent circuit, this stage is the heart of the whole enterprise. A general approach to such macroscopic–microscopic transformation for unsupported systems has been outlined in detail in Macdonald and Franceschetti [1978], and the method is illustrated in, for example, Macdonald, Hooper, and Lehnen [1982] and Macdonald and Cook [1984, 1985].

Here we shall illustrate the method for a simple unsupported situation, that where the equivalent circuit of Figure 2.2.8 applies with only C_∞, R_∞, C_R, and R_R present. Assume that univalent charge of only a single sign, say, positive, is mobile and the partially blocking electrodes are identical with known spacing l. The equations presented in Section 2.2.3 become (per unit area)

$$C_\infty = \epsilon\epsilon_0/l \tag{8}$$

$$R_\infty = l/F\mu_p c_p^0 \tag{9}$$

$$C_R = \left(\epsilon\epsilon_0 F^2 c_p^0/RT\right)^{1/2} \tag{10}$$

and

$$R_R = RT/(F^2 \, c_p^0 k_p^0) \tag{11}$$

Assume that either ϵ, the bulk dielectric constant, is known or that C_∞ has been determined from CNLS fitting of the data. In either case, $\epsilon\epsilon_0$ can be obtained and used in Eq. (10) to obtain from the C_R estimated value an estimate of c_p^0, the bulk concentration of the mobile positive charge carriers. Then this value of c_p^0 may be used in eq. (9) along with the estimated value of R_∞ to obtain an estimate of μ_p, the mobility of the positive charges. Finally, the c_p^0 estimate may be used in Eq. (11) in conjunction with the R_R estimate to calculate k_p^0, the effective reaction rate for the reacting positive charges. Thus from the four macroscopic estimates C_∞, R_∞, C_R, and R_R, one obtains estimates of the four microscopic quantities ϵ, c_p^0, μ_p, and k_p^0. These values and their dependences on controllable variables such as ambient temperature should then finally lead to valuable insight into the electrical behavior of the material–electrode system.

This section was written by J. Ross Macdonald.

CHAPTER FOUR

APPLICATIONS OF IMPEDANCE SPECTROSCOPY

N. Bonanos
B. C. H. Steele
E. P. Butler
William B. Johnson
Wayne L. Worrell
Digby D. Macdonald
Michael C. H. McKubre

4.1 CHARACTERIZATION OF MATERIALS

4.1.1 Microstructural Models for Impedance Spectra of Materials

4.1.1.1 Introduction

In polycrystalline solids, transport properties are strongly affected by microstructure, and impedance spectra usually contain features that can be directly related to microstructure. Nowhere is this more clearly illustrated than in the ceramic electrolytes zirconia and β-alumina. Much of the work done on correlating microstructure and electrical properties has been carried out on these materials since the pioneering work of Bauerle (Bauerle [1969]). One objective of this research effort has been to optimize electrical conductivity for use in solid state electrochemical devices, as described in Section 4.2. We believe that impedance spectroscopy (IS) has an equally important area of application in the study of materials that are not intended for electrical applications, for example, structural ceramics, in which transport properties are incidental to the main application. In the following pages we shall demonstrate that the combined use of IS and electron microscopy is a powerful means of characterizing materials. With the two techniques combined, it is possible to extract conductivity data for minority as well as majority phases with confidence and, in certain cases, to derive microstructural information that is not accessible by electron microscopy alone. As the interpretation of impedance spectra requires models, in this section we discuss models that relate to microstructural features, with a special emphasis on ceramics. Some of the microstructural features that may be modeled are grains and grain boundaries of differing phase composition, suspensions of one phase within another, and porosity.

It is worth remembering that the electrical properties of heterogeneous media have been the subject of much modeling work, stretching back 100 years. Several models in common use today are due to Maxwell [1881] or extensions of these. Many of these models have been reviewed and refined by Meredith and Tobias [1962], while Mitoff [1968] has given a clear account of their scope and usefulness. These articles cover the restricted case where the conductivity or permittivity are real; this normally means dc conductivity or permittivity of loss-free dielectrics, rather than the complete impedance spectrum. AC properties have been discussed by Wimmer, Graham, and Tallan [1974], with special reference to ceramics. The literature devoted to dielectric properties has been reviewed by van Beek [1965]; Dukhin and Shilov [1974] have described models which include the effects of a double layer. Much of the above literature is of relevance to the IS of materials.

The properties of two-phase mixtures can be described terms of bulk-intensive properties such as the complex conductivity ψ, the complex resistivity ρ, permittivity ϵ, and modulus M. To enable comparison of true materials properties, most of the impedance spectra presented in this section have been converted to the complex resistivity and labeled as ρ' and ρ''. For simplicity, spectra of both Z and ρ are described as *impedance spectra*.

A quantity that is most useful in expressing models of this type is the complex conductivity, defined as

$$\psi_i = \sigma_i + j\omega\epsilon_i \tag{1}$$

where σ_i is the dc conductivity and ϵ_i the permittivity of phase i. The role of the model is then to provide a hypothetical microstructure, for which the total complex conductivity ψ_t may be accurately calculated. Having obtained $\psi_t = \psi_t(\omega)$, this may be directly plotted, or converted to some other representation, for example, complex conjugate resistivity ρ_t^*. In general, one can write ψ_t as $\psi_t(\omega) = \sigma_t(\omega) + j\omega\epsilon_t(\omega)$. In many, but not necessarily all cases, there is an equivalent circuit representation for the model. This is very convenient, as the circuit parameters can be estimated from the model and compared to those obtained empirically from impedance spectroscopy. Equation (1) is equivalent to assigning to each phase a Voigt element. Some models for the impedance spectra of two-phase microstructures are equivalent to a circuit consisting of two such elements in series, as shown in Figure 4.1.1. A number of models give rise to this type of behavior, but not all the models predict the same values for the volume-extensive conductive and capacitative components g_1, g_2, c_1, and c_2. The models fall into two types; layer models and effective medium models.

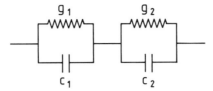

FIGURE 4.1.1. Circuit equivalent of a two-phase microstructure.

4.1.1.2 Layer Models

The earliest model used to describe the electrical properties of a two-phase mixture is the *series layer model* (Maxwell [1881]) shown microstructurally in Figure 4.1.2a. The two phases are assumed to be stacked in layers parallel to the measurement electrodes, with total thicknesses of each phase made proportional to the volume fractions x_1 and x_2. The series layer model is described by the equation

$$\psi_t^{-1} = x_1 \psi_1^{-1} + x_2 \psi_2^{-1} \tag{2}$$

or

$$\rho_t = x_1 \rho_1 + x_2 \rho_2 \tag{2a}$$

which expresses a linear mixing rule for the complex resistivity. The series layer model is equivalent to the circuit of Figure 4.1.1, in which

$$g_1 = \sigma_1/x_1, \qquad g_2 = \sigma_2/x_2 \tag{3}$$
$$c_1 = \epsilon_1/x_1, \qquad c_2 = \epsilon_2/x_2$$

The impedance and modulus spectra for this model consist of two semicircles. The diameters of these are g_1^{-1}, g_2^{-1} in the impedance and c_1^{-1}, c_2^{-1} in the modulus spectra. In practice good resolution cannot be obtained in both Z and M plots. The time constants of the two phases are defined as $\tau_1 = c_1/g_1$ and $\tau_2 = c_2/g_2$. If these time constants differ as a result of differences in c, then the arcs will be well resolved in the impedance spectrum. If they differ as a result of g, they will be resolved in the modulus spectrum (Hodge, Ingram and West [1976]).

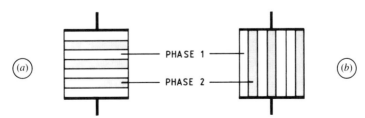

FIGURE 4.1.2. Hypothetical two-phase microstructures described by simple layer models: (*a*) series layer model; (*b*) parallel layer model.

It is interesting to compare the series layer model of Figure 4.1.2a with the corresponding parallel model of Figure 4.1.2b, in which the layers are stacked across the electrodes. For the *parallel layer model,* the complex conductivity rather than the resistivity follows a linear mixing rule,

$$\psi_t = x_1 \psi_1 + x_2 \psi_2 \tag{4}$$

The equivalent circuit would be that of Figure 4.1.3a with values

$$g_1 = x_1 \sigma_1, \qquad g_2 = x_2 \upsilon_2$$
$$c_1 = x_1 \epsilon_1, \qquad c_2 = x_2 \epsilon_2 \tag{5}$$

The behavior of this circuit differs qualitatively from the previous one, because conductances g_1, g_2 and capacitances c_1, c_2 are in parallel. Thus the circuit is equivalent to that of Figure 4.1.3b. This circuit shows only one relaxation. For the microstructure of Figure 4.1.2b the individual relaxations cannot be resolved by any method, graphical, CNLS, or other. Although in principle this model would appear to be as plausible as the series layer model, it fails to describe the behavior of most common ceramic electrolytes.

The two microstructures described illustrate two extreme cases where both or neither of the phases are continuous, both unlikely situations for ceramic microstructures. A third model, originally suggested by Beekmans and Heyne [1976], has been termed the *brick layer model* (van Dijk and Burggraaf [1981], Verkerk, Middlehuis, and Burggraaf [1982]). It is more realistic, treating the microstructure as an array of cubic-shaped grains, separated by flat grain boundaries, as shown in Figure 4.1.4a. The grains have a side D and the grain boundaries a thickness d, where $d \ll D$, and so the volume fraction of the grain boundary phase is $3d/D$. The current flow is assumed to be one-dimensional, and curvature of the current paths at the corners of the grains is neglected. In this case the two paths available to the current are either through grains *and* across grain boundaries or along grain boundaries, as depicted in the exploded diagram shown in Figure 4.1.4b. Depending on the relative magnitudes of σ_{gi} and σ_{gb}, one of the two paths may dominate.

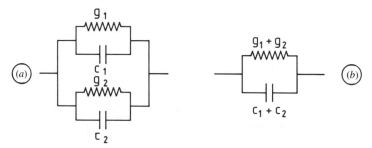

FIGURE 4.1.3. Circuit equivalent of the parallel layer model. Two parallel connected Voigt elements (a) transform to a single Voigt element (b).

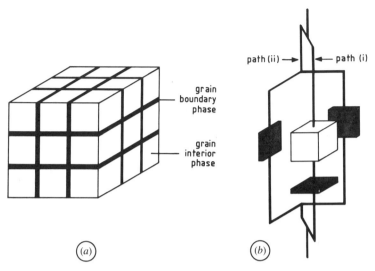

FIGURE 4.1.4. Brick layer model for a two-phase ceramic: (*a*) Overall view, showing array of cubic grains, separated by flat grain boundaries. (*b*) Exploded view of a single cell, showing parallel electrical paths: (i) through grains and grain boundaries, and (ii) along grain boundaries.

Case (i): $\sigma_{gi} \gg \sigma_{gb}$. Conduction along the grain boundaries is negligible, and conduction through the grains and across grain boundaries dominates. The behavior can best be described in terms of complex resistivity:

$$\rho_t = \rho_{gi} + \frac{x_{gb}}{3} \rho_{gb} \tag{6}$$

Case (ii): $\sigma_{gi} \ll \sigma_{gb}$. Conduction along the grain boundaries is dominant. The complex conductivity is given by

$$\psi_t = \psi_{gi} + \frac{2}{3} x_{gb} \psi_{gb} \tag{7}$$

According to Näfe [1984], the two paths (i) and (ii) may be combined into a network describing the polycrystalline properties for all ratios of σ_{gi}/σ_{gb}. Using our own notation, the complex conductivity would then be given by:

$$\psi_t = \left(\psi_{gi}^{-1} + \frac{x_{gb}}{3} \psi_{gb}^{-1} \right)^{-1} + \frac{2}{3} x_{gb} \psi_{gb} \tag{8}$$

Case (i). As described by Van Dijk and Burggraaf [1981], for $\sigma_{gi} \gg \sigma_{gb}$ the brick layer model is equivalent to the series layer model but with a one-third weighting of the grain boundary resistance. This reflects the fact that grain boundaries in only one of the three orientations block the current. The circuit equivalent of the brick layer model is that of Figure 4.1.1 with parameters

$$g_{gi} = \sigma_{gi}, \qquad c_{gi} = \epsilon_{gi}$$
$$g_{gb} = 3\sigma_{gb}/x_{gb}, \qquad c_{gb} = 3\epsilon_{gb}/x_{gb} \qquad (9)$$

An important expression can be derived from these equations in terms of the grain interior and grain boundary permittivities and the volume fractions

$$x_{gb} = 3 \frac{c_{gi}}{c_{gb}} \frac{\epsilon_{gb}}{\epsilon_{gi}} \qquad (10)$$

or, expressed in terms of grain size D and gb thickness d,

$$\frac{d}{D} = \frac{c_{gi}}{c_{gb}} \frac{\epsilon_{gb}}{\epsilon_{gi}} \qquad (11)$$

These relations establish a connection between microstructure and electrical properties allowing one to relate the microscopically observed grain boundary thickness to a thickness estimated from electrical properties. The volume fraction of a grain boundary is difficult to measure by electron microscopy, and it is usually easier to estimate it by assuming $\epsilon_{gb} = \epsilon_{gi}$ and using Eq. (10). In practice, the thickness of a grain boundary phase may easily vary along one boundary by more than the uncertainty introduced by estimating ϵ_{gb}.

Often one refers loosely to g_{gi} and g_{gb} as grain interior and grain boundary conductivities. A more appropriate term would be macroscopic conductivities, as they are corrected for the macroscopic shape of the sample (length/area). For the brick layer model, Eq. (9), and for $d \ll D$, it can be seen that g_{gi} is also the microscopic conductivity of the grain interior. By contrast, g_{gb} is usually 100 to 1000 times higher than σ_{gb} because x_{gb} is small. Thus, we can have $g_{gb} > g_{gi}$ without implying $\sigma_{gb} > \sigma_{gi}$.

Case (ii). When the current is mainly carried along grain boundaries, the brick layer model is equivalent to the parallel layer model (Fig. 4.1.2a), but with a two-thirds weighting on the grain boundary conductance term. A ceramic sample will then show only one arc in the impedance or modulus spectrum and little will be learned about the microstructure.

Thus it is possible on the basis of impedance spectra alone to differentiate between conducting grains with blocking grain boundaries on one hand [case (i)] and poorly conducting grains with highly conducting grain boundaries on the other [case (ii)]. The conductivity equation based on Näfe's [1984] model combines the two extremes in one expression, and is valid at high or low conductivity ratios as it reduces to the expressions given for cases (i) and (ii). We have reservations about the use of this model over the entire σ_{gi}/σ_{gb} range, since it is not clear how, when $\sigma_{gb} \simeq \sigma_{gi}$, this assumption that the current flows via two separate mechanisms can be tenable.

Blocking of Ions—Easy Paths. The brick layer model described in the previous section assumed a continuous grain boundary region separating the individual grains. However, in many instances where $\sigma_{gi} > \sigma_{gb}$ it is found empirically that the activation energies for the two conductivities are equal or very similar. This observation led Bauerle [1969] to suggest that there are regions of the grain boundary where good intergranular contact is established; these are called *easy paths* (Fig. 4.1.5a). The circuit chosen by Bauerle (Fig. 4.1.5b) expresses the idea that migrating oxygen ions are sequentially partly blocked at grain interiors and grain boundaries. Thus g_{ep} is the conduction through constricted ionic paths. Since $\epsilon_{gb} \simeq \epsilon_{gi}$, the existence of a small intergranular contact should not affect c_{gb} noticeably.

A slightly different model, proposed by Schouler [1979], shown in Figure 4.1.5c divides the ionic current into two paths, one of which (g_b, c_b) is blocked capacitatively, while the other (g_a) is not. The proportion of ionic current blocked is then given by

$$\beta = g_b/(g_a + g_b) \tag{12}$$

Circuits (b) and (c) are equivalent, as shown in Chapter 1. Thus, the same ratio β could be expressed in terms of a series model as

$$\beta = r_{gb}/(r_{gi} + r_{gb}) \tag{13}$$

Both models successfully explain the identical activation energies of r_{gi} and r_{gb} observed over a limited range of temperatures. However, it may be convenient to choose between models depending on the expected coverage of grain boundaries by second phase. If the coverage is nearly complete and intergranular contact occurs at a few isolated points, then constriction resistances according to Bauerle's model may be appropriate. But if the grain boundary phase is in the form of isolated islands and the coverage is low, it is difficult to imagine a constriction resistance, and it may be preferable to visualize the situation in terms of a proportion of the ionic charge being blocked.

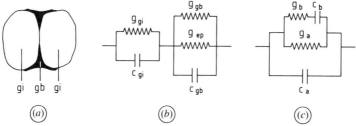

FIGURE 4.1.5. Easy path model for a two-phase ceramic: (*a*) Schematic representation of grains separated by a discontinuous grain boundary phase. (*b*) Series circuit equivalent according to Bauerle [1969]. (*c*) Parallel circuit equivalent according to Schouler [1979].

4.1.1.3 Effective Medium Models

The Maxwell–Wagner Models.   The layer models previously presented have the advantage of clarity, but are derived under slightly unrealistic assumptions concerning the current distributions in heterogeneous media. Alternative models have been developed to take into account the real current distributions, based on a formulation originally devised by Maxwell and now termed the *effective medium technique*. The approach may be summarized as follows. A continuous medium, called the *effective medium*, of conductivity ψ_{eff} has a portion removed and replaced by an equal portion of the heterogeneous system consisting of phases 1 and 2. By applying the constraint that the current distribution in the effective medium is not altered by this operation, an expression is found for ψ_{eff} as a function of ψ_1 and ψ_2. The total conductivity ψ_t is then equal to ψ_{eff}. It is important to note that this method also starts from a hypothetical microstructure for the heterogeneous medium, and Figure 4.1.6 shows the geometrical arrangement for two possible microstructures.

The first exact model of this type was put forward by Maxwell [1881] for the conductivity of a dispersion of spheres of phase 2 in a continuous medium of phase 1. Wagner [1914] showed that the model originally proposed for treating dc conductivity also worked for complex conductivity. The Maxwell–Wagner model gives the following expression for complex conductivity:

$$\psi_t = \psi_1 \frac{2\psi_1 + \psi_2 - 2x_2(\psi_1 - \psi_2)}{2\psi_1 + \psi_2 + x_2(\psi_1 - \psi_2)} \tag{14}$$

Despite the complexity of the resulting expression compared to the layer models, the Maxwell–Wagner model has proven to be extremely versatile in describing the impedance spectra of heterogeneous media. One factor that may have prevented the more widespread use of this model is the absence of an obvious relationship between Eq. (14) and any equivalent circuit.

An embodiment of the Maxwell–Wagner model is shown in Figure 4.1.6*a*, in which the heterogeneous medium consists of a large number of spheres suspended in a continuous medium. The model assumes that any distortion to the electric

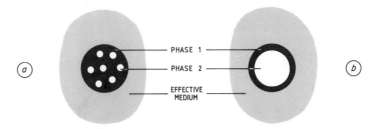

FIGURE 4.1.6. Hypothetical microstructures that may be described by the effective medium model: (*a*) Continuous matrix of phase 1 containing a dilute dispersion of spheres of phase 2. (*b*) A grain boundary shell of phase 1 surrounding a spherical grain of phase 2.

field caused by a particle is local and that neighboring particles see a uniform electric field. This has traditionally been taken to mean that the suspension must be dilute in phase 2.* Thus the model seems to be restricted to matrix–precipitate microstructures. An impedance and modulus simulation for a suspension for $x_2 = 0.25$ is given in Figure 4.1.7a and b, where it is clear that the modulus plot is preferable for resolving the impedance spectra.

A different approach has been taken on this matter by Brailsford and Hohnke [1983], who have applied the effective medium technique to model grain interior–grain boundary phase systems. Their microstructural model, shown in Figure 4.1.6b, consists of a spherical grain of radius r_2 surrounded by a shell of outer radius r_1, which is the grain boundary.* The volume fraction of grain boundary is thus $x_2 = 1 - (r_2/r_1)^3$. Brailsford and Hohnke [1983] derive an equation for ψ_t which in our notation would be

$$\psi_t = \frac{3\psi_2 - 2x_1 (\psi_2 - \psi_1)}{3 + x_1 (\psi_2 - \psi_1)/\psi_1} \tag{15}$$

This expression is none other than Eq. (14) of the Maxwell–Wagner model expressed in terms of the volume fraction of the continuous phase x_1 rather than the discontinuous phase x_2. The equation is very interesting, for, as the authors pointed out, for $x_1 \to 0$ and $\psi_2 \gg \psi_1$, it becomes identical to Eq. (6) for the brick layer model. Furthermore, we would like to add that for $x \to 0$ and $\psi_1 \gg \psi_2$, it reduces to Eq. (7), also an expression appropriate to the brick layer model. The authors appear to have considered this situation but to have arrived at a different conclusion (sec. 3.5 in Brailsford and Hohnke [1983]).

The conditions of low grain boundary phase volume fraction and large disparity in the conductivities of the two phases (particularly $\sigma_{gi} \gg \sigma_{gb}$) assumed are quite realistic as far as ceramic electrolytes are concerned. It is therefore encouraging to see that models derived from two different grain geometries (cubes and spheres) give the same results under these conditions. This suggests that, in practice, the estimation of grain boundary properties from electrical circuit parameters [Eqs. (9), (10), and (11)] may be subject to only a relatively small error as a result of assumptions made about the shape of grains, at least for isotropically shaped grains. This lends confidence to the use of impedance spectroscopy in conjunction with electron microscopy as a tool for the characterization of materials.

Simulated complex resistivity and modulus spectra based on Eq. (14) or (15) for the blocking grain boundary situation (Fig. 4.1.6b) are shown in Figure 4.1.7c and d.

It is evident that for the parameters chosen (continuous phase of lower conductivity) *either* the impedance *or* the modulus spectra resolve the microstructural

---

*The range of volume fraction x_2 over which Eq. (14) is valid has not been accurately specified, and values as low as 0.1 and as high as 0.5 have been proposed. A reasonable estimate might be 0.3, since at volume fractions in the region of 0.35 particles are expected to begin to connect to form continuous paths (percolation limit).

*We have used different subscripts than have Brailsford and Hohnke [1983].

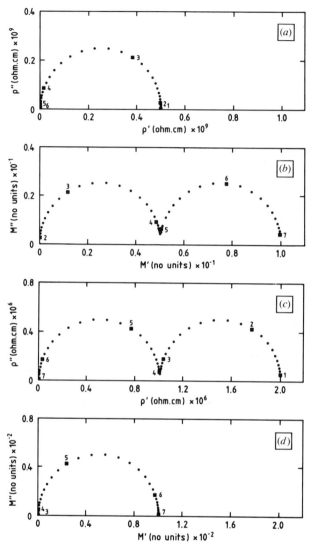

FIGURE 4.1.7. Simulated impedance and modulus spectra for a two-phase microstructure, based on the effective medium model. Values of the input parameters are given in Table 4.1.1. (*a, b*) Spectra for a matrix of phase 1 containing 25% by volume of spheres of phase 2. Resolution is achieved in the modulus spectrum (*b*) but not the impedance spectrum (*a*). (*c, d*) Spectra for a spherical grain of phase 2 surrounded by a grain boundary shell of phase 1. The ratio of shell thickness to sphere radius is 10^{-3}. Resolution is achieved in the impedance spectrum (*c*) but not the modulus spectrum (*d*).

components. For an appreciable volume fraction of high conductivity suspended phase (Fig. 4.1.6*a*), the impedance spectrum is of limited value and the modulus spectrum is preferable. For the situation with a low volume fraction of continuous grain boundary phase (Fig. 4.1.6*b*) the impedance or complex resistivity spectra give good resolution.

Brailsford and Hohnke [1983] make two important observations regarding porosity for the case $\sigma_{gi} \ll \sigma_{gb}$. If the volume fraction of pores is x_p, then for intragranular pores,

$$g_{gi} = \frac{\sigma_{gi}}{1 + \frac{3}{2}x_p} \tag{16}$$

The same amount of porosity distributed intergranularly has a larger effect on the grain boundary conductivity [Eq. (17)], thus highlighting the importance of the spatial distribution of pores:

$$g_{gb} = \frac{3\,\sigma_{gb}}{x_{gb} + \frac{3}{2}x_p} \tag{17}$$

Comparing the last two equations to Eq. (9) shows that the presence of pores modifies the impedance plots (complex conductivity plots) by altering the diameters of the grain interior arc for intragranular porosity and the grain boundary arc for intergranular porosity. Nevertheless, in neither case do pores introduce a new arc or other feature. Thus, when stating that the effect of pores can be seen on the impedance spectrum one should not imply that from an empirical impedance spectrum the degree or type of porosity can be established. This deduction can only be made if the electrical properties σ, ϵ are known beforehand *both* for grains *and* for grain boundary phases, a situation which is unlikely.

Looking at the simulated impedance and modulus plots of Figure 4.1.7 suggests to one that despite being slightly involved, Eq. (14) generates spectra that are similar to those of simple RC circuits. Bonanos and Lilley [1981] showed that the Maxwell–Wagner model, subsequently used by Brailsford and Hohnke [1983], is equivalent to the two-element circuit of Figure 4.1.1. They derived a set of equations giving parameters g_1, g_2, c_1, and c_2 in terms of σ_1, σ_2, ϵ_1, and ϵ_2 and the volume fraction of the discontinuous phase x_2:

$$g_1 = \frac{\sigma_1 A_\epsilon - \epsilon_1 A_\sigma}{\sigma_1 B_\epsilon - \epsilon_1 B_\sigma}\,\sigma_1, \qquad g_2 = \frac{\sigma_1 A_\epsilon - \epsilon_1 A_\sigma}{A_\epsilon B_\sigma - B_\epsilon A_\sigma}\,A_\sigma$$

$$c_1 = \frac{\sigma_1 A_\epsilon - \epsilon_1 A_\sigma}{\sigma_1 B_\epsilon - \epsilon_1 B_\sigma}\,\epsilon_1, \qquad c_2 = \frac{\sigma_1 A_\epsilon - \epsilon_1 A_\sigma}{A_\epsilon B_\sigma - B_\epsilon A_\sigma}\,A_\epsilon \tag{18}$$

where

$$\begin{aligned}
A_\sigma &= 2\sigma_1 + \sigma_2 - 2x_2(\sigma_1 - \sigma_2) \\
B_\sigma &= 2\sigma_1 + \sigma_2 + x_2(\sigma_1 - \sigma_2) \\
A_\epsilon &= 2\epsilon_1 + \epsilon_2 - 2x_2(\epsilon_1 - \epsilon_2) \\
B_\epsilon &= 2\epsilon_1 + \epsilon_2 + x_2(\epsilon_1 - \epsilon_2)
\end{aligned} \tag{19}$$

These equations describe the formal identity between the effective medium model of Maxwell and Wagner and the two-element circuit; they are analogous to

Eq. (9) for the brick layer model and they are valid at all frequencies. So given a particular microstructural arrangement, the circuit parameters may be calculated directly. When an impedance spectrum is required, this may be more easily obtained by simulation of the circuit, rather than by computation of Eq. (14). In Table 4.1.1 we have calculated the circuit parameters for the model microstructures simulated in Figure 4.1.7. For the matrix precipitate microstructure, where $x_1 = 3x_2$ and $\sigma_1 \ll \sigma_2$, we have obtained $c_1 = c_2$ and $g_1 \ll g_2$. For the grain interior-grain boundary microstructure, where $x_1 = 300x_2$ and $\sigma_1 \ll \sigma_2$, we obtained $c_1 \ll c_2$ and $\sigma_1 = \sigma_2$. This clarifies how the relaxations are resolved in the modulus and impedance plots, thus providing some further justification for the assertions made earlier on the basis of the series layer model.

The Fricke Model for Two-Phase Dispersions. Expressions similar to those of Maxwell have been derived by Fricke [1953] for ellipsoidal particles. The expressions contain form factors ϕ which depend on the ellipsoidal axial ratio and on their orientation with respect to the electric field. For ellipsoids oriented randomly in the field, the corresponding equation is

$$\psi_t = \psi_2 + \frac{(\psi_1 - \psi_2)(1 - x_2)}{1 + (x_2/3) \sum_{n=1}^{3} (\psi_1 - \psi_2)/(\phi_n\psi_1 + \psi_2)} \tag{20}$$

where

ψ_t = complex conductivity of the dispersion
ψ_1, ψ_2 = complex conductivities of the matrix and the dispersed phase, respectively
x_2 = volume fraction of the dispersed phase
$\phi_{1,2,3}$ = form factors which depend on the axial ratios of the ellipsoidal particles defined by the semiaxes a, b, c, where $\geqslant b \geqslant c$ (Table 4.1.2)

The case of perfectly oriented ellipsoids has also been treated by Fricke [1953], but the above expression is probably the most useful as it describes a realistic

TABLE 4.1.1. Equivalent Circuit Parameters Calculated Using Eqs. (18) and (19) for Microstructures Simulated in Figure 4.1.7

Volume Fraction x_2	g_1	g_2	c_1	c_1	Ratio g_1/g_2	Ratio c_1/c_2
	Per ohm/cm		F/cm			
0.25	2×10^{-9}	10^{-6}	1.8×10^{-13}	1.8×10^{-13}	2×10^{-3}	1
0.997	10^{-6}	10^{-6}	8.8×10^{-10}	8.8×10^{-13}	1	997

Note: Conductivities σ_1, σ_2 are 10^{-9} and $10^{-6}/\Omega$-cm, respectively. Permittivities ϵ_1, ϵ_2 are both equal to 8.85×10^{-13} F/cm, corresponding to a dielectric constant of 10.

TABLE 4.1.2. Selection of Form Factors According to Fricke [1953] for Use in Eq. (20)

a/b	b/c	ϕ_1	ϕ_2	ϕ_3	Particle Shape
1	1	2.00	2.00	2.00	Spherical
1	2	4.79	1.42	1.42	
1	4	12.2	1.16	1.16	Spheroidal
1	6	21.6	1.09	1.09	
2	2	7.9	2.50	0.66	Ellipsoidal
2	6	20.7	6.9	0.209	

Note: These factors have been calculated from the axial ratios of the ellipsoidal particles a, b, c, where $a \geqslant b \geqslant c$.

microstructure. The model is normally used to describe a situation analogous to that of Figure 4.1.6*a*. Simulated Z and M spectra are shown in Figure 4.1.8. The parameters used for this simulation are given in Table 4.1.3.

For the particular parameters chosen, there are two readily resolvable arcs in the modulus spectrum Figure 4.1.8*b*. The low-frequency arc corresponds to the low-conductivity continuous phase and is apparently a perfect semicircle with its center on the real axis. The high-frequency arc corresponds to the suspended phase and is composed of three relaxations corresponding to the three possible orienta-

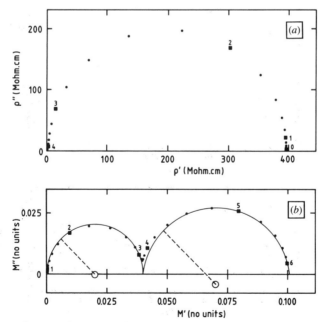

FIGURE 4.1.8. Simulated impedance and modulus spectra for a two-phase microstructure comprising a matrix of phase 1 with 25% by volume of randomly oriented ellipsoids of phase 2. (*a*) Impedance spectrum showing only one arc. (*b*) Modulus spectrum resolving two arcs, one being nonideal due to anisotropy of the ellipsoids. Values of input parameters are given in Table 4.1.3.

TABLE 4.1.3. Parameters Used in the Simulation of Impedance and Modulus Spectra (Fig. 4.1.8a, b) for a Matrix of Phase 1 Containing Randomly Oriented Ellipsoids of Phase 2 According to the Model Developed by Fricke [1953]

Conductivities: $\sigma_1 = 10^{-9}/\Omega$-cm, $\sigma_2 = 10^{-6}/\Omega$-cm
Permittivities: $\epsilon_1 = \epsilon_2 = 8.85 \times 10^{-13}$ F/cm
Ratios of ellipsoid semiaxes: $a/b = 2$, $b/c = 2$
Form factors: $\phi_1 = 7.90$, $\phi_2 = 2.50$, $\phi_3 = 0.66$
Volume fraction of suspended phase: $x_2 = 0.25$

tions of the ellipsoids. In Figure 4.1.8 these are not resolved, but cause the arc to be nonideal. An attempt to fit a circular arc graphically would result in the center being below the real axis. In the above example the impedance plot (Fig. 4.1.8a) shows only the near-ideal low-frequency arc, with a very small distortion at high frequency. Although the shape of the impedance plot is insensitive to the microstructural details, the right-hand intercept is affected by the presence of a second phase, and is governed by Eq. (20) for the special case when ψ is real (dc).

The dispersion can be represented by the circuit shown in Figure 4.1.9a or by its electrical equivalent shown in Figure 4.1.9b, consisting of four series-connected RC elements. For ellipsoids of higher symmetry (spheroids, spheres) the number of elements decreases, so that for spheres, where orientation is immaterial, Eq. (20) reduces to the Maxwell–Wagner model.

Meredith and Tobias [1962] have used the low-volume fraction approximation of Fricke's model as a basis for more elaborate models for higher-volume fractions. These models, however, are only valid for dc conductivity and have therefore not been considered here.

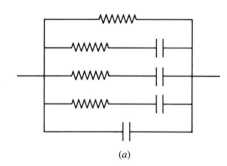

(a)

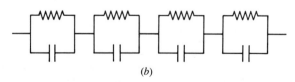

(b)

FIGURE 4.1.9. Two equivalent circuits, for a matrix of phase 1 containing randomly oriented ellipsoids of phase 2 according to model proposed by Fricke [1953]: (a) parallel circuit; (b) series circuit.

The above examples and those of Figure 4.1.7 for the Maxwell–Wagner model illustrate the way microstructural variables can affect the impedance spectra. Furthermore, Fricke's extension of this model provides one example where purely microstructural factors can produce nonideal arcs. This should not, however, be taken to mean that nonideal behavior necessarily implies heterogeneity. The assumption of σ and ϵ being frequency-independent is not always justified theoretically and rarely verified in practice (Jonscher [1975c, 1983]). An explanation has also been given for frequency-dependent conductivity of nonmicrostructural origin, observed in even the well-behaved ceramic electrolytes such as zirconia (Abélard and Baumard [1982]). But these departures from ideality are not in general large compared to the microstructural effects discussed here.

By means of examples we have illustrated the importance of the correct choice of representation: impedance or modulus (or other). In the majority of cases, where grain interior–grain boundary effects are dominant, the impedance representation is the best starting point. In others, the modulus is more useful. We think the importance of the modulus spectrum has, in general, been underrated and suspect that it may also be useful in representing simulated data based on models of homogeneous materials.

4.1.2 Experimental Techniques

4.1.2.1 Introduction
A wide range of materials can be usefully characterized by IS, namely electrical and structural ceramics, magnetic ferrites, semiconductors, synthetic membranes, polymeric materials, and protective paint films. Measurement techniques used to characterize materials are generally simpler than those used for electrode processes. Impedance spectra which accurately reflect microstructural features are usually independent of applied potential (both ac amplitude and dc bias) up to potentials of 1 V or more. Consequently, it is unnecessary to fix the potential of electrodes in contact with the sample, as in potentiostatic experiments, and two-electrode symmetrical cells are most commonly used.

Considerations of Frequency and Impedance Range. As shown in Section 4.1.3., the frequency range to be chosen depends on the relaxation frequencies f_n of phases present in the sample under study and on the microstructure. The highest relaxation frequency is normally that of the grain interiors and is given by $f_{gi} = \sigma_{gi}/(2\pi\epsilon_{gi})$, corresponding to the apex of the grain interior arc in an impedance or modulus plot. Conductivities vary enormously from one material to another, and with temperature for any one material. By contrast, the dielectric constants of most compounds, apart from ferroelectrics, lie within the range of 5 to 100. Taking 30 as a typical value allows us to illustrate the relationship between conductivity and frequency in Figure 4.1.10. The frequency range over which impedance spectra can most accurately be obtained is shown in the upper shaded region. At frequencies above a few megehertz problems of lead length become very serious, whereas at low frequencies the restriction comes from the time taken to perform one cycle of

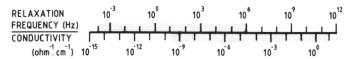

FIGURE 4.1.10. Relationship between relaxation frequency and conductivity and for a solid having a dielectric constant of 30.

measurement. Values of bulk conductivity normally encountered in IS of materials are shown in the lower shaded region. The most highly conducting solid electrolytes, such as $RbAg4I_5$ and related compounds, have conductivities in the region of $1/\Omega$-cm. On the low conductivity side the cutoff depends on the sensitivity of current measurement, with the equipment described in this section (high impedance adaptor) conductivities of less than $10^{-9}/\Omega/cm$ can be measured. In solid electrolytes at moderate temperatures (e.g., yttria-stabilized zirconia at 300°C) grain boundary phases and electrode double layers show up in the impedance spectra at frequencies that correspond to conductivities of the order of $\sim 10^{-12}/\Omega$-cm. However, these regions are very thin, and so their overall contribution to the impedance is often comparable to that of the bulk.

4.1.2.2 Sample Preparation—Electrodes

Bulk properties such as conductivity are calculated using the length over area l/A of each sample; hence, wherever possible, samples should be cut so as to have two parallel faces and a well-defined cross section. For measurement of grain interior or grain boundary properties, surface finish is not critical, as this only affects the electrode impedance (Armstrong et al. [1973]). If the grain boundary and electrode arcs partly overlap, the resolution can sometimes be improved by polishing.

Electrodes are normally applied to the sample by painting, vacuum evaporation, or sputtering. For cation conductors such as β-alumina, reversible electrodes can be made by pressing on alkali metal foils, but this operation requires special precautions such as dry handling. Electrodes are usually made of the precious metals silver, platinum, or gold or of graphite. The types of electrodes commonly used are described below.

Types of Electrodes. Silver paints or "dags" provide good general-purpose electrodes for use at moderate temperatures ($\sim 600°C$). Platinum-painted electrodes are widely used in the IS of ceramics. These require baking at temperatures in the range 600–1300°C, depending on the type used. Some platinum paints contain an inorganic flux or glass frit, which may be undesirable, particularly for the study of electrode impedance. Fluxless platinum paints developed for oxygen sensors require baking at $\sim 1300°C$, which limits the range of materials to which they may be applied. Gold paints are available but not often used in IS studies, as evaporated films can easily be prepared. Water-based carbon dags can be used when painted electrodes are required and the sample is sensitive to organic solvents.

Silver and gold electrodes may be easily applied by vacuum evaporation, but platinum poses difficulties because of its lower vapor pressure. Most metals, including stainless steel, may be deposited by RF sputtering; if required, sputtering can be preceded by gas discharge etching of the surface. Typical film thicknesses are 200–1000 Å. This process produces electrodes of high purity and reproducibility. Prior to electrode attachment, the samples must be thoroughly cleaned, preferably in a refluxing solvent. After cleaning the samples should be baked at 400–500°C to remove traces of organic compounds. Failure to prepare the surfaces properly results in loss of adhesion or in blistering when samples are heated in the course of measurement. The use of thin sputtered Pt film electrodes can pose problems at temperatures over 600°C. Here the appearance and electrode characteristics can slowly change due to a gradual loss of electrode material. The electrode characteristics of gold and platinum electrodes are similar in the temperature region of 0–500°C, that is, relatively blocking to oxygen, but silver gives a lower electrode impedance.

4.1.2.3 A Measurement System for IS of Materials

Presently available commercial measurement instruments do not cover the range of frequency and conductivity shown in Figure 4.1.10. The system described below is based on a SOLARTRON 1174 digital frequency response analyzer (FRA), chosen because of its exceptionally wide frequency range ($10^{-4}–10^{6}$ Hz). The impedance range of the system is $10^{2}–10^{9}$ Ω and the stray capacitance is less than 0.5 pF. Because of this it can be used for materials other than solid electrolytes, for example, poorly conducting structural ceramics, semiconducting ferrites, or organic coatings for corrosion protection of metals. Other systems based on FRAs or network analyzers have been described previously by Morse [1974], Engstrom and Wang [1980], and Staudt and Schön [1981]. A versatile microprocessor-based system has been recently developed by Boukamp [1984].

A schematic diagram of the system is given in Figure 4.1.11. The sample is held in a temperature-controlled rig which is connected to the SOLARTRON FRA via a high-impedance adaptor. All three of these instruments are interfaced to a microcomputer (Research Machines Ltd. 380-Z) which controls the measurement and plots the results in real time. Facilities exist for storing and retrieving files of impedance data and for outputting these to a printer or plotter. Data can be displayed in any of the usual formalisms Z, Y, M, ϵ, $Z''(f)$, and so on.

Principle of Operation. The FRA generates a sinusoidal voltage of specified frequency and amplitude which is applied to the sample under test and measures the in-phase and 90° out-of-phase components of the voltage applied to two inputs designated ''X'' and ''Y.'' The results are expressed as two complex voltages $V_x = V_x' + jV_x''$ and $V_y = V_y' + jV_y''$ and are transmitted to the computer along with the measurement frequency f. Since the inputs of the FRA both measure voltage, its use for the measurement of impedance requires some external components. As long as the measured impedances are not too great, the voltage divider configuration shown in Figure 4.1.12 may be used. In this configuration input ''X'' is used to monitor the generator output, while input ''Y'' measures the potential at the midpoint of the voltage divider. The unknown impedance is then given by

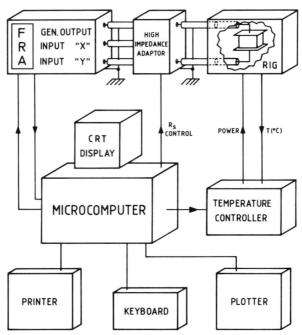

FIGURE 4.1.11. Schematic diagram of measurement system for the characterization of materials by IS.

$$Z_u = \frac{V_y}{V_x - V_y} Z_s \tag{21}$$

Although simple, this configuration suffers from two disadvantages.

1. The inputs of the FRA have a finite impedance (1 Ω in parallel with 50 pF), and the impedance of input "Y" is connected across the sample. When using IS for the characterization of materials, measured impedances are often larger than 1 MΩ, and capacitances of the order of 10 pF; hence errors are introduced. Up to a point these may be calculated out, but a maximum practical limit of about 1 MΩ measured impedance is found for this configuration.

2. The voltage V_u dropped by the unknown is not the generator output V_{out}, but is given instead by $V_u = V_{out} Z_u/(Z_s + Z_u)$, and hence V_u varies with frequency. Thus it is necessary to change the value of the standard resistor as

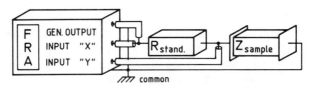

FIGURE 4.1.12. Impedance adaptor for frequency response analyzer (FRA) based on voltage divider network. (After Morse [1974])

the frequency is changed, but the switchgear required for this adds to the already existing capacitance across the sample.

These problems can be partly reduced by modifying the above circuit. The inputs "X" and "Y" can be "buffered" by a very fast unity-gain integrated amplifier such as the LH0033 chip. This device has an input capacitance of 7 pF, an impedance of $> 10^9$ Ω, and a negligible phase shift in frequency range of 1 MHz or below. This refinement is strongly recommended considering the low cost of the devices compared to the FRA.

By the time the stray capacitance of cables, rig and so on are included, the total stray capacitance is around 25 pF, that is, larger than that of a typical sample. In the above configuration, the position of standard and unknown may be reversed, so that the stray capacitance appears across the standard. If the standard impedance is kept very small, the effect of the stray capacitance is negligible. Ratios of Z_u / Z_s of up to 10^4 have been used in conjunction with a high-gain amplifier to compensate for the low level of signal from the standard impedance. A system of dielectric measurements based on this configuration has been described by Morse [1974].

In the study of electrode processes in solid electrolytes it is necessary to restrict the applied voltage to a few millivolts. In such a situation, the loss of signal strength resulting from a low value of the standard resistor would pose serious problems. In designing an impedance adaptor for our system we have adopted a different approach and eliminated the problems of stray capacitance and of variable voltage by using an active current-to-voltage converter as shown in Figure 4.1.13.

Our system uses the fast operational amplifier LH0032CG with the standard impedance placed in its feedback loop. Configured in this way, the amplifier delivers an output voltage such that the point marked "T" in the diagram is essentially at ground potential. (The voltage applied to the sample and the output of the operational amplifier will have opposite signs). As point "T" is not actually connected to ground, it is referred to as a *virtual ground*. As a consequence of this arrangement:

1. The voltage applied to the sample is constant with frequency and equal to the output of the FRA; it is measured at input "X."
2. The voltage across the standard is the output voltage of the amplifier and is measured at input "Y."
3. Since points "X" and "Y" are driven, stray capacitances from these points to earth do not affect the measurement.

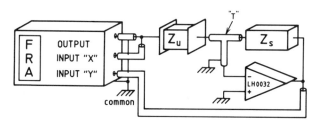

FIGURE 4.1.13. Impedance adaptor for frequency response analyzer (FRA) incorporating fast operational amplifier.

4. Since point "T" is held at earth potential, any stray capacitance from "T" to earth does not accumulate charge and plays no part in the measurement.

5. As the effect of stray capacitances has been eliminated, ordinary screening techniques may be used. The advantage of this is considerable. With a suitably designed rig one may screen cables to within 1 cm of the sample, thus reducing the susceptibility of the system to electrical noise as well as reducing the stray capacitance. With this system measurements can be made at low applied potentials, typically 10 mV across 1 MΩ of unknown.

In the configuration of Figure 4.1.13 the unknown impedance is given by the following relation:

$$Z_u = \frac{V_x}{V_y} Z_s \qquad (22)$$

The simplest situation arises when Z_s is real (i.e., a standard resistor), but it has been found that a parallel combination of standard resistor and standard capacitor gives a standard impedance that better matches that of the sample over the entire frequency range and so avoids the need to change resistors throughout the measurement. In that case Z_s should be calculated as $(G_s + j\omega C_s)^{-1}$ at each measurement frequency and then inserted into Eq. (22). In the system described, the standard resistors used are in the range 100 Ω to 100 MΩ (decade values); the standard capacitor is fixed at 25 pF.

Test Procedures. To verify the correct operation of the system the following tests are incorporated into the automated measurement sequence:

1. The voltage V_x is tested. If its magnitude falls below 50% of the desired V_{out}, the operator is warned and the measurement is rejected. This is to check for short circuits in cables, rigs, or cells.

2. The voltage V_y is tested. If its magnitude exceeds 10 V, the measurement is rejected and the operator is warned. Since V_y is the output of an operational amplifier, one must ensure that it is below the saturation level, typically 10–12 V, depending on the power supply voltage.

3. The ratio V_y/V_x is tested. If this falls below 0.1 the standard resistor is incremented. (This test is not usually necessary when a standard capacitor is used, but only when Z_s is real.)

A flowchart of the measurement procedure is shown in Figure 4.1.14.

Constructional Details. The operational amplifier and associated components are built into a small metal box. The connection between generator output and input "X" is made inside this box. The power supply lines are decoupled as specified by the manufacturer. Compensating capacitors of 30 pF are used as specified by the manufacturer for unity-gain amplification. All lead lengths are kept as short as possible, preferably 1 m or less.

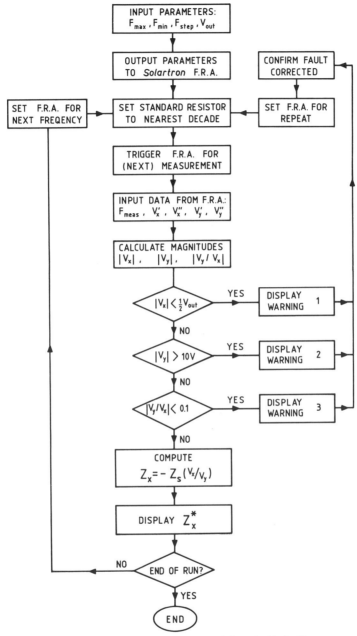

FIGURE 4.1.14. Flowchart of the measurement procedure followed with the IS measurement system shown in Figure 4.1.11.

Performance and Errors. One way of assessing the performance of the measurement system described is to construct RC circuits from components of known values and compare the measured impedance spectra with those calculated from the component values. Figure 4.1.15 shows such a comparison for a circuit equivalent

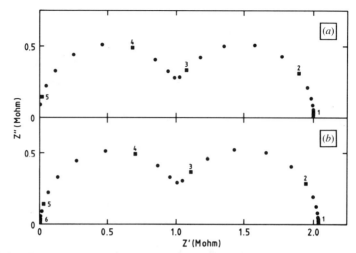

FIGURE 4.1.15. Demonstration of the performance of the high-impedance adaptor by comparison of measured and simulated spectra obtained for a two-element Voigt circuit (Fig. 4.1.1): (*a*) measured impedance spectrum; (*b*) simulated impedance spectrum.

to a two-phase material. This is not the most rigorous test of the system, but gives a good indication as to its usefulness for spectroscopy of materials. According to this criterion, the performance of the system is very satisfactory. From measurements on standard components it is also found that the stray capacitance introduced by the system is, as expected, very small (less than 0.5 pF). Many of the spectra shown in Section 4.1.3 were obtained on this system, and in several cases, where they were compared with measurements on manual bridges, they were found to be in close agreement.

At the highest decade of frequency of the FRA (10^5–10^6 Hz) an error does appear and produces the following effect: In the impedance spectrum the measured points are shifted downward by an amount proportional to frequency. This is particularly obvious for measured impedances on the low side of the scale (100 Ω to 1 kΩ). This effect is shown schematically in Figure 4.1.16*a* for the case of simple RC combination. In the modulus spectrum, the error corresponds to an upward shift of points, also proportional to frequency, as shown in Figure 4.1.16*b* (which should have been an arc).

Both of these errors can be attributed to phase shift in the current-to-voltage converter of Figure 4.1.13. An analysis of this circuit, taking account of the nonideal behavior of the operational amplifier (T. Goldrick [1982], unpublished work) has shown that for certain simplifying assumptions, the apparent measured impedance Z_m is related to the unknown impedance Z_u as follows:

$$Z_m = Z_u + j\frac{f}{f_o}(Z_s + Z_u) \tag{23}$$

where f is the measurement frequency and f_o the frequency at which the open-loop gain of the amplifier becomes unity. Attempts to relate the value of f_o estimated

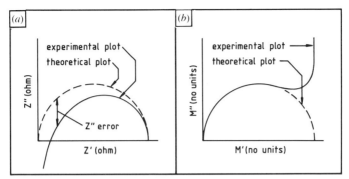

FIGURE 4.1.16. Effect of error introduced by the high-impedance adaptor at the top decade of frequency: (a) Error introduced into the impedance spectrum of a parallel RC element. (b) Error introduced into the modulus spectrum of a parallel RC element.

from plots such as Figure 4.1.16a to that measured directly on the operational amplifier have broadly substantiated this analysis. For example, f_o was estimated from an impedance plot as 12 MHz, while measurement of the open-loop gain and extrapolation to unity gave f_o as 18 MHz, in rough agreement. Thus we believe that we can explain the origin of the high-frequency error, but not with sufficient confidence to "calculate it out."

For the special case when $Z_u = Z_s$, Eq. (23) simplifies to

$$Z_m = Z_u\left(1 + j\frac{2f}{f_o}\right) \qquad (24)$$

Thus it can be shown that at a maximum measurement frequency of 1 MHz the error in the imaginary part of impedance is considerable, but that in the real part is small. Both errors rapidly decrease with falling frequency. In many practical situations, the phase error cannot be observed on the impedance plot. This is because at high measurement frequencies the impedance tends to zero anyway, and an error of this magnitude cannot be resolved. This happy coincidence does not apply to the modulus plot, and for the collection of this type of data it may be preferable to use another configuration, or possibly a different measurement system altogether, such as a bridge, or the Hewlett–Packard 4192A impedance analyzer.

Measurement Rig. Important considerations in the design of a rig for IS of materials are (i) the ability to provide a controlled temperature and gaseous environment for the sample under test and (ii) the provision of the best possible electrical path to the sample.

The importance of temperature control is apparent if one considers that for a sample having a conductivity activation energy of 1 eV, an error of 1° at 300°C would give rise to a 4% error in conductivity. At 100°C the same temperature error would produce a 9% error in conductivity. Therefore, temperature control to approximately 0.25°C is required, and this can be achieved without difficulty with present-day three-term temperature controllers.

A typical measurement rig that meets the above requirements consists of a furnace and an electrode assembly, with the furnace constructed from a tube of alumina, noninductively wound with Nichrome or Kanthal heating wire. The furnace is controlled by a remote-setpoint three-term controller with digital input (e.g., CRL 507).

A control thermocouple near the windings is provided, but it was found that better temperature control could be achieved by using a thermocouple in the electrode assembly for control. At the ends of the alumina tubes are fitted standard vacuum flanges to which the electrode assembly may be easily attached.

The electrode assembly, shown in Figure 4.1.17 is based on an alumina "shoe" that slides over an alumina lumina supporting tube. It is fitted with alumina spacers and a disc supporting a piece of platinum foil attached using high-temperature cement. A harness of Nichrome wire and ordinary laboratory springs provides a load of approximately 20 g on the contacts. Contact is made with the sample via a platinum foil on one side and a loop of platinum wire on the other. The leads are carried in two ceramic sheaths, one of which fits inside the support tube. The sheaths are coated on the outer side with platinum paint to provide electrical screening. The cold ends of the tubes are bonded to the metal base of the assembly with silver-loaded epoxy cement to provide good grounding. With this arrangement, the total unscreened length of lead is kept as low as 1 cm per lead. The rig can be operated at temperatures of up to 1000°C and frequencies from dc to several megahertz.

For measurements below 500°C alternative designs based on copper or stainless steel may be preferable because of their superior screening and temperature control (Strutt, Weightman, and Lilley [1976]).

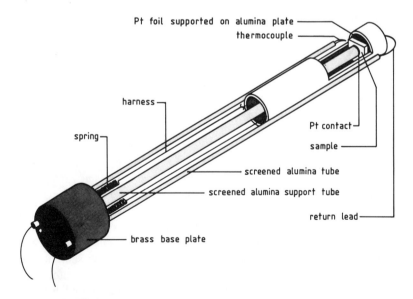

FIGURE 4.1.17. Rig used for the characterization of ceramic samples.

4.1.3 Interpretation of the Impedance Spectra of Materials

4.1.3.1 Introduction

This section covers the interpretation of experimental impedance spectra in terms of the two-phase models described in Section 4.1.1. Examples are given from the work of the authors and their colleagues, illustrating the application of these models to ceramics. The two-dimensional complex plane representation was chosen, as used in the source publications, because it permits graphical estimation of the parameters of the circuit equivalent.

The relationship between microstructural models and circuits shows its real merit when used to correlate the individual parameters of the circuit equivalent of a material with changes in the external conditions or in the microstructure—in other words, when applied to problems of materials science.

In ceramics where it is possible to resolve the resistances due to the grain interiors and the grain boundaries this has facilitated the study of some important processes namely sintering, grain growth, and solid state precipitation. Most of the examples given relate to the family of zirconia ceramics because these materials give well-resolved spectra of microstructural origin. These materials also exhibit a range of interesting phenomena, such has polymorphism, ionic conductivity, and transformation toughening and are technologically important for electrical and structural applications. Their structure is briefly described below.

Microstructural Aspects of Zirconia Ceramics. The compound ZrO_2 is polymorphic and can have one of three crystal structures: at ambient temperatures ZrO_2 is monoclinic (*m*) but transforms to tetragonal (*t*), and then to cubic (*c*) at high temperatures (1170 and 2370°C, respectively). On cooling, the transformation sequence is reversed, and the structure normally reverts to monoclinic. It has not proven possible to fabricate ceramics from pure ZrO_2 because of the disruptive effect of the $t \rightarrow m$ transformation. This effect can, however, be suppressed by alloying ZrO_2 with other oxides such as CaO, MgO, and Y_2O_3 with which it forms solid solutions, collectively called *stabilized zirconias* (Subbarao [1981], Subbarao and Maiti [1984]). At stabilizer concentrations of around 10% molar the solid solutions have a cubic structure similar to the high-temperature cubic phase; these materials are termed *fully stabilized* (FSZ). Solid solutions containing 2–6 mole % Y_2O_3, however, retain the tetragonal structure provided the grain size is sufficiently small, typically around 500 nm (Gupta et al. [1977]). These ceramics can usually be sintered at lower temperatures than cubic zirconia and have exceptional mechanical strength and toughness; they are referred to as *tetragonal zirconia polycrystals* (TZPs). At intermediate levels of stabilizer, ceramics are termed *partially stabilized* (PSZ) consisting of a matrix of *c* containing a dispersion of *t* and/ or *m* precipitates (Kobayashi, Kayajima, and Masuki [1981]. If a significant volume fraction of metastable *t* phase is present, these precipitates give the ceramic good mechanical properties by a process known as *transformation toughening*. Briefly, this effect is due to the ability of the precipitates to transform from *t* to *m* under the influence of a propagating crack (Garvie, Hannink, and Pascoe [1975]).

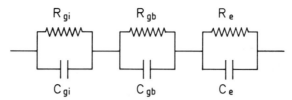

FIGURE 4.1.18. Circuit equivalent for a ceramic electrolyte according to Bauerle [1969] and modeling the impedance of the grain interiors (gi), grain boundaries (gb) and electrode (e).

In ceramics optimally aged for mechanical strength and toughness the precipitates are mostly below a critical size (approximately 100 nm), above which they spontaneously transform to monoclinic on cooling.

When ZrO_2 is alloyed with oxides of elements of lower valence, these elements occupy the Zr^{4+} sites, and vacancies are introduced at oxygen lattice sites. Migration of these vacancies gives zirconia ceramics their well-known ionic conductivity at high temperatures (Subbarao and Maiti [1984]) and leads to their use in electrochemical cells such as fuel cells or oxygen monitors (Steele [1976], Steele et al. [1981]). Principally FSZ has been used for electrical application, while PSZ and Y-TZP have been developed for structural applications. Recently it has, however, become apparent that for moderate temperature devices, PSZ and TZP may be better suited (Bonanos, Slotwinski, Steele, and Butler [1984a,b]).

Bauerle's Circuit Equivalent. The application of IS to the characterization of ceramic ionic conductors started after Bauerle [1969] showed that for zirconia with platinum electrodes the individual polarizations of grain interiors, grain boundaries, and electrode–electrolyte interfaces could be resolved in the admittance plane. He presented a circuit equivalent for this arrangement, which has since proven to be typical of most solid electrolytes (see also Section 1.4). In this circuit the elements corresponding to grain interiors (gi), grain boundaries (gb), and electrode (e) connected in series are shown in Figure 4.1.18.* The estimation of the circuit parameters was made slightly complicated by Bauerle's choice of the admittance plane; many subsequent workers have therefore preferred to work in the impedance plane, where a more direct relationship exists between the spectrum and the circuit (Armstrong et al. [1974], Schouler et al. [1981]). A comparison of an admittance and impedance spectrum for a sample of ZrO_2: mole % Y_2O_3 at 240°C is given in Figure 4.1.19$_{a,b}$; the spectra were obtained using the system described in Section 4.1.2. The temperature was chosen to give the largest possible coverage of the grain interior and grain boundary arcs within the available frequency range. A simulated impedance plot is shown in Figure 4.1.19c using circuit parameters estimated from the impedance plot (Fig. 4.1.19b); these circuit parameters are given in Table 4.1.4. The level of agreement between experiment and simulation is quite satisfactory for the grain interior and grain boundary arcs both in terms of shape and distribution of frequencies on the arcs, supporting the view that circuit equivalents are a meaningful way of representing the data.

---

*The circuit shown in Figure 4.1.18 contains one extra capacitor for the grain interior element not included by Bauerle because the grain interior relaxation lay above the frequency range he had covered.

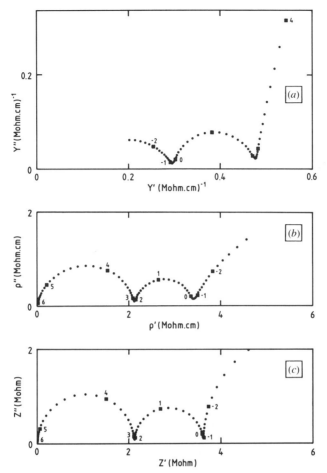

FIGURE 4.1.19. Comparison of admittance and impedance spectra for a zirconia solid electrolyte (ZrO_2:6 mole % Y_2O_3) at 240°C: (*a*) Experimental admittance spectrum. (*b*) Experimental impedance spectrum. (*c*) Simulated impedance spectrum, using the circuit of Figure 4.1.18 and parameter values given in Table 4.1.4.

4.1.3.2 Characterization of Grain Boundaries by IS

Understanding the origin of polarizations in ceramic electrolytes is important not only in maximizing their conductivity, but also in aiding microstructural interpretation, with the object of optimizing other properties such as strength. Fabrication of ceramics often involves additions of small amounts of impurities that may form liquids at the sintering temperature. In zirconia ceramics, trace amounts of silica and alumina fulfill this function, even when they are not added deliberately. On cooling, various aluminosilicate amorphous or crystalline phases form at the grain boundaries and can have a deleterious effect on the ionic conductivity (Bauerle [1969], Schouler et al. [1973], Schouler [1979], Kleitz et al. [1981], Bernard [1981]). It has, however, also been noted that small additions of Al_2O_3 to ZrO_2 at levels above the solubility limit can reduce the grain boundary resistance, and

TABLE 4.1.4. Circuit Parameters Graphically Obtained from the Impedance Spectrum of Figure 4.1.19*b* and Used to Simulate the Spectrum of Figure 4.1.19*c*

C_{gi} = 4.8 pF	C_{gb} = 1.7 nF	C_e = 2.0 μF
R_{gi} = 2.1 MΩ	R_{gb} = 1.5 MΩ	R_e = 5.0 MΩ

it has been suggested that this is accomplished by scavenging of SiO_2 to form discrete nonconducting particles of mullite ($Al_6Si_2O_{13}$) (Drennan and Butler [1973]). Subsequent grain boundary movement during grain growth leaves these particles in the interior of grains, where their influence on the ionic conductivity is minimal. This effect is to be expected on the basis of Eqs. (16) and (17), which describe the effect of voids or high-resistivity inclusions on the impedance spectrum of ceramics.

Microstructure–Grain Boundary Property Correlations. The way in which grain boundary structure can affect the impedance spectrum is illustrated in the following example by comparing 3 mole % Y_2O_3 tetragonal ceramic (TZP) developed for structural applications and a 6 mole % mainly cubic material developed for oxygen sensor applications and optimized for oxygen ion conductivity.

A typical transmission electron micrograph of the tetragonal ceramic is shown in Figure 4.1.20*a*. Clearly visible features are the fine grains (~ 600 nm) and the thin grain boundary film, marked by triangles in the micrograph. Selected area diffraction patterns show these areas to be amorphous, while energy-dispersive x-ray microanalysis reveals the presence of silica, alumina, and yttria at levels of 65, 20, and 70 wt %, respectively (Butler and Bonanos [1985]). These films have been carefully examined by transmission electron microscopy (TEM) to determine whether they are continuous and appear to be so. However, although high-resolution TEM is a powerful method of investigation, it samples only a very small part of the grain boundary area, and it is difficult to be sure that small grain boundary defects have not been missed. It is not straightforward to measure the thickness of thin-grain-boundary films with accuracy because of problems of contrast and viewing angle, but on the basis of the micrographs one would estimate the thickness of the grain boundary phase as 10 nm or less.

A transmission electron micrograph of the 6 mole % material is shown in Figure 4.1.20*b*; the grain size is found to be approximately 10 μm. Pockets of second phase can be clearly seen along the grain boundaries; their lenticular shape indicates that they originated from nonwetting liquid phases present at high temperatures. Both crystalline and amorphous grain boundary phases are found when these area are subjected to selected area electron diffraction analysis, but all the grain boundary phases appear to be discrete, allowing limited intergranular contact. Once again, however, definitive verification of the absence of grain boundary phases from the interlens regions is not possible; these phases could simply be too thin to detect.

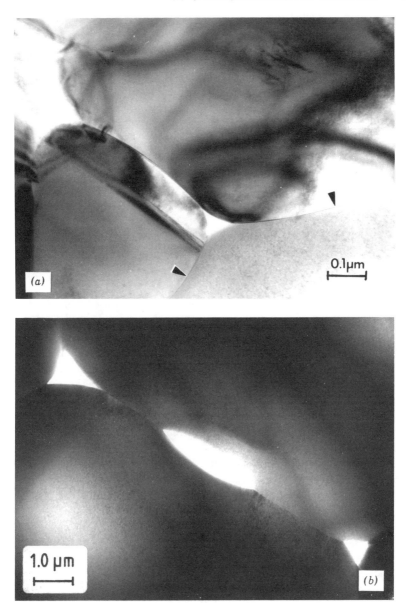

FIGURE 4.1.20. Transmission electron micrographs of two zirconia ceramics: (*a*) Tetragonal zirconia ceramic (ZrO_2 : 3 mole % Y_2O_3) showing thin, continuous grain boundary phase (arrowed). (*b*) Partially stabilized ceramic (ZrO_2 : 6 mole % Y_2O_3) showing discrete, lenticular grain boundary phase.

Impedance spectra of the two zirconia ceramics obtained at 300°C are shown in Figure 4.1.21. They both follow the Bauerle type of behavior. The spectrum of the 3 mole % material, Figure 4.1.21*a*, indicates that the resistivity is dominated by the grain boundaries. The spectrum of the 6 mole % material, shown in Figure

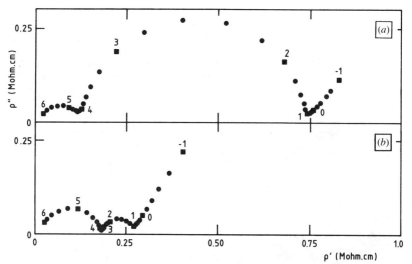

FIGURE 4.1.21. Impedance spectra for two zirconia ceramics, obtained at 300°C using sputtered platinum electrodes: (a) Tetragonal zirconia ceramic (ZrO_2 : 3 mole % Y_2O_3) with large grain boundary arc due to continuous grain boundary phase. (b) Partially stabilized ceramic (ZrO_2 : 6 mole % Y_2O_3) with small grain boundary arc due to discrete grain boundary phase. (Courtesy of Silicates Industriels.)

4.1.21b, for the same temperature reveals a relatively small grain boundary contribution to the total resistivity, consistent with the fact that this material has been optimized for ionic conductivity. It would appear that in this case the grain boundary resistance has been minimized by making the grain boundary phases discrete. We emphasize that the observation of an impedance spectrum in which $r_{gb} < r_{gi}$ does not imply $\rho_{gb} < \rho_{gi}$ or that the ionic current is carried *along* the grain boundaries.

Further indirect information about the topology of the grain boundary phases can be obtained from the temperature dependence of the quantities r_{gi} and r_{gb}. In crystalline extrinsic ionic conductors the conductivity is thermally activated:

$$\sigma = (\sigma_0/T) \exp(-\Delta H_m/kT) \tag{25}$$

where H_m is the activation enthalpy for ionic migration. The Arrhenius plots of the ionic conductivity or resistivity linearize this equation and their slope give $\Delta H_m/k$. Arrhenius plots of r_{gi} and r_{gb} for both ceramics (in Fig. 4.1.22) are linear over the temperature range examined (200–500°C). In the tetragonal ceramic (Fig. 4.1.22a) the slopes are visibly different (with r_{gb} higher), as expected on the basis of the brick layer model without easy paths as described in Section 4.1.1. By contrast, the slopes for the mainly cubic ceramic (Fig. 4.1.22b) are quite similar, suggesting that a partial blocking model is more appropriate to this material. Estimated activation enthalpies are given in Table 4.1.5 for the zirconia ceramics discussed and for a third PSZ ceramic intended for electrical applications. The conclusions based on the analysis of the impedance spectra are in agreement with TEM observations in assigning the models appropriate to the two ceramics.

The treatment of the capacitive elements of the circuit equivalent differs ac-

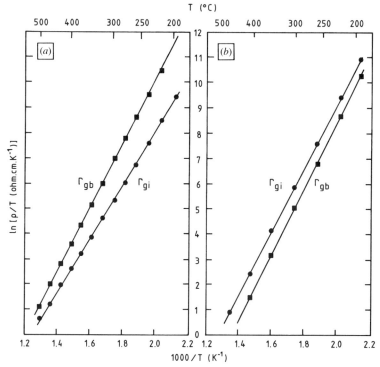

FIGURE 4.1.22. Arrhenius plots of the grain interior and grain boundary resistivities for two zirconia ceramics: (*a*) Tetragonal zirconia ceramic (ZrO_2 : 3 mole % Y_2O_3) in which the lines have different slopes, as expected from the brick layer model. (*b*) Partially stabilized ceramic (ZrO_2 : 6 mole% Y_2O_3) in which the slopes of the lines are similar, as expected for discrete grain boundary phase. (Courtesy of Silicates Industriels.)

TABLE 4.1.5. Activation Energies for Conduction, Estimated for Three ZrO_2 : Y_2O_3 Ceramics of Different Composition and Structure

Sample	Composition	Structure	ΔH_{gi} (eV)	ΔH_{gb} (eV)
Y-TZP	3.0 m% Y_2O_3	Tetragonal	0.92	1.09
Y-PSZ	4.7 m% Y_2O_3	Cubic + tetragonal	1.07	1.15
Y-PSZ	6.0 m% Y_2O_3	Cubic + tetragonal	1.07	1.12

cording to which model applies. For the "bricklayer" tetragonal ceramic, the grain boundary capacitance may be used to calculate the grain boundary thickness using Eq. (11). Ideally this would require knowledge of the dielectric constants of the two phases, but in the absence of this information one may assume $\epsilon_{gi} = \epsilon_{gb}$, and the error introduced by this assumption should not be larger than the spread in the distribution of grain size, that is, about a factor of two. Thus for the tetragonal ceramic we calculate a grain boundary thickness of ~4 nm. This value is compatible with the value estimate of ~ 10 nm from the electron micrograph.

For the mainly cubic 6 mole % ceramic this type of calculation would be inappropriate, as the grain boundary phases are not continuous. Instead, Eq. (13) describing the partially blocked situation can be used to calculate a blocking coefficient β. In this case a value of $\beta = 0.3$ is found, implying grain boundary coverage of 30% by second phase. This would seem to be roughly in accord with the appearance of the boundaries in Figure 4.1.20 and in other micrographs examined.

Thus, for two different zirconia ceramics the techniques of TEM and IS, although both subject to errors and requiring interpretation, give results that are similar in terms of the continuity and the dimensions of the grain boundary phases.

Materials Lacking a Grain Boundary Impedance Arc. An obvious example of such materials are single crystals, and for some time fully stabilized crystals of ZrO_2 have been grown as artificial gems by a process known as *skull melting*. Single crystals of $ZrO_2:Y_2O_3$ containing 2.2 and 3.4 mole % Y_2O_3 were characterized by x-ray diffraction and TEM (Bonanos and Butler [1985]). Transmission electron micrographs of the 2.2 mole % crystal (Fig. 4.1.23) showed striated areas where there had been a reorientation of the crystals known as *twinning*. Careful electron diffraction analysis showed the presence of monoclinic and tetragonal phases. The morphology of the latter was similar to that of a nontransformable tetragonal phase *t*, observed by Lanteri et al. [1983] in similar crystals and originally identified by Scott [1975]. In the 3.4 mole % crystal only the *t* phase was observed.

A comparison of the impedance spectra of the two types of zirconia is made in Figure 4.1.24. The spectrum of the 2.2 mole % crystal (Fig. 4.1.24a) shows a large, depressed grain interior arc and an electrode arc. It is assumed that the

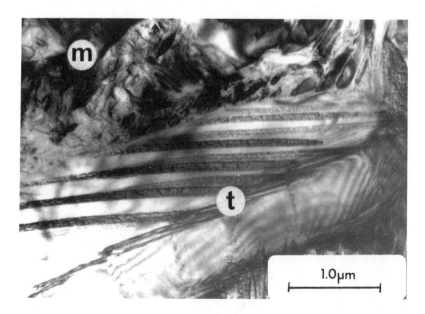

FIGURE 4.1.23. Transmission electron micrograph of a single crystal ($ZrO_2:2.2$ mole% Y_2O_3), showing separate regions of monoclinic (m) and tetragonal (t) phases. The micrograph was obtained at 1 MV. (Courtesy of Silicates Industriels.)

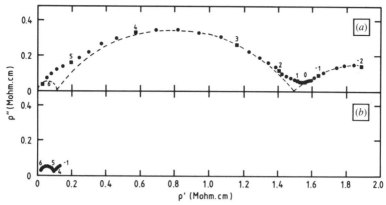

FIGURE 4.1.24. Comparison of the impedance spectra obtained at 300°C for two zirconia–yttria single crystals. (*a*) Tetragonal + monoclinic single crystal (ZrO_2:2.2 mole % Y_2O_3) showing large bulk arc. (*b*) Fully tetragonal single crystal (ZrO_2:3.4 mole % Y_2O_3) showing small bulk arc.

depressed arc is due to the combination of tetragonal and monoclinic phases identified by TEM; indeed, it appears to be composed of two poorly resolved arcs (dashed lines). The 3.4 mole % crystal displays one high-frequency arc due to the t phase. The resistivity of 0.10 MΩ-cm agrees well with the value of 0.13 MΩ-cm found in the 3 mole % Y-TZP previously mentioned. The two plots illustrate the sensitivity of the impedance spectrum to phase composition in zirconia ceramics and in doing so lend confidence to its use as a basis for investigating complex microstructures.

Although the three-arc response is typical of most ceramics, situations have been found where the grain boundary arc is absent in polycrystalline ceramics. One example has been reported in the ZrO_2:(Y_2O_3 + MgO) system (Slotwinski, Bonanos, and Butler [1985]). Figure 4.1.25 shows the impedance spectrum of such a ceramic, aged for 10 h at 1400°C and measured at 300°C. The observed arc can be identified as a bulk property from the fact that it passes through the origin: this assertion is supported by the capacitance associated with it (4 pF/cm).

As this spectrum is unusual, it is suspected to be caused by some unique feature in the grain boundary microstructure. A transmission electron micrograph of the

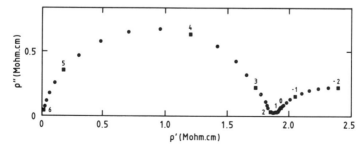

FIGURE 4.1.25. Impedance spectrum for a partially stabilized zirconia ceramic of composition ZrO_2:(7.0 mole % MgO + 1.5 mole % Y_2O_3), obtained at 300°C, not resolving a grain boundary arc. (Reprinted from R. K. Slotwinski, N. Bonanos, and E. P. Butler, *J. Mat. Sci. Lett.* 4, 641-644, 1985, courtesy of Chapman & Hall.)

grain boundaries (Fig. 4.1.26) shows two grains, labeled A and B, and the intervening grain boundary phase. Unlike other zirconia-based systems, in $ZrO_2 : (Y_2O_3 + MgO)$ there are finely twinned crystallites along substantial lengths of grain boundary. Selective area electron diffraction patterns of this phase showed a tetragonal rather than monoclinic symmetry, resembling the patterns of the t phase (Lanteri et al. [1983]), which has a relatively high ionic conductivity (Bonanos and Butler [1985]). The authors have therefore suggested that t-phase "colonies" along most of the grain boundary area form "short circuits" and effectively eliminate the grain boundary impedance. Further work on this system is necessary to verify this hypothesis.

Another example of a ceramic lacking a grain boundary response is found in very dense $Bi_2O_3 : Er_2O_3$ ceramics (C. P. Tavares and Bonanos 1984 unpublished work). Bi_2O_3 has a high-temperature phase with the highest known oxygen ion conductivity. Efforts to stabilize this phase by doping, for example, with Y_2O_3, have been successful, albeit with some loss of conductivity compared to that of Bi_2O_3 extrapolated from high temperatures (Subbarao and Maiti [1984]). Figure 4.1.27 shows the impedance spectrum of a pellet of $Bi_2O_3 / 17$ mole % Er_2O_3 measured at 300°C. The absence of a grain boundary arc can be tentatively explained by the absence of grain boundary phases altogether. Unlike zirconia, bismuth oxide is very reactive and has considerable solid solubility with silica and alumina, two common constituents of grain boundary phases in ceramics. Thus it is possible that bismuth oxide segregated at grain boundaries has a "self-cleaning" effect.

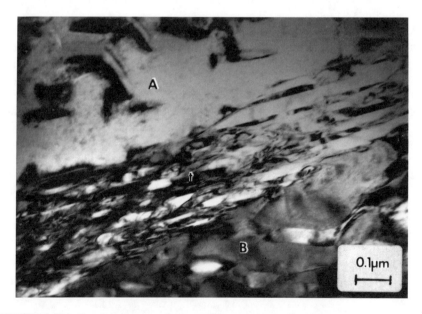

FIGURE 4.1.26. Transmission electron micrograph of partially stabilized zirconia ceramic $ZrO_2 : (7.0$ mole % $MgO + 1.5$ mole % $Y_2O_3)$ showing two grains (labeled A, B) and the intervening tetragonal grain boundary phase (arrowed). (Reprinted from R. K. Slotwinski, N. Bonanos, and E. P. Butler, *J. Mat. Sci. Lett.* 4, 641-644, 1985, courtesy of Chapman & Hall.)

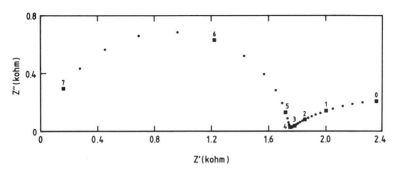

FIGURE 4.1.27. Impedance spectrum obtained at 300°C for a pellet of Bi_2O_3 : 17 mole % Er_2O_3, showing no grain boundary arc. (The pellet was supplied by Dr. C. P. Tavares of Basic Volume Ltd., London, England.)

Indirect support for this explanation is provided by the appearance of the sample, which is almost translucent.

It has been possible in the ZrO_2 and Bi_2O_3 systems so far discussed to provide interpretations of the grain boundary impedance spectra in terms of the presence of second phases with electrical properties differing from those of the grain interiors. However even ultrapure-zirconia-based ceramics may still exhibit "grain boundary" arcs, and it is necessary to recognize that the electrical behavior of grain–grain interfacial contact regions can differ from that of the interior even in the absence of second phases. For example, dopants such as Y_2O_3 may segregate to the interfacial region to provide a relatively high-resistance layer. In addition, ceramic grain boundaries will usually be charged, leading to a depletion (or accumulation) space charge region adjacent to the interfacial region which would be expected to exhibit electrical behavior dissimilar to that of the grain interior. The nonohmic electrical behavior of grain boundaries in ceramic oxide semiconductors is used in technological devices such as varistors and PTC thermistors (Levinson [1981]). The presence of interfacial segregation and space charges in ionic conductors has also been involved (Heyne [1983], Burggraaf et al. [1985], Steele and Butler [1985]) to interpret the appearance of "grain boundary" arcs in nominally very pure ceramics. There will be many situations in ceramic materials where second phases, segregation, and space charges will all be present, producing a rich variety of electrical behavior. For such situations as many techniques as possible should be used to unravel the complexities of the interfacial properties.

4.1.3.3 Characterization of Two-Phase Dispersions by IS

Two-phase dispersions are common in engineering materials and are usually formed by solid state precipitation. Such microstructures are common in metallic alloys, especially in steel and aluminium, and in ceramic alloys, such as zirconia-toughened alumina and partially stabilized zirconia. In ceramic ionic conductors the impedance spectra can be understood in terms of the effective medium models described in Section 4.1.1. Since the treatment differs depending on how well resolved are the time constants of the phases, the two cases are dealt with separately.

Extraction of the Conductivities in an Alkali Halide System. Alkali halide crystals have been studied as model systems for the study of solid state precipitation as well as for point-defect studies. In the $NaCl:CdCl_2$ system single crystals can be grown containing an ordered precipitate phase $CdNa_6Cl_8$ (Suzuki phase). An electron micrograph of a NaCl crystal containing cubic precipitates of Suzuki phase is shown in Figure 4.1.28. This was taken using a 1-MV electron microscope with cryogenic stage to prevent electron damage to the crystals (Guererro, Butler, and Pratt [1979]). The crystal was grown at the University of Sussex by the Bridgman–Stockbarger technique.

Electrical measurements on the crystals were reported by Bonanos and Lilley [1981]. The impedance spectrum showed only one arc, due to the bulk conductivity, and an electrode impedance. Significantly, there was no grain-boundary-type arc that could be attributed to polarization of the $NaCl:CdNa_6Cl_8$ interface.

By contrast, the modulus spectrum shown in Figure 4.1.29 reveals two overlapping arcs. Consideration of the models described in Section 4.1.1. and the variation with Suzuki-phase content of the relative sizes of the arcs led to the conclusion that these were due to the matrix and precipitate phases, respectively. The ionic conductivity and hence relaxation frequency of Suzuki phase is expected to be higher than that of the matrix on the basis of a higher dopant cation vacancy

FIGURE 4.1.28. Transmission electron micrograph of a $NaCl:CdCl_2$ single crystal with precipitates of the Suzuki phase $CdNa_6Cl_8$. The micrograph was taken by A. L. Guererro and E. P. Butler at a voltage of 1 MV, at liquid helium temperature. (Reprinted with permission from N. Bonanos and E. Lilley, Conductivity Relaxations in Single Crystals of Sodium Chloride Containing Suzuki Phase Precipitates, *J. Phys. Chem. Solids* **42**, 943–952. Copyright 1981 Pergamon Journals Ltd.)

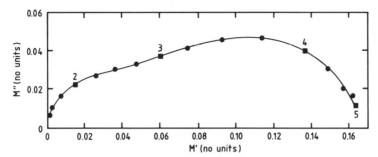

FIGURE 4.1.29. Modulus spectrum obtained at 150°C for a single crystal of NaCl : CdCl$_2$, of which a micrograph is shown in Figure 4.1.28. (Reprinted with permission from N. Bonanos and E. Lilley, Conductivity Relaxations in Single Crystals of Sodium Chloride Containing Suzuki Phase Precipitates, *J. Phys. Chem. Solids* **42**, 943–952. Copyright 1981 Pergamon Journals Ltd.)

concentration. Although the precipitates were not spherical, it proved possible to interpret the modulus spectra in terms of the Maxwell–Wagner effective medium model (Section 4.1.1.). The estimated parameters of the circuit equivalent and the volume fraction of precipitate phase were used to obtain microscopic conductivities of the two phases over a range of temperatures. The conductivity of the Suzuki phase agreed with that measured earlier on polycrystalline samples prepared by rapid quenching from the melt (Bonanos and Lilley [1980]).

Structure–Property Correlations in Polyphase Ceramics. In polyphase ceramics it is not usually possible to identify each individual phase by its own arc in an impedance or modulus spectrum because of the proximity of the time constants. Even so, the impedance spectrum in conjunction with a full microstructural characterization can be used to glean information that is otherwise inaccessible—for example, regarding the conductivities of individual phases.

As mentioned earlier, the grains in PSZ consist of a matrix of cubic zirconia containing a fine dispersion of tetragonal particles, slightly elongated and equally distributed along three crystallographically equivalent axes (Garvie, Hannink, and Pascoe [1975]). A typical microstructure of a ceramic partially stabilized with (Y$_2$O$_3$ + MgO) is shown in Figure 4.1.30. The appearance of this system suggests that the random ellipsoidal effective medium model proposed by Fricke [1953], discussed in Section 4.1.1. should provide an accurate description of this system. The model predicts that the macroscopic electrical properties of such a system will be isotropic, thus implying that the effect of various orientations of the grains within the ceramic may be ignored. From the circuit equivalent derived for this model (Fig. 4.1.19b) one would expect up to four relaxations arising from the polyphase structure in addition to the usual grain–grain boundary relaxations (the exact number would depend on the symmetry of the shape of the precipitates).

Further aging produces a transformation from tetragonal to monoclinic without significant change in the shape, size, or chemical composition of the precipitates, as they are constrained in the cubic matrix. Since the *c*, *t*, and *m* phases have different conductivities, impedance spectroscopy should be a sensitive technique

FIGURE 4.1.30. Transmission electron micrograph of grain interior of a partially stabilized zirconia ceramic ZrO_2:(7.0 mole % MgO + 1.5 mole % Y_2O_3), aged for 40 h at 1400°C. Bright areas correspond to the tetragonal phase.

for following microstructural changes during aging (Bonanos, Slotwinski, Steele, and Butler [1984a]).

For ZrO_2 stabilized with 8 mole % CaO, XRD and TEM showed that the $t \rightarrow m$ transformation was complete in materials aged for approximately 20 h at 1400°C. From analysis of x-ray diffraction (XRD) data the volume fraction of t-ZrO_2 was estimated as 0.3. Electron microscopy revealed that the particles were ellipsoidal with axial ratios a/b and b/c close to 2. The corresponding form factors are found in Table 4.1.2 viz: $\phi_1 = 7.9$, $\phi_2 = 2.5$, $\phi_3 = 0.66$. Thus the parameters in Eq. (20) are known, and the only unknowns are the complex conductivities.

Impedance spectra are shown in Figure 4.1.31 for Ca-PSZ aged for various times at 1400°C. In contrast with the expectations, distinct relaxations for the matrix and dispersed phases are not observed, in these spectra, or in the corresponding modulus spectra (not shown). Instead the material shows a Bauerle-type spectrum. The failure of the impedance spectra to resolve matrix and precipitate relaxations in this system means that the parameters of a circuit such as that of Figure 4.1.9b are not accessible and that the grain interiors must be approximated by a single RC element. The variation of the resistivities with aging time is shown in Figure 4.1.32; r_{gi} shows a "bathtub"-shaped plot, with three well-defined stages of behavior; these can be interpreted in terms of the sequence of events occurring during aging. The as-received material, furnace-cooled from 1800°C, is deficient in t phase and has a nonequilibrium composition. The initial fall in r_{gi} (stage I) can be explained in terms of the increase in volume fraction of t phase on the assumption that $\sigma_t > \sigma_c$. Considerations of the Fricke model show that purely

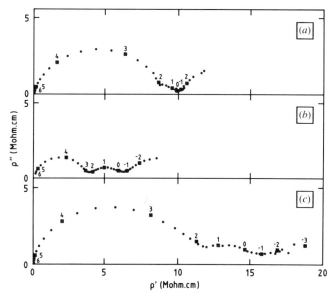

FIGURE 4.1.31. Impedance spectra obtained at 300°C for partially stabilized zirconia ceramics (ZrO$_2$:8 mole % CaO): (*a*) as fired; (*b*) aged for 15 h at 1400°C; (*c*) aged for 30 h at 1400°C. (Reprinted from N. Bonanos, R. K. Slotwinski, B. C. H. Steele, and E. P. Butler, *J. Mat. Sci.*, **19**, 785–793, 1984, courtesy of Chapman & Hall.)

morphological changes in the precipitates could not account for such a large effect. Stage II is regarded as period of precipitate growth at constant volume fraction. According to the model, particle size should not affect the conductivity, and this is indeed observed. Compositional changes during this period can be ruled out because of the effect they would have on the impedance spectrum. The large increase in resistivity accompanying the transformation of the precipitates from tetragonal to monoclinic (stage III to stage III transition) demonstrates conclusively that $\sigma_t \gg \sigma_m$, as determined from Figure 4.1.7*a*. Since the *t*-to-*m* transformation occurs without change in chemical composition, the conductivity of the matrix, σ_c, must remain the same in stages II and III. This constraint has been applied to a numerical analysis of the conductivity changes during the stage-II-to-stage III transformation to obtain sets of σ_c, σ_t, σ_m such that when inserted into Eq. (20), the pairs σ_t, σ_t and σ_c, σ_m combine to give the stage II and stage III conductivities, respectively. This analysis has shown that physically realizable solutions can be obtained only for $\sigma_t > \sigma_c > \sigma_m$, which is surprising in view of the lower concentration of dopant possessed by the *t* phase. Meanwhile, the grain boundary resistivity r_{gb} (also shown in Figure 4.1.32) increases slowly with aging, due to the formation of monoclinic zirconia at grain boundaries through an entirely different process.

The above work demonstrates that impedance spectroscopy can be usefully applied to complex engineering ceramics with polyphase microstructures. Tetragonal calcia-stabilized polycrystals have not so far been prepared, so a direct measurement of their properties could not have been made. It also illustrates the necessity

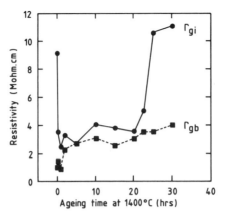

FIGURE 4.1.32. Grain interior and grain boundary resistivities obtained at 300°C for partially stabilized zirconia ceramics (ZrO_2 : 8 mole % CaO) as a function of aging time at 1400°C. (Reprinted from N. Bonanos, R. K. Slotwinski, B. C. H. Steele, and E. P. Butler, *J. Mat. Sci.*, **19**, 785–793, 1984, courtesy of Chapman & Hall.)

of having a suitable microstructural model with which to interpret the experimental impedance spectra. The evolution of the circuit equivalent with aging has been studied in a number of other PSZ systems, namely ZrO_2 doped with CaO + MgO, CaO + Y_2O_3, MgO + Y_2O_3, and Y_2O_3 alone. The last three systems show only stage I and II behavior even for long aging times: there is agreement with XRD and TEM results, which show difficulty in inducing the *t*-to-*m* transformation. The unusual property of the MgO + Y_2O_3 system in having no grain boundary resistance has been mentioned previously in Section 4.1.3.2.

4.1.3.4   Assessment of IS as a Characterization Technique

Effect of Grain Anisotropy in β-Alumina Ceramics. In the examples of zirconia ceramics discussed so far the shape and conductivity of the grains was approximately isotropic. Sodium β-alumina ceramics, however, consist of elongated grains, whose sodium ion conductivity is highly anisotropic, with sodium ions migrating only along specific crystallographic planes.

An IS study by Powers and Mitoff [1975] distinguished between grain interior and grain boundary conductivities with activation energies of approximately 0.2 and 0.4 eV, respectively. This result was confirmed by others, notably Lilley and Strutt [1979], who studied the impedance spectrum up to several megahertz and over the temperature range −135 to 400°C using a combination of manual bridges. Low temperatures were required in order to observe the grain interior arc because of the high sodium ion conductivity, at ambient temperature, of $5 \times 10^{-3}/\Omega$-cm (cf. Fig. 4.1.10). The impedance spectrum at −135°C (Fig. 4.1.33*a*) reveals a large grain boundary resistance.

Significantly, Arrhenius plots of the two conductivities (Fig. 4.1.33*b*) display two regions: a low-temperature region where the two activation energies are identical at 0.18 eV and a high-temperature region where the grain boundary activation energy rises to approximately 0.45 eV. The two lines cross over at 220°C, where

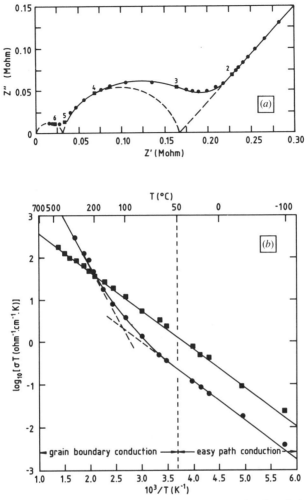

FIGURE 4.1.33. Application of impedance spectroscopy to a study of sodium β-alumina by Strutt and Lilley [1979]: (a) Impedance spectrum obtained at $-135\,^\circ$C using evaporated gold blocking electrodes. (b) Arrhenius plot of σ_{gi} and σ_{gb} showing transition between grain boundary and easy path conduction. (Reprinted from E. Lilley and J. E. Strutt, *Phys. Stat. Sol.* (a) **54**, 639–650, 1979, courtesy of Akademie-Verlag.)

$r_{gi} = r_{gb}.$* The first region is explained in terms of an easy-path model (Fig. 4.1.34a). The second region is reached when conduction through the grain boundary regions (having a higher activation energy) exceeds the conduction across the regions of intergranular contact. These results convincingly demonstrate the transition between easy-path and true grain boundary conduction in one and the same material, thus supporting the validity of the models reviewed in Section 4.1.1.

---

*As pointed out previously, this does not imply that $\sigma_{gi} = \sigma_{gb}$.

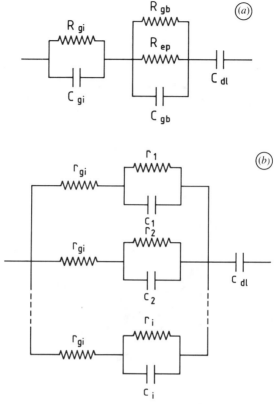

FIGURE 4.1.34. Circuits proposed for modeling the impedance spectrum of polycrystalline sodium β-alumina: (*a*) Easy path model according to Lilley and Strutt [1979] (*b*) Multielement model according to De Jonghe [1979].

Activation energies obtained in this analysis are in good agreement with previously determined values (Whittingham and Huggins [1971], Powers and Mitoff [1975]).

At about the same time, De Jonghe [1979] studied the correlation between TEM and IS in sodium β-alumina and was led to the conclusion that the simple two-RC network in general use was inadequate, since it failed to take account of the varying cross section and orientation of the grains in β-alumina ceramics. As an alternative he proposed a circuit having a series–parallel configuration (Fig. 4.1.34*b*). De Jonghe's model was based on careful electron microscopic observations of boundaries of grains at various angles of coincidence and on simulation of the effect of blocking on the dc conductivity. The effect of inhomogeneity was incorporated into this model by allowing the grain boundary related properties r_i, c_i to vary along the branches, while r_{gi} was held constant. The resistances r_{gi} and r_{gb} were assigned activation energies of 0.17 and 0.35 eV, respectively. In one simulation the admittance of 100 branches was summed and the resulting spectra were reanalyzed in terms of the series model. The above procedure was carried out at several temperatures, and the resulting values of r_{gb} and c_{gb} were compared to those obtained by summing the r_i, c_i branch admittances, ignoring r_{gi}. An example of a simulated impedance spectrum was not shown, nor was the exact method of analysis but, the r_{gi}, r_{gb} estimates were reported to deviate from the true values, at all but the lowest temperatures. Thus r_{gi} was overestimated by up to three times; the apparent capacitance also increased with temperature. Hence, De Jonghe concluded that IS was unsuitable for extracting microstructural parameters of any fundamental importance except possibly at the low temperatures and that it was only capable of producing "simple engineering correlations." He also stressed the difficulty of interpreting c_{gb} in microstructural terms when discontinuous grain boundary phases are present, a point with which we agree.

De Jonghe's simulation was only intended as an illustration of the effects of inhomogeneity and anisotropy, not as a technique to be used for parameter estimation. Nevertheless, it was on the basis of this simulation that he criticized the use of series models and, by inference, the application of IS techniques to other microstructural property evaluations. It is therefore worth examining to what extent the predictions of his model are validated in the light of the contemporary work of Lilley and Strutt [1979]. These authors went to great lengths to ensure that they could obtain reproducible impedance spectra and reported values of capacitative parameters at various temperatures (Table 4.1.6). They found that the grain boundary capacitance remained constant at around 800 pF over the whole temperature range. According to De Johnge's model, a variation of a factor of two or more would have been expected; this should have been easy to detect had it occurred. In our opinion this key observation undermines the plausibility of De Jonghe's series–parallel circuit and points toward the series circuit as a satisfactory model for β-alumina ceramic electrolytes.

In the context of the above discussion it is pertinent to ask as to whether the grain boundary resistance has any meaning outside the impedance spectrum and whether this resistance can be detected by a nonspectroscopic method. One possible approach would be to perform dc measurements on ceramics of the same chemical purity but of different grain size and to fit the total material resistivity so obtained ($r_{gi} + r_{gb}$) to one of the microstructural models of Section 4.1.1, for example, Eq. (6). Such an estimation of grain size would be difficult for β-alumina because of the extreme grain anisotropy.

Recently Powers [1984] has employed an ingenious interpretation of the dc resistivity which does not require explicit knowledge of the grain size. This de-

TABLE 4.1.6. Values of Capacitances in Circuit Equivalents of β-Alumina at Various Temperatures, as Determined by Lilley and Strutt [1979]

Temperature (K)	C_{gi} (F)	C_{gb} (F)	C_{el} (F)
138	4×10^{-12}	7×10^{-10}	1.5×10^{-7}
149	—	8×10^{-10}	1.5×10^{-7}
371	—	8×10^{-10}	1.0×10^{-7}
652	—	—	0.8×10^{-7}

pends on constructing, for a series of ceramic samples, a plot of the total dc resistivity of the material at a given temperature vs. the resistivity at a reference temperature (300K). Powers showed that this plot should be linear, with a slope related to the grain boundary activation energy, and he verified linearity for a range of samples (Fig. 4.1.35). By making a hypothesis regarding the relative magnitudes of r_{gi} and r_{gb} at the reference temperature, this plot can be used to find the corresponding ratio at the measurement temperature. Initially the hypothesis may be based on single-crystal data or guessed; subsequently the values of r_{gi} and r_{gb} are refined by adjustment so as to minimize the variance in the fit of the Arrhenius plot. Powers concluded that, contrary to the predictions of De Jonghe's multielement model, "contamination" of ΔH_{gb} with ΔH_{gi} was not significant and hence that r_{gi} and r_{gb} were indeed separable.

Effect of Conduction along Grain Boundaries. A further criticism of the series model is that it ignores the effect of conduction *along* grain boundaries. This view has been put forward by Näfe [1984], who has proposed an alternative, series-parallel model, reviewed in Section 4.1.1. In this section the application of the above model to ceramic electrolytes is critically discussed.

The analysis proposed by Näfe starts with the assumption that the grain interior and grain boundary phases have conductivities described by the Arrhenius relation, with distinct activation energies and preexponential factors. Using the series–parallel model, an analytical expression was obtained for the total dc conductivity σ_t as a function of temperature and was fitted to a large collection of dc conductivity data published for solid electrolytes of fluorite structure, namely, $ZrO_2 : Y_2O_3$, $ZrO_2 : CaO$, $ThO_2 : Y_2O_3$, and $ThO_2 : CaO$.

Each fit yielded estimates of the activation energies ΔH_{gi}, ΔH_{gb}, the preexponentials a_{gi}, a_{gb}, and the ratio of grain boundary thickness to grain size, d/D. In

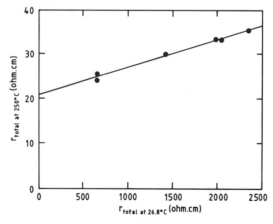

FIGURE 4.1.35. Resistivity distribution plot for a series of sodium β-alumina ceramic samples, according to Powers [1984]. (Reprinted from R. W. Powers, *J. Mat. Sci.* **19,** 753–760, 1984, courtesy of Chapman & Hall.)

all cases it was found that $\Delta H_{gi} > \Delta H_{gb}$; moreover, values of the ratio d/D agreed with those estimated by TEM. It was claimed that the above analysis carried no assumptions regarding the dominant mechanism of conduction. The activation energies obtained by Näfe for the $ZrO_2 : Y_2O_3$ system are given in Table 4.1.7 and may be compared to our own results on the same system given in Table 4.1.5.

While the success in estimating d/D gives prima facie support for Näfe's analysis, the result $\Delta H_{gi} > \Delta H_{gb}$ conflicts with a large body of data obtained by IS of ceramic electrolytes, showing that grain boundary phases are mainly high resistivity phases with high activation energies for conduction (see Section 4.1.1). In particular, the disagreement in the conductivity parameters of Tables 4.1.5 and 4.1.7 is too great be explained by variations in composition of samples or their method of preparation. In view of this, we have sought to test the Näfe model by extending it to ac behavior and examining its predictions regarding the type of impedance spectra expected of $ZrO_2 : Y_2O_3$ ceramic electrolytes.

First, to examine the dc behavior, we have reconstructed the Arrhenius plot of the total dc conductivity σ_t (Fig. 4.1.36), calculated using Näfe's own equation and the parameters given in Table 4.1.7. Lines for the grain interior and grain boundary specific conductivities are also drawn. From the varying slope of the reconstructed plot it is clear that, according to this model, at either end of the temperature scale, the dc conductivity is dominated by the grain boundaries, while at the midrange the conductivity is almost entirely determined by the grain interior conductivity. Transitions are expected to occur at temperatures of roughly 600 and 1500K.

Although Näfe's equation refers purely to dc conductivity, the more general block model (Fig. 4.1.4) and the corresponding equation [(Eq. (8)], may be used to simulate the whole impedance spectrum. This model includes conduction both *along* and *across* grain boundaries.

Impedance spectra were thus obtained using the conductivities σ_{gi}, σ_{gb} calculated from Table 4.1.7. The permittivities ϵ_{gi}, ϵ_{gb} were taken to be 3.3×10^{-12} pF/cm, corresponding to a dielectric constant of 37.3. Simulations carried out at two selected temperatures, 500K and 2000K, are shown in Figure 4.1.37. The low-temperature simulation (Fig. 4.1.37a) shows a single arc. In our own terminology of Section 4.1.1, this would be described as case II, where $\sigma_{gb} \gg \sigma_{gi}$ and the grain boundaries "short out" the grains. By contrast, the simulated high-temperature spectrum (Fig. 4.1.37b) shows two arcs, corresponding to case I, where $\sigma_{gi} \gg \sigma_{gb}$. As the calculated conductivities are high, of the order of 1 Ω/cm, the spectrum covers frequencies in the range of 10^7 to 10^{12} Hz.

TABLE 4.1.7. Arrhenius Parameters for Conduction in the $ZrO_2 : Y_2O_3$ System as Estimated by Näfe [1984]

ΔH_{gi} (eV)	ΔH_{gb} (eV)	a_{gi}	a_{gb}	d/D
1.14	0.38	5.5×10^5	27	5×10^{-3}

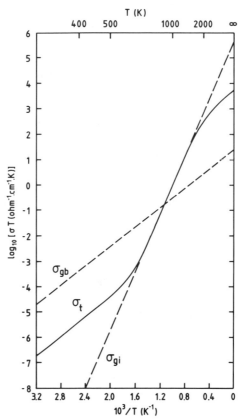

FIGURE 4.1.36. Arrhenius plot reconstructed from data published by Näfe [1984] for a $ZrO_2 : Y_2O_3$ ceramic.

To summarize the above results, our calculations, based on an ac extension of the Näfe model, predict that zirconia ceramics at moderate temperatures should give impedance spectra with a single arc, but that at high temperatures two arcs could, in principle, be observed with high-frequency measuring equipment. These predictions are at odds with experimental data on zirconia and other ceramic electrolytes. Evidence presented earlier shows that microstructural features are best resolved at lower temperatures and that grain boundary blocking becomes, if anything, less severe at elevated temperatures (Bernard [1981]). Thus, we consider that treatments based only on dc properties, such as the above, do not adequately describe the relationship between microstructure and conductivity. These conclusions do not necessarily imply a failure of the microstructural model proposed by Näfe. They do, however, cast serious doubts over the validity of the computer fit used and hence over the view that ionic conduction *along* grain boundaries plays a significant role in the electrical properties of zirconia ceramics.

Conclusions. In view of the work of Lilley and Strutt [1979] and Powers [1984] and of our own work on zirconia ceramics cited previously, we draw the following

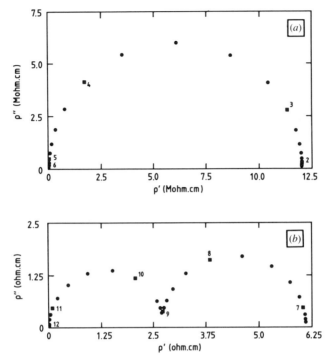

FIGURE 4.1.37. Impedance spectra for a $ZrO_2 : Y_2O_3$ ceramic simulated using the model described by Näfe [1984], extended to ac: (*a*) Impedance spectrum at 500K, showing a single arc. (*b*) Impedance spectrum at 2000K, resolving two arcs.

conclusions regarding the applicability of IS to microstructural-property correlations:

1. The behavior of grain boundary capacitance cited by De Jonghe as supportive of a multielement model is not borne out by a careful IS study of β-alumina.

2. There is independent evidence for the separability of r_{gi} and r_{gb} from dc measurements on β-alumina ceramics.

3. In any case, the objections raised for β-alumina do not apply to cubic ceramics of uniform, isotropic grain size.

4. Conductivity along grain boundaries is not a factor affecting the properties of common ceramic electrolytes.

5. For materials with continuous grain boundary phases, series models do assist in microstructural interpretation.

Acknowledgment

The authors wish to thank R. K. Slotwinski for providing some of the data presented here and for producing the diagrams, Dr. E. Lilley for providing data on

β-alumina ceramics, and Dr. C. P. Tavares for supplying a sample of Bi_2O_3 : Er. They are also grateful for many useful discussions with other members of the Wolfson Unit for Solid State Ionics at Imperial College, London. Much of the work in developing the measurement system described in Section 4.1.2 is credited to C. D. Waters.

This section was written by N. Bonanos, B. C. H. Steele, and E. P. Butler

4.2 SOLID STATE DEVICES

In this section examples of several different applications of impedance spectroscopy (IS) will be presented. Four different devices have been chosen: solid electrolyte chemical sensors, secondary (rechargeable) batteries, photoelectrochemical devices, and semiconductor–insulator–electrolyte sensors. In each subsection one selected application of IS will be briefly summarized to indicate the utility of this technique in determining the parameters important to that device. These sections are not intended to provide an extensive review of the area, but rather to show the power of the technique to solid state researchers. Thus this section is not a description of every device to which IS has been applied. Devices not discussed here which have been studied using IS include, among others, ion selective membranes (Sandifer and Buck [1974], Buck [1980, 1982]) and high-temperature steam electrolyzers (Schouler et al. [1981]).

Before beginning a detailed discussion it will be helpful to indicate the advantages and limitations that IS has in general application. Particularly desirable features of IS inherent to each specific applications will be discussed in the subsections below. The most important advantage is the ability to determine all of the time constants associated with a given interface in one experiment. That means that it is possible to determine diffusive, electrochemical, and chemical rate constants for a process from a single impedance spectrum. Further, the impedance is measured with a small ac signal, and a dc bias voltage can be superimposed with the ac signal so that the impedance and the rate information can be determined under various conditions. Such potential control is particularly important for electrochemical systems because the applied potential influences the rate of electron transfer at the interface. By measuring the impedance in such systems as a function of applied potential (i.e., dc bias), it is possible to determine the importance of the electrochemical reaction step to the overall rate of the reaction.

Although the equipment necessary to measure impedance spectra is readily available from many different suppliers, it remains expensive ($15,000–20,000 in 1985 dollars) even before the purchase of the nearly mandatory computer required for control and data analysis. Another disadvantage of IS is that very careful cell design is required to minimize stray capacitances and inductances. In addition to requiring three-electrode cell arrangements, as with all electrochemical systems, lead effects, including length, shielding, and the nature of all electrical contacts leading to and from the sample, must be considered. Ideally, the impedance of the cell should be measured under the actual experimental conditions but in the absence of the sample. These results can be used to verify or correct the experimental results. However, it is usually sufficient to minimize the stray impedances so that their values are negligible in comparison to the sample impedances.

A final drawback of the technique is the cumbersome data analysis which is required to obtain the desired physical quantities from the impedance spectra. A model electrical circuit which approximates the physical process being examined must be formulated. The model parameters are then obtained by determining parameter values which give the best fit to the impedance data. Finally, the model must be correlated with the physical system to establish the reliability of the model and to establish that the model values determined from the fit are physically reasonable. If not, the model may have to be modified and the entire analysis process repeated. In Chapter 3 there is a detailed discussion of this entire procedure. Clearly the required analysis is not always straightforward and is usually quite involved.

4.2.1   Electrolyte–Insulator–Semiconductor (EIS) Sensors

Electrolyte–insulator–semiconductor (EIS) sensors are one of a larger class of chemically sensitive electronic devices (Zemel in Janata and Huber [1985]) which meld integrated circuit technology with traditional chemical sensor technology. The EIS device is composed of a doped semiconductor, normally Si, acting as a substrate for a thin insulating layer, normally an oxide or nitride, which can be immersed in an electrolyte containing a fixed concentration of an ionic species to be measured. General reviews of the construction [Huber in Janata and Huber [1985]), thermodynamics (Janata in Janata and Huber [1985]), and operation (Abe, Esashi, and Matsuo [1979], Lauks and Zemel [1979], Bergveld and De Rooij [1981]) of such devices are available. Only a brief overview of the area will be given here so that those unfamiliar with these devices will be able to appreciate the application of IS described below.

A schematic of a typical EIS sensor is shown in Figure 4.2.1a. The operation of the sensor can easily be understood by considering the solid state analog of it, the metal oxide semiconductor (MOS) capacitor shown in Figure 4.2.1b. In the MOS device the capacitance is controlled by applying an external voltage between the gate and substrate. When there is a negative voltage relative to a p-type substrate the capacitance will be large because the holes in the substrate will be attracted to the insulator–semiconductor interface, giving rise to a wider region of dielectric material through which charge is separated. If the gate voltage is increased toward zero, the space charge layer in the capacitor will become narrower so the capacitance will decrease. As the voltage goes positive, the space charge layer in the semiconductor becomes narrower until it eventually disappears, and electrons build up at the surface of the semiconductor, forming what is normally called an *inversion layer*. This process is shown schematically in Figure 4.2.1c. For an n-type substrate, the capacitance curve is inverted (Fig. 4.2.1d), as when $V_G < 0$, an inversion layer forms (holes at surface of n-type Si), and when $V_G > 0$, there is a wider space charge layer.

The EIS functions in exactly the same fashion except the gate is formed by a

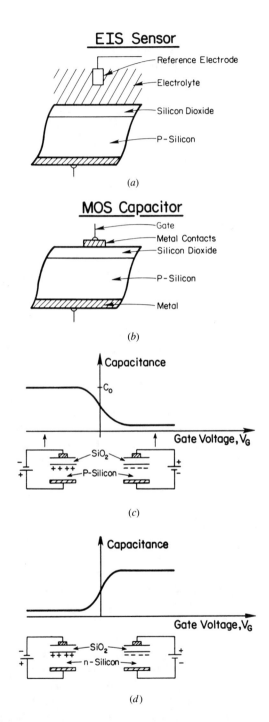

FIGURE 4.2.1. A schematic of a typical (*a*) EIS sensor and (*b*) an analogous MOS capacitor. (*c*) The capacitance of a MOS capacitor as a function of gate voltage for a *p*-type substrate, and (*d*) the equivalent response for an *n*-type substrate.

reference electrode in solution rather than a metal contact. The ability of the EIS to respond to ions in solution results from a modification of the electric charge distribution at the insulator–liquid and/or the insulator–semiconductor interface. Thus, at a given value of the reference potential the capacitance of the device will change depending on the ionic concentration in the solution. For example, for the simple device shown in Figure 4.2.1a, it has been shown that the capacitance will respond to a change in pH in the solution (Bergveld [1970], Siu and Cobbold [1979], Leroy et al. [1982], Bousse [1982], Bousse and Bergveld [1983], among others).

One real advantage of these sensors lies in the fact that an ion-selective membrane can act as a gate directly on a field effect transistor (FET) (Janata and Huber [1985]). These ion-selective field effect transistors (ISFET, shown schematically in Fig. 4.2.2a) again are the analog of a solid state device, the metal oxide semiconductor field effect transistor (MOSFET—Fig. 2.2.2b).

A MOSFET operates by controlling the concentration of charge carriers in p-type substrate between two n-type regions, called the source and the drain. When the gate is negatively biased with respect to the substrate, the region between the two n-type regions below the gate has no free electrons and the conductivity between the source and drain is very low. But as the voltage is increased until it becomes positive, at some point an inversion layer will form so there will be electrons available to form a channel. The conductivity between the source and drain will then increase. If there is a drain voltage supplied in this case, then a drain current can exist. At a given gate voltage, as the drain voltage is increased the drain current will saturate because the inversion layer is no longer of uniform thickness and becomes pinched off at the drain end. The transfer curve shown in Figure 4.2.2c results. As before, the ISFET operates in the same fashion except that an ion-selective membrane and reference electrode operate as a gate.

Although the impedance characteristics of the MOS devices are reasonably well understood (Nicollian and Goetzberger [1967], Nicollian and Brew [1982]), IS has not been applied nearly as widely to the EIS or ISFET devices. In this section the IS results of one of the simplest EIS devices, the Si–SiO₂–electrolyte pH sensor (Barabash and Cobbold [1982]), Bousse and Bergveld [1983], Diot et al. [1985]), will be used to illustrate the relative advantages of the technique.

An equivalent circuit for an EIS device has been derived (Bousse and Bergveld [1983], adapted by Diot et al. [1985]). From left to right in Figure 4.2.3a, it consists of the reference electrode impedance Z_{ref}; the electrolyte solution resistance R_s; the electrolyte–insulator interface impedance, which is composed of the double-layer capacitance C_{dl}, a diffusion impedance associated with the ionic species in solution (hydrogen ions for pH sensor), Z_w, and the SiO₂–electrolyte interface capacitance C_a; the insulator capacitance C_i; and the semiconductor–insulator interface impedance, which is composed of the space charge capacitance in the semiconductor, C_{sc}, as well as a capacitance and resistance C_{it} and R_{it}, respectively, associated with the interface states at this interface. At high frequency the electrolyte–insulator impedance is small with respect to C_i, as is the impedance of the reference electrode, so the equivalent circuit reduces to that shown in Figure 4.2.3b, where C_p represents the combined response of the C_{sc} and C_{it}.

ISFET

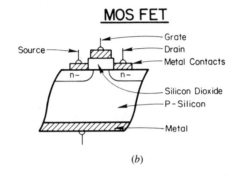

Reference Electrode
Electrolyte
Ion Sensitive Membrane
Source — Drain
Encapsulant
n– n–
Silicon Dioxide
P–Silicon

(a)

MOS FET

Grate
Source — Drain
Metal Contacts
n– n–
Silicon Dioxide
P–Silicon
Metal

(b)

Drain Current, I_D

Increasing V_G

Drain Voltage, V_D

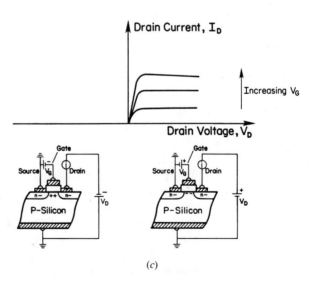

(c)

FIGURE 4.2.2. A schematic of a typical (a) ISFET device and (b) the analogous MOSFET device. (c) The current in the drain circuit as a function of the drain voltage at various different gate voltages. When the gate voltage is negative (left portion of Figure) the n–p–n junction will not conduct and there will be negligible drain current.

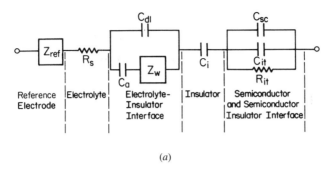

(a)

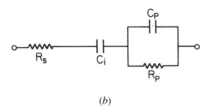

(b)

FIGURE 4.2.3. The equivalent circuit of EIS sensor shown in Figure 4.2.1a (after Bousse and Bergveld [1982]). (b) The reduced equivalent circuit for a MIS where the reference electrode–electrolyte interface impedances are small. Here R_P and C_P are the equivalent R and C elements associated with the combined impedance of the semiconductor and semiconductor–insulator interface (After Diot et al. [1985])

In the EIS structure the value of C_p depends upon the biasing of the device as described in Figure 4.2.1. In forward bias, C_p will be large with respect to C_i because an accumulation layer exists in the semiconductor (Fig. 4.2.1d, $V_G < 0$). In this case, the measured capacitance at high frequency, which is a series combination of C_{sc} and C_i, reduces to $C_i (1/C_i + 1/C_{sc} \cong 1/C_i)$, which is independent of applied potential. By evaluating the relative voltage in the space charge layer and the oxide (Diot et al. [1985], Sze [1985]), it can be shown that

$$E - E_{fb} = eN_D\epsilon_0\epsilon_i[(C_i/C)^2 - 1]/2C_i^2 \tag{1}$$

where E is the applied potential, E_{fb} is the flatband potential, e is the charge on an electron, N_D is the doping level, ϵ_0 and ϵ_i are the vacuum permittivity and relative dielectric constant, respectively, C_i is the insulator layer capacitance, and C is the total measured capacitance. The value of C varies with applied potential because the semiconductor capacitance, C_{SC}, depends on E through the surface voltage ψ_s (see below). Thus, the values of N_D and E_{fb} can be determined from the slope and intercept, respectively, of the linear portion of a plot of E vs. $(C_i/C)^2 - 1$. An example of such a result is shown in Figure 4.2.4.

Additional information about the semiconductor can be obtained from the interface capacitance C_{it}, which arises because each interface state stores a charge. A surface potential ψ_s can be defined as the potential at the semiconductor–insu-

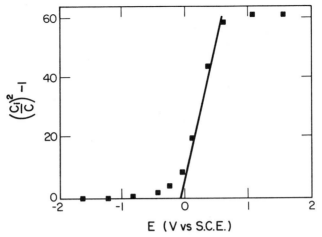

FIGURE 4.2.4. A plot of $(C_i/C)^2 - 1$ for n-type Si/SiO_2 EIS, where C_i is the insulator capacitance derived from high-frequency data and C is the total capacitance of the system. The oxide thickness is 94 nm and pH = 2.5. The values of the doping density N_d and the flat-band potential E_{fb} calculated from the linear portion of the curve are $2.2 \times 10^{20}/m^3$ and -0.06 V, respectively. (Diot et al. [1985])

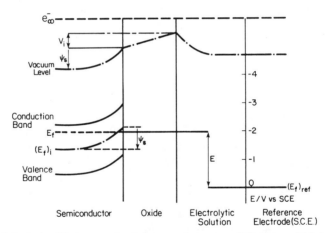

FIGURE 4.2.5. A simplified energy band diagram of an n-type EIS device which has an applied voltage E such that the semiconductor oxide interface is in the inversion regime [shown by crossing of E_f and $(E_f)_i$] leading to a buildup of holes at the interface. Two energy scales are shown, one referenced to an electron at infinity (e_∞^-) and the other referenced to the saturated calomel electrode (SCE). The surface potential ψ_s, the voltage drop across the insulator, V_i, and the Fermi levels of the reference electrode, the semiconductor under an applied voltage E, and the intrinsic semiconductor $[(E_f)_{ref}, E_f,$ and $(E_f)_i$ respectively] are all shown. (After Diot et al. [1986])

lator interface which causes the center of the band gap of the semiconductor [the Fermi level of the intrinsic material, $(E_f)_i$] to shift to a new value (Fig. 4.2.5). This surface potential arises whenever the applied potential causes charge to build up at the interface. For example, for an n-type material when $E = E_f - (E_f)_{ref}$ is very much less than zero, $(E_f)_i$ will cross E_f as shown in Figure 4.2.5, leading to an accumulation of holes at the interface, that is, inversion as described above for the MOS devices. Now ψ_s can be calculated from the capacitance data described above by means of (Nicollian and Goetzberger [1967])

$$\psi_s = \int_{E_{fb}}^{E} \left(1 - CC_i^{-1}\right) dE \qquad (2)$$

Thus, the important electrical characteristics of the semiconductor can be determined.

Further information about the interface states can also be extracted from the impedance data. From measured values of the total conductance G and capacitance C, the interface conductance can be calculated at any given potential after correcting for the solution resistance according to (Diot et al. [1985])

$$\frac{G_p}{\omega} = \frac{C_i^2\, G}{\omega\left[\left(G/\omega\right)^2 + \left(C_i - C\right)^2\right]} \qquad (3)$$

The insulator capacitance can be determined as described above. Thus, at any given reference potential the surface conductance G_p can be determined as a function of frequency. An example for typical results (Diot et al. [1985]) of a Si–SiO$_2$–electrolyte EIS is shown in Figure 4.2.6. Alternatively, G_p/ω can be calculated

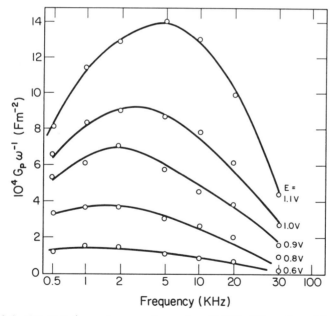

FIGURE 4.2.6. Plot of G_P/ω vs. frequency for a p-type Si/SiO$_2$ EIS at pH = 4.0 with doping density $N_A = 5.0 \times 10^{21}$/m^3 and oxide thickness of 60 nm. Here G_P is the total equivalent conductance associated with the semiconductor–semiconductor interface as described in Figure 4.2.3b. (Diot et al. [1985])

as the reference potential is swept and then converted to the representation in Figure 4.2.6.

Using these data, it is possible to calculate the number of interface states N_{it} at the semiconductor–insulator interface. In addition, a characteristic time constant τ_{it} associated with these states can be calculated. It is the time constant associated with the exponential decay of the interface states back to equilibrium after a perturbation. Qualitatively, when a small ac potential applied to the system swings in one direction, the electrons will be promoted from the interface states into unoccupied states in the silicon band and then demoted back to them as it swings in the other sense. The characteristic time for the electrons (in an n-type material) to decay back to the equilibrium configuration is τ_{it}.

The interface conductance can be understood, then, simply as the energy loss associated with this RC circuit such that $R_{it}C_{it} = \tau_{it}$ or $G_{it} = C_{it}/\tau_{it}$. The loss will be a maximum when the applied frequency reaches resonance with the characteristic time. In that case, G_p/ω will reach a maximum value. For the case of weak inversion in the semiconductor, the maximum will occur when $\omega\tau_{it} = 1$ and $(G_p/\omega)_{max}$ will equal $C_{it}/2$. The number of interface states N_{it} is given by C_{it}/e.

Using the techniques above, the effect of pH in the electrolyte has been examined on the Si–SiO$_2$–electrolyte EIS to ascertain the nature of the interaction between the hydrogen ions and the device (Bousse and Bergveld [1983]; Diot et al. [1985]). One immediate advantage of IS appears in the ability to measure very-low-interface state concentrations, lower than 10^{15} m^{-2}(eV)$^{-1}$ (Diot et al. [1985]). Plots of the N_{it} (calculated as described above) vs. the surface potential [calculated from Eq. (2)] have also been made as a function of pH and show that the effect of pH is very small (Fig. 4.2.7). This implies that the SiO$_2$–electrolyte interface is the one responsible for the change in potential of the device, not the SiO$_2$–Si interface (Diot et al. [1983]). Based upon a similar conclusion, a theoretical model which depends only on the sensitivity of the SiO$_2$ to the ionic concentration in the electrolyte has been used to successfully model the Si–SiO$_2$–electrolyte capacitance as a function of pH (Bousse and Bergveld [1983]). In addition, the general shape of Figure 4.2.7 is identical to that observed in MOS devices, thus reinforcing the contention that MOS and EIS devices function in exactly the same manner. Measurements of τ_{it} (Diot et al. [1985]) are also consistent with this supposition.

To reiterate, IS can be used to determine the important semiconductor electrical characteristics [E_{fb} and N_d (or N_a)], the insulator characteristics (C_i), and the nature of the semiconductor–insulator interface states (N_{it} and ψ_s). The technique is quite sensitive, allowing interface state concentration measurements below 10^{15} m^{-2}(eV)$^{-1}$. Results from several different studies (Barabash and Cobbold [1982]; Diot et al. [1985]) verify that the EIS device behavior is identical to that of the MOS device except that the metal gate is replaced by an electrolyte containing a reference electrode. Finally, and most importantly to device operation, in the Si–SiO$_2$–electrolyte device the electrolyte–insulator interface is shown (Diot et al. [1985]; Bousse and Bergveld, [1983]) to be the one that responds to changes in pH rather that the Si–SiO$_2$ interface. A major advantage of IS is its ability to gather such detailed interfacial information which is not easily accessible with other measurement techniques.

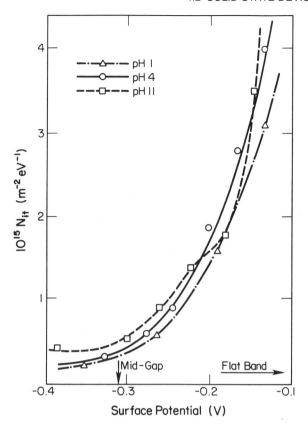

FIGURE 4.2.7. The number of interface states N_{it} as a function of surface potential ψ_s at three different pH values for a p-type Si/SiO$_2$ EIS (doping density $N_A = 4 \times 10^{-21}/m^3$ and oxide thickness of 92.5 nm). (Diot et al. [1985]).

4.2.2 Solid Electrolyte Chemical Sensors

The role of IS in the development and characterization of solid electrolyte chemical sensors (SECSs) is rapidly expanding. SECSs are electrochemical cells designed to measure the concentration or pressure of chemical species in gases or fluids. IS is emerging as an extremely useful technique to investigate the critical parameters which determine the electrolyte and electrode performances in these sensors.

The most successful SECSs are those which use zirconia-based electrolytes to measure oxygen concentrations. The three most common applications of these electrolytes are to measure oxygen concentrations of steel melts and in combustion gas environments and to control the air–fuel ratio in automobile engines. In the latter two applications, there is increasing interest in lowering the sensor temperature below 600°C, the current minimum temperature of operation because of low ionic conductivity and slow charge-transfer reactions at electrode–electrolyte interfaces.

An excellent example of the advantages and limitations of IS is the recent use of this technique to examine the effect of various electrode materials on the properties of zirconia-based oxygen sensors at temperatures below 600°C. (Matsui [1980], Badwal [1983], Mizusaki et al. [1983], Badwal et al. [1984]). The most common electrode material is platinum. However, the charge-transfer reaction (I) at the electrode–electrolyte interface is restricted to regions at or near lines of three-phase (gas–electrode–electrolyte) contact:

$$\tfrac{1}{2}\,O_2\,(\text{gas}) + 2e^-\,(\text{electrode}) = O^{-2}\,(\text{electrolyte}) \tag{I}$$

Because of this, a finely dispersed, porous electrode structure is formed on the electrolyte surface to maximize the regions of three-phase contact. However, an optimum pore structure is very difficult to maintain due to electrode sintering upon exposure to elevated temperatures.

Several authors (e.g., Matsui [1981], Badwal [1983], Mizusaki et al. [1983], Badwal et al. [1984]) have used IS to investigate the effects of different electrode materials and their pretreatment temperatures upon oxygen sensor performance at low temperatures. It is particularly interesting to compare the properties of gold and silver electrodes with the commonly used platinum electrodes. A typical impedance spectra of a zirconia-based oxygen sensor at 500°C is characterized by two semicircles, as shown in Figure 4.2.8. (Matsui [1981]). The semicircle in the low-frequency range shows a characteristic distortion depending largely on the electrode material and preparation. The intersection of the low-frequency semicircle at the extreme right side of the abscissa (3000 Ω) is determined by the resistance arising from the oxygen electrode reaction (I) and is represented in the equiv-

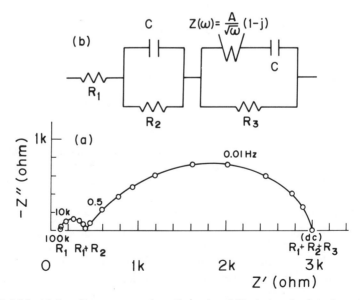

FIGURE 4.2.8. (*a*) Impedance response for a Pt/yttria-stabilized zirconia electrolyte with additives/Pt cell at 500°C and (*b*) the corresponding equivalent circuit. (Matsui [1981])

alent circuit by R_3. The values of R_1 and R_2 in Figure 4.2.8 represent the bulk and grain boundary resistance, respectively. As shown in Figure 4.2.8, the centers of the semicircle are usually below the real axis. In general this may result from two factors as described in Section 1.3, a constant-phase element such as that arising either from diffusion (the Warburg impedance) or from a distribution of time constants around an ideal value.

Oxygen sensor electrodes can experience temperatures as high as 900°C during cell preparation because of the necessity to remove organic impurities in the platinum paste electrodes and/or to ensure adherent platinum films on the zirconia electrolyte. The exposure time and temperature can affect and significantly increase the electrode impedance due to a reduction in the three-phase contact regions for reaction (I), which is caused by sintering of the finely dispersed, porous electrodes at high temperatures. Thus, oxygen cells with similar platinum paste electrodes but having different exposure times and temperatures will exhibit different complex-impedance spectra and electrode resistances. The IS data can therefore be used to optimize the sintering times and temperature to provide an electrode with better properties.

The electrode preparation technique is another important factor in determining the electrode resistance. For example, the difference between porous platinum electrodes prepared from a platinum paste (A) and a sputtering technique (B) is shown in Figure 4.2.9 (Mizusaki et al. [1983]), which shows only the low-frequency part of the complex-impedance spectrum. Although both cells were held at 900°C for 50 h in 1 atm oxygen, the resistance of the sputtered platinum electrode is less than that of the one prepared from platinum paste. However, the resistance of the oxide ($U_{0.5}Sc_{0.5}O_{2+x}$) electrode (C) is about an order of magnitude less than that of the platinum electrodes. These oxide electrodes significantly de-

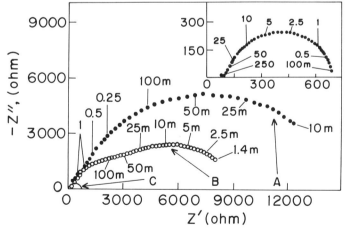

FIGURE 4.2.9. Complex impedance response for (a) 6082 Pt paste electrodes, (b) sputtered Pt electrodes 0.9 μm thick, and (c) $U_{0.5}Sc_{0.5}O_{2+x}$ electrodes (also inset on enlarged scale) at 600°C in 100% oxygen. All cells were given a prior heat treatment at 900°C for 50 h. Numbers on the arcs are frequencies in hertz (Badwal [1983])

crease the electrode resistance by increasing the interfacial area for charge-transfer reaction (I). Both oxygen ions and electrons are mobile in these electrodes (Badwal et al. [1984]), and reaction (I) can occur over the entire gas–electrode interfacial area. Scanning electron micrographs of the three electrodes shown in Figure 4.2.9, taken before and after heating at 900°C, clearly indicate substantial sintering of the platinum electrodes, while only small morphological changes are observed with the oxide electrodes (Badwal [1983]).

The impedance spectra (only the low-frequency region) for three noble metals (Ag, Au, Pt) electrodes are shown in Figure 4.2.10 (Badwal et al. [1984]). The results clearly indicate a significant difference between the silver and the gold electrode resistance in an oxygen sensor cell at 600°C. Although the resistance of the silver electrode is only slightly smaller than that of the platinum electrode, the resistance of the latter electrode significantly increases upon exposure to high temperatures. These sintering effects are not as severe for the silver electrodes because the appreciable solubility of oxygen in silver enables reaction (I) to occur over the entire electrode–electrolyte interfacial area rather than only at or near the three-phase contact region, as in the case with the platinum and gold electrodes.

Figure 4.2.10 clearly indicates that silver is a better electrode material in low-temperature oxygen sensors. However, significant volatility and microstructural

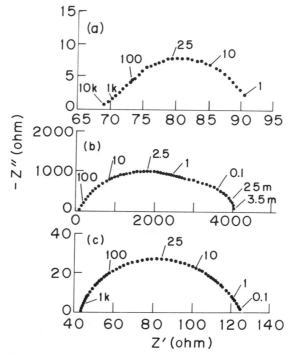

FIGURE 4.2.10. Complex impedance response at 600°C in 1 atm of oxygen after heating at 600°C for 50 h for the cells (*a*) Ag/yttria-stabilized zirconia/Au, (*b*) Au/yttria-stabilized zirconia/Au, and (*c*) Pt/yytria-stabilized zirconia/Pt. The numbers on the arcs are frequencies in hertz. [Badwal et al. [1984]]

changes of silver can occur at elevated temperatures, particularly at or above 900°C (Badwal et al. [1984]). Silver has been combined with platinum to form a Pt–Ag electrode, which possibly could exhibit the advantages of both metals. As shown in Figure 4.2.11 (Matsui [1981]), a Pt–Ag electrode (circles with centers) does have a significantly lower resistance than that of the platinum one (open circles). The impedance spectra shown in Figure 4.2.11 and zirconia-cell results at 300°C indicate that the Pt–Ag electrode could be a very useful electrode in a low-temperature oxygen sensor (Matsui [1981]).

The use of impedance spectra to determine the optimum electrode materials and preparation procedures for low-temperature oxygen sensors is only one example of the application of this technique in solid electrolyte sensors. For example, impedance spectra have already been used to examine the properties of zirconia stabilizers such as yttria and calcia in low-temperature zirconia electrolyte oxygen sensors (Badwal [1983]). The use of this technique in the development and characterization of other solid state sensors should increase significantly in the next few years.

4.2.3 Solid State Batteries

Recently, there has been increased interest in solid state batteries based upon insertion reactions in which a metal reacts reversibly with a cathode material by entering into the relatively open spaces—for example, planes or tunnels—of the cathode material (e.g., Johnson and Worrell [1982], Whittingham [1978], Rouxel [1979], Scrosati [1981]). Impedance spectroscopy has been applied to some of these electrochemical cells, which generally have the form of cell (A):

$$M/M^+ \text{ electrolyte, solid or liquid}/I \qquad\qquad (A)$$

where M is a metallic or a metallic alloy anode, often lithium or sodium, and I is an insertion electrode. The electrolyte in cell (A) can be either a liquid organic electrolyte or a solid electrolyte. In this section the lithium insertion reactions in

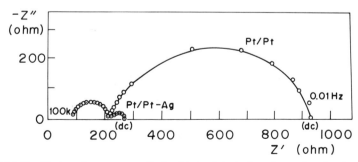

FIGURE 4.2.11. The impedance response for two kinds of electrodes using a tube of yttria-stabilized zirconia as the electrolyte. The arcs in the high-frequency range coincide, but the low-frequency arcs show a significantly lower resistance for the Pt/Pt-Ag electrode. (Matsui [1981])

tungsten and vanadium oxide bronzes will be used to illustrate the application of IS to the battery field.

Using cell (B), the kinetics of the insertion reaction of lithium into single-crystal sodium tungsten bronze has recently been investigated (Raistrick [1983]):

$$\text{Li}/1 \text{ M LiAsF}_6 \text{ in propylene carbonate}/\text{Na}_x\text{WO}_3 \qquad (B)$$

The complex plots were seen to follow the simple equivalent Randle's electrical circuit (Fig. 4.2.12) in the absence of specific adsorption (Ho et al. [1980]). The exchange current density I_0 can be determined from the interfacial reaction resistance θ according to

$$\theta = RT/nI_0F \qquad (4)$$

where n is the number of electrons transferred at the interface, F is Faraday's constant, and R and T are the ideal gas constant and the absolute temperature, respectively. The interfacial reaction resistance can be determined from a curve-fitting analysis of the circuit model in Figure 4.2.12, which includes this resistance as one of the components. Often in battery studies, the interfacial charge transfer resistance is in parallel with the double-layer capacitance C_{DL}, which gives rise to a semicircle in the impedance plane at higher frequencies. The exact frequency range over which this semicircle occurs will depend on the values of θ and C_{DL}, but it will often fall in the range of $10\text{--}10^4$ Hz.

The diffusion coefficient of lithium, D_{Li}, can be calculated from

$$Z_w = (1 - j) \frac{V_m(dE/dy)}{nFs(2\omega D)^{1/2}} \qquad (5)$$

where dE/dy is the slope of the open circuit voltage vs. composition curve, s is the surface area, and V_m is the molar volume. A typical impedance plot, shown in Figure 4.2.13, shows that at low frequencies (less than 1 Hz) there is a linear response at an angle of 45° to the real axis, indicative of a diffusion process. Because the diffusion in the solid tungsten bronze is so much slower than any other

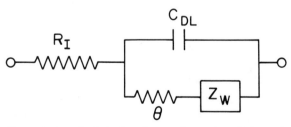

FIGURE 4.2.12. The equivalent circuit for insertion reactions, where C_{dl} is the double-layer capacitance, R_l is the electrolyte resistance, Z_w is the Warburg diffusional impedance, and θ is the interfacial charge-transfer resistance. (After Raistrick [1983])

diffusive process in the rest of the cell, this portion of the response results from that diffusive process.

The process between 1 Hz and 5 kHz is a depressed semicircle which was interpreted to result from the non-Warburg impedance elements in Figure 4.2.12. Thus, the solution resistance R_i, the charge transfer resistance θ, and the double-layer capacitance C_{DL} can all be easily determined. For example, from Figure 4.2.13, $R_i \simeq 45\ \Omega$ and $\theta \simeq 105\ \Omega$ (150 − 45). Using Eq. (4), I_0 was determined at four different potentials over a range of 50°K (Fig. 4.2.14). The results show

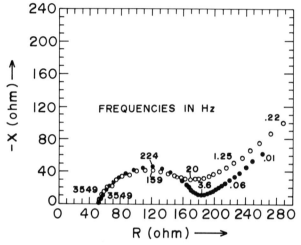

FIGURE 4.2.13. Impedance data for cell (B) with $x = 0.64$. Open circles are at $E = 1.85$ V vs. Li; closed circle are at $E = 2.00$ V vs. lithium. (Raistrick, [1983])

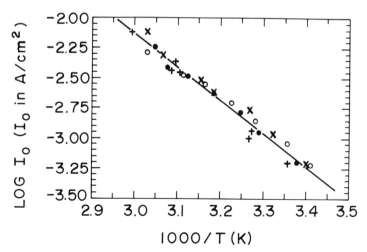

FIGURE 4.2.14. Log of the exchange current density vs. $1/T$ for different compositions of $Li_x Na_{0.64}$ WO_3. Plusses: $E = 2.25$ V, open circles: $E = 2.14$ V, closed circles: $E = 2.00$ V; Crosses: $E = 1.85$ V (all vs. lithium). (Raistrick [1983])

that over this temperature range the reaction kinetics have a single activation energy of 50 kJ/mole and that they are independent of potential (and thus electron concentration in the cathode). Therefore, Raistrick [1983] suggested that the interfacial reaction is associated with the ion transfer rather than the electron transfer.

Several features of this study are particularly noteworthy. First, with just one IS experiment information about the electrode–electrolyte interface reaction *and* diffusion in the electrolyte or the electrode can be determined. In addition, the correction for solution resistance is not necessary during the experiment since it can be immediately determined from the IS results. One common source of error in electrochemical experiments is thus immediately removed.

Determining the electrode reaction kinetics and the presence (and/or magnitude) of electrode or electrolyte diffusion processes while controlling the voltage is particularly important in studies of battery systems because the state of charge in any given battery system is a direct function of the voltage. By measuring the impedance spectra at different voltages it is possible to examine the importance of the reaction and diffusion steps at any given level of charge of the battery. Major differences in the spectra at different voltages will indicate changes in the kinetics of the cell and will permit a deeper understanding of the discharge process and lead to improved battery designs that enhance performance.

In this particular case, the interpretation of the IS results were quite simple due to the form of the results. Yet Raistrick still correlated the results with previous diffusivity measurement to ensure that there was no misinterpretation. Correlation of the parameters obtained from a given model associated with the impedance results and the physical parameters of the system obtained using other experimental approaches is essential in the data analysis procedure. Without such an analysis, the chosen equivalent circuit and thus the very nature of the interpretation is open to question.

4.2.4 Photoelectrochemical Solar Cells

Photoelectrochemical solar cells (PESCs) are devices which harness light energy and convert it into electrical or chemical energy by means of an electrochemical reaction at an interface. A general review of the electrochemistry of these devices can be found in most electrochemistry texts (e.g., Bard and Faulkner [1980]), but a cursory description will be given here for those unfamiliar with these devices. Most PESCs are composed of a semiconductor–electrolyte interface with an appropriate redox couple in solution. For an *n*-type semiconductor, when light with energy greater than the band gap strikes the interface, photons are absorbed and electron–hole pairs are created in the semiconductor. Some of these electron–hole pairs will simply recombine in the bulk, dissipating their energy thermally by the creation of phonons, by photon emission and so on. However, some proportion of the holes created at the interface will be available to oxidize the reduced species in solution, liberating an electron in the semiconductor which can flow in the external circuit. This photocurrent is absent in the dark where the concentration of holes is very low, so no reaction with the species in the solution is possible. The behavior of p-type semiconductors under irradiation is analogous; however, in this case electrons assist a reduction process in the solution and a current is produced by holes in the semiconductor.

In practice, the electrochemical behavior of semiconductor–electrolyte interfaces is far more complex than that described above (for a good review, see Boddy [1965]). One of the complications arises because the semiconductor surface at the electrolyte–semiconductor interface is not equivalent to that in the bulk. In particular, the energy states localized at the surface for holes and/or electrons are different than those present in the bulk. These surface states may arise in several ways—for example, through pretreatment (etching, polishing, etc.) of the semiconductor surface before immersion in the electrolyte. The surface states can be deterimental to the PESC efficiency if they increase the recombination of the electron–hole pairs in the semiconductor, thus reducing the number of holes (electrons for p-type material) available for chemical reaction with the redox species in solution.

Another complication arises because the semiconductor may chemically or electrochemically react with the electrolyte after immersion, leaving a layer on the surface of the semiconductor which has different electrical or electrochemical characteristics (e.g., an insulating layer) from the semiconductor. Because the photocurrent under illumination is very sensitive to the semiconductor–electrolyte interface, these surface perturbations not only change the electrochemical behavior but they can, in extreme cases, completely inhibit the photoresponse.

Impedance spectroscopy offers an excellent tool to examine the existence of surface states or other modifications of the ideal semiconductor–electrolyte interface. The general response of such interfaces was reviewed as early as 1965 (Boddy [1965]), but detailed studies of the response of PESCs are fairly recent. Dutoit et al. [1975] found that the capacitance of these interfaces at a given dc potential was dependent on the measuring frequency for CdSe, CdS and TiO_2 in several different aqueous and nonaqueous electrolytes. Tomkiewicz [1979] and McCann and Badwal [1982] have made more thorough investigations of the impedance response of several different technologically important semiconductor–electrolyte interfaces. The capacitance of a semiconductor–electrolyte junction has also been measured as a function of incident wavelength and used to characterize energy levels in semiconductors (Haak and Tench [1984], Haak, Ogden, and Tench [1982]). One particular study (Shen, Tomkiewicz, and Cahen [1986]) will be examined in more detail here to illustrate the kinds of effects that can be resolved using IS.

The impedance response of n-$CuInSe_2$ in polyiodide solutions has recently been used to understand the behavior of this material in a PESC (Shen, Tomkiewicz, and Cahen [1986]). Typical current potential response curves for n-$CuInSe_2$–polyiodide solutions are given in Figure 4.2.15 (Shen, Tomkiewicz, and Cahen [1986]) in which the effect of various pretreatments are shown. Polishing + etching or polishing + etching + oxidation significantly improves the photoresponse over simple polishing. For example, from Figure 4.2.15 at -0.2 V vs. Pt the photocurrent increases by approximately a factor of two after each additional pretreatment. IS was used in combination with electroreflectance (Shen, Siripala, and Tomkiewicz [1986]) to understand this behavior.

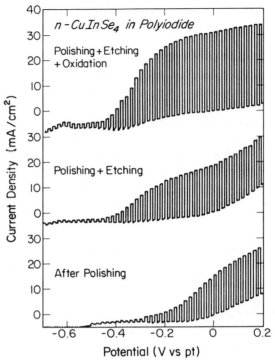

FIGURE 4.2.15. The effect of surface preparation on the current–potential response curves of n-CuInSe$_2$ in a solution of 6M KI + 0.1M InI$_3$ + 0.0125M I$_2$ at pH 6.0. The square-wave response results from using a chopped white light source of intensity 100 mW/cm^2. Etching was in a 2% Br$_2$–methanol solution for 60 s; oxidation was for 2 h at 150°C. (Shen, Tomkiewicz, and Cahen [1986]) Reprinted by permission of the publisher, The Electrochemical Society, Inc.

The impedance response obtained for polished + etched crystals and polished + etched + oxidized crystals (Fig. 4.2.16a and b, respectively) show qualitatively different behavior, the principal one being the addition of at least one more time constant in the oxidized case as manifested by (at least) one additional peak in the imaginary part in Figure 4.2.16b. Such behavior is reasonable since there is an additional interface between the oxide and semiconductor.

Using the equivalent circuits shown in Figure 4.2.16a and b, the high-frequency data were analyzed to determine the capacitance associated with the two fastest time constants in the polished + etched material and the fastest time constant in polished + etched + oxidized material. The low-frequency data were not analyzed because their physical interpretation was not clear. For the data shown in Figure 4.2.16a and b, the fastest time constant has been associated with the space charge layer in the semiconductor (C_{SC} in Fig. 4.2.16a and C_1 in Fig. 4.2.16b). In the polished + etched material, the next fastest is that associated with surface states on the semiconductor–electrolyte interface. It was assumed that the surface states are characterized by one time constant which does not significantly overlap with the time constants of any other states. It has been pointed out (McCann and Badwal

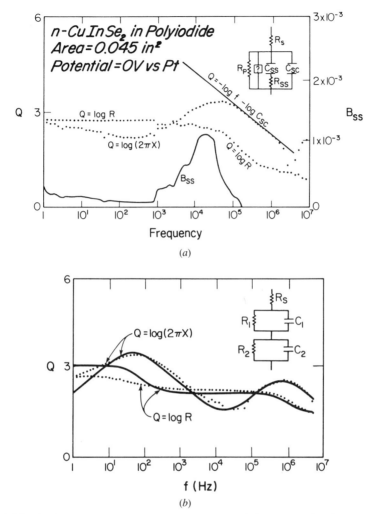

FIGURE 4.2.16. The impedance response and equivalent circuit for n-CuInSe$_2$ in the same solution as described in Figure 4.2.15 for: (a) Polished + etched sample. B_{ss} is the imaginary portion of the measured admittance less ωC_{sc}. The value of C_{sc} is calculated from the linear part of the high-frequency portion of the imaginary response of the impedance. The low-frequency response was not analyzed. (b) Polished + etched + oxidized sample. The solid line is a theoretical fit assuming the equivalent circuit shown. (Shen, Tomkiewicz, and Cahen [1986]) Reprinted by permission of the publisher, The Electrochemical Society, Inc.

[1982]) that should there be overlap of states with time constants close to one another, the time constants become essentially continuous, and a frequency-dependent resistance and capacitance must be used to model the interface. Here, though, the essential features of the interface appear to be adequately described without resorting to such elements.

In the case of the polished + etched + oxidized sample, C_1 was associated with the space charge layer capacitance. No further data were used. Thus, C_1 is representative of the change in capacitance of the space charge layer from the presence of the oxide layer.

Analyzing the effect of applied potential on the capacitance arising from the surface states, C_{ss}, in the polished + etched material (Fig. 4.2.17) led to the conclusion that there were two surface states, one centered 0.17 eV below the conduction band $[-0.69 - (-.72)$ in Fig. 4.2.17] and the other at 0.45 eV below the conduction band. Assuming a Gaussian distribution of surface states (Tomkiewicz [1979]), the area density of both states was calculated to be less than 1% of a monolayer. Thus, it was concluded that one major effect of etching was to remove most of the surface states. This conclusion is consistent with electro-reflectance results (Shen, Siripala, and Tomkiewicz [1986]) on the same system which show that in unetched samples the surface states pin the Fermi level, while after etching the surface states are nearly completely removed.

To determine the effect of oxidation, a Mott–Schottky plot of the space charge capacitance before and after oxidation was compared. In these plots, which were originally derived for a metal–semiconductor interface (Schottky [1939, 1942], Mott [1939]) but hold equally well for the metal–electrolyte interface, a linear relationship is predicted between the applied potential and one over the square of the capacitance arising from the space charge layer in the semiconductor. The slope is inversely proportional to the effective donor or acceptor concentration in the semiconductor. For the semiconductor–electrolyte interface (Bard and Faulkner [1980]),

$$\frac{1}{C_{sc}^{2}} = \frac{2}{e\epsilon_{o}\epsilon N_{d}}\left(-\Delta\phi - \frac{kT}{e}\right) \tag{6}$$

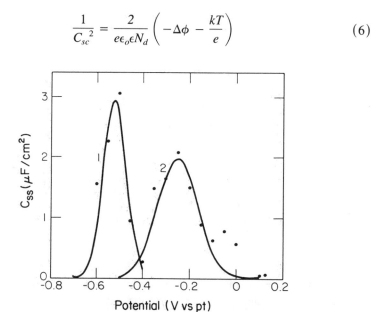

FIGURE 4.2.17. The variation of the capacitance associated with the surface states C_{ss} as a function of potential. The solid lines are a theoretical fit to two Gaussian line shapes as described in text. (Shen, Tomkiewicz, and Cahen [1986]) Reprinted by permission of the publisher, The Electrochemical Society, Inc.

where $\Delta\phi$ is the difference between the applied potential and the flatband potential $E - E_{fb}$, C_{sc} is the space charge capacitance, ϵ is the dielectric constant, ϵ_o is the permittivity of free space, k is Boltzmann's constant, T is absolute temperature, and N_d is the concentration of donors or acceptors. For the n-CuInSe$_2$ electrode the Mott–Schottky plot (Fig. 4.2.18) shows that the polish + etch + oxidation procedure does not change the flatband potential, but the effective doping level decreases by nearly one order of magnitude from that observed in the polished + etched material.

Several conclusions can be drawn from these data. First, the oxidation produced a layer that does not alter the electrical characteristics of the semiconductor since the flat-band potential did not change. Second, the oxide layer decreases the doping level, thus increasing the width of the space charge layer. This wider layer in turn leads to higher photocurrent because most of the light is absorbed within the space charge layer so that recombination of charge carriers in the bulk is reduced. Finally, by applying a simple model of photoresponse, Tomkiewicz [1980] deter-

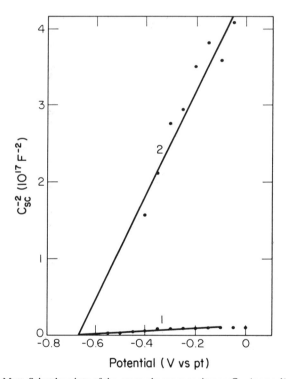

FIGURE 4.2.18. Mott–Schottky plots of the space charge capacitance C_{SC} (curve 1) as derived from data like those shown in Figure 4.2.9a and the capacitance associated with the high-frequency response, C_1, (curve 2) derived from data like those shown in Figure 4.2.9b. The flat-band potential is the same in both cases (0.69 V), but the doping level, as calculated from the slope of the lines, is an order of magnitude lower for curve 2 (polished + etched + oxidized sample) than for curve 1 (polished + etched sample). (Shen, Tomkiewicz, and Cahen [1986]). Reprinted by permission of the publisher, The Electrochemical Society, Inc.

mined that surface recombination arises from the surface state at 0.17 eV below the conduction band. Electroreflectance measurements (Shen, Siripala, and Tomkiewicz [1986]) are consistent with this conclusion. Thus, the improved response caused by etching can be explained by the decrease in density of these surface states observed in the impedance results after etching.

Several key features of this study should be emphasized. IS clearly can be used to successfully model a semiconductor–electrolyte interface in a PESC. The ability to probe the physics of this interface using IS while controlling the applied potential can allow significant insight into the important parameters of the device. In particular, the surface states at the semiconductor electrolyte interface may be determined, as can their relative importance after several different pretreatments or in different cell configurations. The electrical characteristics of the interface— for example, the flat-band potential and the space charge capacitance, can also be determined.

The work described above also shows that it is not always necessary to analyze the entire frequency spectrum (i.e., determine the complete equivalent circuit) of a cell in order to obtain significant insights into its operation if it is possible to associate a particular region of the spectrum with a meaningful physical quantity. In the case of the semiconductor–electrolyte interface described above, a strong theoretical background describing the expected behavior along with other experimental findings (electroreflectance, current–potential curves) on the system permitted such a limited but meaningful analysis. Further, a more detailed analysis of the results would probably have led to a more complete description of the operation of the PESC. This fact was recognized by the authors themselves (Shen, Tomkiewicz, and Cahen [1986]). Although a limited analysis may allow significant insights, it should be made with extreme caution, as the results could lead to erroneous conclusions. A complete detailed analysis of the entire frequency spectrum is far preferable and leads to a more complete understanding of the device operation.

This section was written by William B. Johnson and Wayne L. Worrell.

4.3   CORROSION OF MATERIALS

4.3.1   Introduction

Corrosion is defined as the spontaneous degradation of a reactive material by an aggressive environment and, at least in the case of metals in condensed media, it occurs by the simultaneous occurrence of at least one anodic (metal oxidation) and one cathodic (e.g., reduction of dissolved oxygen) reaction. Because these partial reactions are charge-transfer processes, corrosion phenomena are essentially electrochemical in nature. Accordingly, it is not surprising that electrochemical techniques have been used extensively in the study of corrosion phenomena, both to determine the corrosion rate and to define degradation mechanisms.

Of all of the electrochemical techniques that are available, impedance spectroscopy promises to be the most valuable because of its ability, in a single experiment, to detect interfacial relaxations covering a wide range of relaxation times. The application of this technique in corrosion science became possible in the practical sense only within the past decade, with the advent of techniques for measuring transfer functions at subhertz frequencies (see Section 3). Instruments and techniques are now available for measuring interfacial impedances at frequencies down to the 10^{-3}–10^{-4} Hz region, where relaxations involving adsorbed intermediates and diffusing species appear.

Over this same period, considerable development has taken place in the theoretical treatment of the impedance properties of corroding interfaces (Macdonald and McKubre [1981]). These theoretical developments have been especially important, since they serve to enhance the quantitative nature of the technique. Indeed, impedance spectroscopy has emerged as probably the most powerful technique currently available for identifying corrosion reaction mechanisms, and methods are now being developed to extract kinetic parameters (rate constants, transfer coefficients) for multistep reaction schemes.

In this section, we review the application of impedance spectroscopy to the study of corrosion phenomena. Emphasis is placed on illustrating how the method is applied to identify the different processes that occur at a corroding interface. We also review the use of impedance measurements for measuring corrosion rate, since this was the initial application of the technique in corrosion science and engineering. The use of impedance spectroscopy to analyze other cause an effect phenomena of interest in corrosion science, including electrochemical–hydrodynamic, fracture, and electrochemical–mechanical processes, is also discussed.

4.3.2 Fundamentals

The response of any physical system to a perturbation of arbitrary form may be described by a transfer function

$$H(s) = \overline{V}(s)/\overline{I}(s) \tag{1}$$

where s is the Laplace frequency and $\overline{V}(s)$ and $\overline{I}(s)$ are the Laplace transforms of the time-dependent voltage and current, respectively (Goldman [1950]). In terms of the steady state sinusoidal frequency domain, the transfer function becomes

$$H(j\omega) = \frac{F[V(t)]}{F[I(t)]} = \frac{V(j\omega)}{I(j\omega)} \tag{2}$$

where F signifies the Fourier transform and $V(j\omega)$ and $I(j\omega)$ are the sinusoidal voltage and current, respectively. Provided that the system is linear, that causality is obeyed, and that the interface is stable over the time of sampling (see later), the transfer function may be identified as an impedance $Z(j\omega)$. Because they are vector quantities, $H(j\omega)$ and $Z(j\omega)$ are complex numbers containing both magnitude and phase information. From a theoretical viewpoint, the impedance (or, more generally, the transfer function) is one of the most important quantities that can be measured in electrochemistry and corrosion science. This is because, if it is sampled over an infinite bandwidth, it contains all the information that can be obtained from the system by purely electrical means.

An important requirement for a valid impedance function is that the system be linear. Theoretically, this implies that the real and imaginary components transform correctly according to the Kramers–Kronig relationships (discussed later in this section). Practically, linearity is indicated by the impedance being independent of the magnitude of the perturbation, a condition that is easily (although seldom) tested experimentally.

4.3.3 Measurement of Corrosion Rate

The Stern–Geary equation provides a direct relationship between the steady state corrosion current and the ''dc'' resistance across the interface (Stern and Geary [1957])

$$i_{corr} = \left[\frac{\beta_a \beta_c}{2.303 \, (\beta_a + \beta_c)} \right] \left(\frac{1}{R_p} \right) \tag{3}$$

where β_a and β_c are the Tafel constants for the anodic and cathodic partial reactions, respectively, and R_p is the polarization resistance (Mansfeld [1976]). Because corroding interfaces are inherently reactive by nature, owing to the presence of capacitive, psuedoinductive, and diffusional impedance terms, it is evident that the polarization resistance is given only by the difference of the measured impedance at sufficiently low and high frequencies:

$$R_p = \left| Z(j\omega) \right|_{\omega \to 0} - \left| Z(j\omega) \right|_{\omega \to \infty} \tag{4}$$

Measurement of the series resistance at the high-frequency limit normally presents few problems, because $Z(j\omega)$ becomes nonreactive at frequencies as low as 10 kHz, in most cases. On the other hand, in the low-frequency region, reactance is commonly observed at frequencies in the vicinity of 10^{-3} Hz, so that special precautions must be adopted to obtain reliable data (Syrett and Macdonald [1979]). The need for these precautions is independent of the form of the perturbation applied to the interface. Accordingly, they apply equally well to the use of potential or current steps and triangular and sinusoidal voltage perturbations in the measurement of the polarization resistance, as well as to the determination of ''steady state'' current–voltage curves. Practically, therefore, it is necessary to use a sufficiently low-frequency (sinusoidal perturbation), low-voltage scan rate (small-amplitude cyclic voltametry) or to wait a long enough time (potential or current step perturbation) before acquiring the response data (e.g., current) for calculating the polarization resistance.

The problem of acquiring impedances at sufficiently low frequencies is amply demonstrated by the data (Syrett and Macdonald [1979]) for 90 : 10 Cu:Ni alloy corroding in flowing seawater (Fig. 4.3.1). Thus, for an exposure time of 22 h, the impedance function can be defined over the entire bandwidth, and an accurate

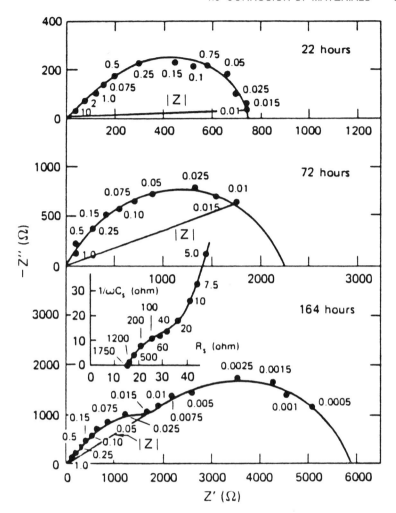

FIGURE 4.3.1. Complex plane impedance diagrams for 90 : 10 Cu : Ni alloy in flowing seawater as a function of exposure time. Flow velocity = 1.62 m/sec, [O_2] = 0.045 mg/liter, specimen area = 11.05 cm^2, T = 26°C; exposure time = 50 h. (From B. C. Syrett and D. D. Macdonald, The Validity of Electrochemical Methods for Measuring Corrosion Rates of Copper–Nickel Alloys in Seawater. Reprinted with permission from *Corrosion*, Vol. 35, No. 11, 1979, NACE, Houston, TX.) Numbers next to each point to frequency in hertz.

value for R_p may be obtained by probing the interface at frequencies above 0.01 Hz. On the other hand, at much longer exposure times, frequencies as low as 0.0005 Hz are not sufficient to completely define the interfacial impedance, and considerable extrapolation is required to acquire a value for R_p.

It is important to emphasize again that, because time domain functions can be synthesized as linear combinations of sinusoidal (sine and cosine) components (Fourier synthesis), this problem is not limited to impedance spectroscopy. Thus, failure to use a sufficiently low sweep rate in the case of small-amplitude cyclic voltametry (SACV) will also introduce significant error (Fig. 4.3.2), depending upon which resistance is considered as being the parameter of interest (R_d or R_{app}, Fig. 4.3.3; Macdonald [1978a]). Interestingly, our experience in using a variety of electrochemical monitoring techniques indicates that SACV is superior, in many respects, to impedance spectroscopy for determining the polarization resistance. Thus, quite reliable values for R_p for systems as reactive as that shown in Figure 4.3.1 generally can be obtained using a single voltage sweep rate of 0.1 mV/s, which is quite accessible using standard electrochemical instrumentation.

Because most impedance measurements are made sequentially at discrete frequencies, the total data acquisition time can be expressed as

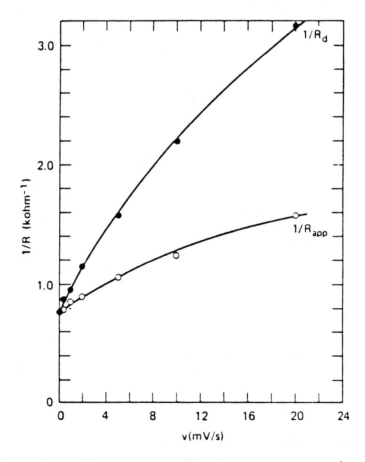

FIGURE 4.3.2. Plots of $1/R_d$ and $1/R_{app}$ as measured using SACV for 90:10 Cu:Ni in flowing seawater. Flow velocity = 1.62 m/s, $[O_2]$ = 0.045 mg/liter, T = 26°C, exposure time = 50 h. (From D. D. Macdonald, An Impedance Interpretation of Small Amplitude Cyclic Voltammetry: I. Theoretical Analysis for a Resistive–Capacitive System, *J. Electrochem. Soc.* **125**, 1443–1449, 1978. Reprinted by permission of the publisher, The Electrochemical Society, Inc.)

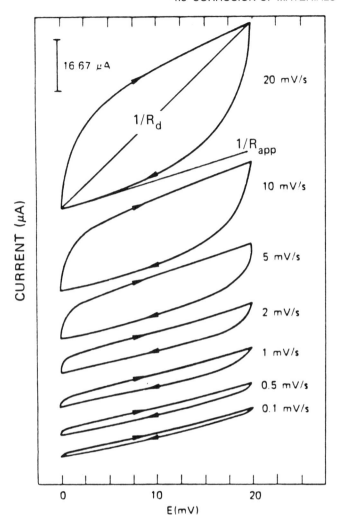

FIGURE 4.3.3. Small-amplitude cyclic voltamograms for 90:10 Cu:Ni alloy in flowing seawater. Experimental conditions are as listed in Fig. 4.3.1. (From D. D. Macdonald, An Impedance Interpretation of Small Amplitude Cyclic Voltammetry: I. Theoretical Analysis for a Resistive–Capacitive System, *J. Electrochem. Soc.* **125**, 1443–1449, 1978. Reprinted by permission of the publisher, The Electrochemical Society, Inc.)

$$T = \sum_i \frac{n_i}{f_i} \tag{5}$$

where n_i is the number of cycles at frequency f_i. The minimum acquisition time is obtained by setting $n_i = 1$. Therefore,

$$T_{\min} = \sum_i \frac{1}{f_i} \tag{6}$$

and it is apparent that the minimum acquisition time is dominated by the low-frequency components. For example, the impedance data shown in Figure 4.3.1, for an exposure period of 164 h, required an acquisition time of more than 1 h. This contrasts with an acquisition time of 100 s required to obtain a reliable value for R_p using SACV with a voltage sweep rate of 0.1 mV/s and a peak-to-peak amplitude for the triangular voltage excitation of 5 mV. Because SACV does not readily yield the mechanistic information afforded by impedance spectroscopy, the two methods are best regarded as being complementary in nature.

According to Fourier's theorem, all small-amplitude techniques must yield identical results (i.e., the same interfacial impedance), regardless of the form of the excitation. This is clearly the case for the system discussed above, as shown in Figure 4.3.4. In this figure, polarization resistance data, obtained using the impedance spectroscopic, potential step, and SACV techniques, are plotted as a function of time for two copper–nickel alloys exposed to flowing seawater (Syrett and Macdonald [1979]). The fact that the polarization resistance data are independent of the technique used for their measurement implies that the experimenter has the freedom to tailor a perturbation for the measurement of interfacial impedance in order to achieve some desired experimental goal. One implementation of this concept is the application of a large number of sine-wave voltage signals simultaneously, so that the total data acquisition time is determined only by the lowest frequency, and not by the summation embodied in Eq. (5). These "structured noise" techniques are now being actively developed for corrosion-monitoring purposes.

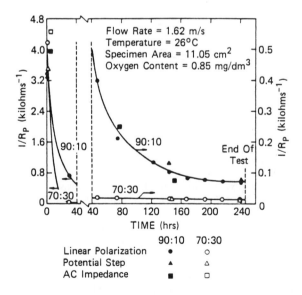

FIGURE 4.3.4. Corrosion rate (as I/R_p) vs. time for 90:10 Cu:Ni and 70:30 Cu:Ni in flowing seawater having an oxygen content of 0.85 mg/dm³. (From B. C. Syrett and D. D. Macdonald, The Validity of Electrochemical Methods for Measuring Corrosion Rates of Copper–Nickel Alloys in Seawater. Reprinted with permission from *Corrosion*, Vol. 35, No. 11, 1979, NACE, Houston, TX.)

The structured noise method stem from the elegant work of Smith and coworkers (Smith [1966]), who developed a multifrequency technique for ac polarography. Subsequently, structured noise techniques have been used in corrosion studies by Smyrl and coworkers (Smyrl [1985a,b], Smyrl and Stephenson [1985]) and by Pound and Macdonald [1985]. In all cases, the perturbation applied to the system is of the form

$$V(t) = \sum_i a_i \sin(\omega_i t + \phi_i) \tag{7}$$

where a_i is the amplitude, ω_i is the frequency, and ϕ_i is the phase. If these parameters are selected in a completely random fashion, the signal is referred to as *white noise*. However, because corroding interfaces are inherently nonlinear, considerable advantages exist in choosing values for a_i, ω_i, and ϕ_i such that certain experimental problems are avoided. For example, nonlinearity produces harmonics of $2\omega_i$, $3\omega_i$, . . . $n\omega_i$ in response to a perturbation at the fundamental frequency ω_i. Because the amplitude of a harmonic decreases rapidly with increasing n, harmonic intrusion may be avoided by ensuring that $\omega_j \neq n\omega_i$ ($n = 2, 3, . . .$) or may at least be minimized by requiring that $n > 3$. Also, the power applied to the interface, which is proportional to the square of the amplitude of each component, may be tailored by choosing appropriate values for a_i. Regardless of the exact form of the perturbation employed, the impedance data are extracted from the perturbation and the response by Fourier or Laplace transformation (Pound and Macdonald [1985], Smyrl [1985a,b]). As an example of this technique, we show the data of Pound and Macdonald [1985] for carbon steel in acidified brine (Fig. 4.3.5). The structured noise data are compared with those obtained using a frequency-by-frequency correlation technique (FRA). Clearly, the structured noise data are considerably more scattered than are those obtained by the correlation method, but that is compensated for by the reduction in the data acquisition time.

4.3.4 Harmonic Analysis

The derivation leading to the Stern–Geary relationship [Eq. (3)] assumes that the corroding electrode responds linearly to the imposed electrical perturbation; that

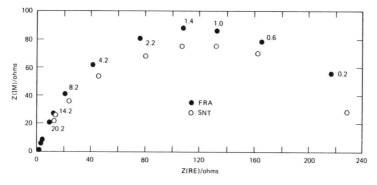

FIGURE 4.3.5. Nyquist plot of impedance data for 1018 steel in HCl-acidified 3% NaCl (pH = 3). Exposure time = 28 h, E_{corr} = −0.682 V (SCE).

is, doubling the perturbing voltage amplitude results in a double current response (but an unchanged impedance). Since physical variables in all physically realizable systems must have a finite first derivative, it is always possible to achieve linear conditions by applying a perturbation of limitingly small amplitude.

The nonlinearity of the current–voltage relationship in corroding systems provides an opportunity to determine corrosion rates without the need to measure indepedently the Tafel constants. The reason is that the electrical perturbation, which is imposed on the system at a frequency of f, in a nonlinear system results in a response at $2f$, $3f$, $4f$, and so on, in addition to a dc component (McKubre [1983], Morring and Kies [1977], McKubre and Macdonald [1984], Bertocci [1979], Bertocci and Mullen [1981, Kruger [1903]). Neither the fundamental response (f_0) nor the total power response ($\sum_{h=0}^{\infty} hf$) can be analyzed to determine uniquely the corrosion rate (as opposed to the polarization resistance). Nevertheless, an analysis of the harmonic responses can be used to determine the unknown parameters in Eq. (3) and thus to measure corrosion rates in systems for which the Tafel coefficients are not known or at potentials removed from the free corrosion potential V_{fc}, as, for example, under conditions of an applied cathodic protection potential.

The origin of the harmonic response is shown schematically in Figure 4.3.6a for an input voltage sine wave at frequency f, superimposed on a current–voltage curve of the form

$$I = I_{fc}\left\{\exp\left[\beta_a(V - V_{fc})\right] - \exp\left[-\beta_c(V - V_{fc})\right]\right\} \qquad (8)$$

where β_a and β_c are the forward anodic and reverse cathodic Tafel coefficients, respectively, and I_{fc} is the free corrosion current flux, defined at the free corrosion potential (V_{fc}) as

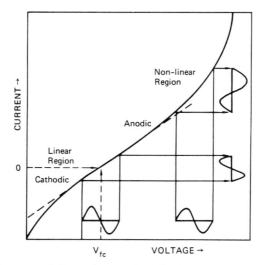

FIGURE 4.3.6a. The source of a harmonic response: Reflection of an input sine wave on a nonlinear current–voltage curve.

$$I_{fc} = I_a = -I_c \qquad (\text{at } V_{fc}) \tag{9}$$

Figure 4.3.6a shows a perturbing voltage sinusoid reflected about the dc current voltage response curve at V_{fc} and at some positive bias. In the linear region, this reflection results in an undistorted current response, with V/I being a constant (equal to the dc corrosion resistance).

The expected response in the time domain is shown schematically in Figure 4.3.6b. The output is generally shifted in phase with respect to the input due to reactive terms associated with diffusional and capacitive processes. The extent of the distortion in the nonlinear region can be quantified by performing a Fourier series analysis. As indicated in Figure 4.3.6c, when a sinusoidal perturbation of moderate amplitude is applied to a corroding electrode, the response will consist of a component at the same frequency (generally shifted in phase with respect to the input), as well as terms at integral multiples of the input frequency (the harmonics). Unless the input excitation is symmetric about V_{fc}, then the output also will show a dc offset that we term the *zero'th harmonic*. An offset is shown even for a symmetric perturbation if the I/V curves is not symmetrical—this is the basis of the faradic rectification effect. These harmonic response terms contain information sufficient to completely specify the current–voltage curve, in principle, at any dc voltage and thus to monitor the instantaneous corrosion rate even in the presence of an applied cathodic polarization.

The analysis of the harmonic response of a system to a sinusoidal current or voltage perturbation has received periodic attention in the electrochemical literature since the pioneering work of Warburg [1899] and Kruger [1903]. This effect

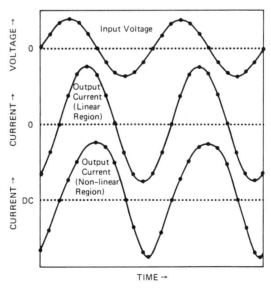

FIGURE 4.3.6b. The source of a harmonic response: Time domain representation of input and output waveforms.

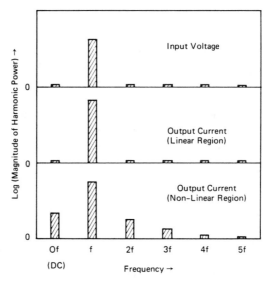

FIGURE 4.3.6c. The source of a harmonic response: Frequency domain representation of input and output waveforms.

has been studied as faradic rectification (Oldham [1957], Barker, Faircloth, and Gardener [1958], Barker [1958], Delahay, Senda, and Weis [1960], Iami and Delahay [1962], Bauer [1964] and faradic distortion (Delahay [1954], Breyer and Bauer [1964], Smith [1966]), and the results of this form of analysis have been applied to the development of ac polarography (Morring and Kies [1977]). More recently, Bertocci [1979], Bertocci and Mullen [1981], and others (Chin and Venkatesh [1979] have investigated the effect of large-amplitude perturbations in increasing the corrosion rates of electrical conduit materials (so-called ac corrosion).

The theroetical treatments referenced above all suffer from a major deficiency. The nonlinear term of interest in corrosion (the electron transfer process) is contained within a circuit comprising other linear (electrolyte resistance) and nonlinear (double-layer capacitance and diffusional impedance) terms. Since the voltage dropped across nonlinear circuit elements cannot be considered to linearly superimpose, we cannot use the equivalent circuit method to isolate the impedance terms of interest. Properly, one must solve for the system as a whole, including diffusional and double-layer terms, and identify the harmonic components associated with the faradic process of interest.

The simplified theoretical treatment presented here is similar in form to that described previously (McKubre [1983], Bertocci [1979], Bertocci and Mullen [1981], Devay and Meszaros [1980], and Devay [1982], Gill, Callow, and Scantlebury [1983], Hladky, Callow, and Dawson [1980], Rangarajan [1975], Ramamurthy and Rangarajan [1977], Rao and Mishra [1977], Callow, Richardson, and Dawson [1976], Devay and Meszaros [1980]).

We are interested in the current response of an electrode to a voltage perturbation of the form

$$V = V_0 + v \sin (\omega t) \qquad (10)$$

Substituting Eq. (10) into Eq. (8) yields

$$I = I_{fc} \left\langle \exp \left\{ \beta_a [\eta + v \sin (\omega t)] \right\} - \exp \left\{ -\beta_c [\eta + v \sin (\omega t)] \right\} \right\rangle \qquad (11)$$

where

$$\eta = V_0 - V_{fc} \qquad (12)$$

One can make the substitution (Abramowitz and Stegun [1965], Bauer [1964])

$$\exp \left[z \sin (x) \right] = J_0(z) + 2 \sum_{k=0}^{\infty} (-1)^k J_{2k+1} (z) \sin \left[(2k + 1)x \right]$$
$$+ 2 \sum_{k=1}^{\infty} (-1)^k J_{2k}(z) \cos (kx) \qquad (13)$$

where $J_n(z)$ is a modified Bessel function of order n. The value of $J_n(z)$ can be calculated by means of the expression,

$$J_n(z) = (z/2)^n \sum_{k=0}^{\infty} \left[(z/2)^{2k} / k! (n + k)! \right] \qquad (14)$$

The first term in Eq. (13) represents the expected dc response (zero'th harmonic or faradic rectification component) attributable to an ac perturbation. The second terms gives the odd-order harmonic response, and the last term gives the even harmonics. In the limit as $v \to 0$, all response functions except the fundamental disappear, and for $\eta = 0$ we obtain the familiar expression for the Stern–Geary (1957) relationship

$$\frac{dI_{fc}}{dV} = \frac{1}{R_p} = \frac{2.303 I_{fc}(\beta_a + \beta_c)}{\beta_a \beta_c} \qquad (15)$$

Under all other conditions, the faradic current must be represented by a Fourier series of harmonic responses, as indicated by Eqs. (12)–(14).

Substituting Eqs. (13) and (14) into Eq. (11), we obtain an equation of the form

$$I/I_{fc} = \left[\exp (\beta_a \eta) \right] ({}_0C^+ + {}_1C^+ + {}_2C^+ + {}_3C^+ + \cdots) \qquad (16)$$
$$- \left[\exp (-\beta_c \eta) \right] ({}_0C^- + {}_1C^- + {}_2C^- + {}_3C^- + \cdots)$$

where the harmonic series of constants $C^\pm$ are $\exp [\pm v \sin (\omega t)]$ evaluated according to Eq. (13) and presubscripts are used to denote the harmonic number.

Figures 4.3.7 and 4.3.8 demonstrate the influence of various corrosion parameters on the expected harmonic response, calculated from Eq. (16). Figure 4.3.7

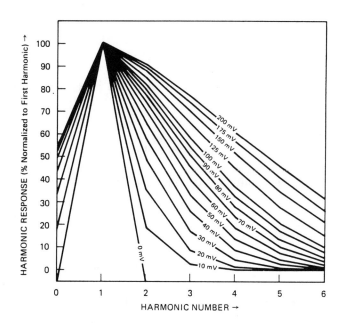

FIGURE 4.3.7. The effect of ac amplitude on the magnitude of the harmonic response for a two-electron process.

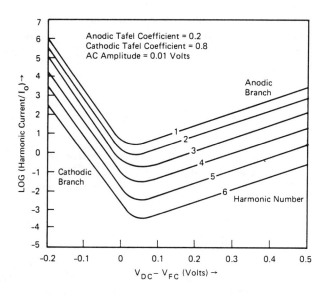

FIGURE 4.3.8. The effect of dc applied potential on the harmonic response for an asymmetric corrosion process.

shows the effect of the ac amplitude (v) on the magnitude of the response at large overvoltages; the responses are normalized by that at the first harmonic. Clearly, the application of perturbation levels as low as 50 mV can result in significant powers of the harmonic response at 0f, 2f, and 3f.

The effect of dc potential and Tafel coefficient on the harmonic response is shown in Figure 4.3.8. The power of the current response normalized by the free corrosion current (I_{fc}), shown on a log-linear scale, parallels that of the dc current response with constant ratio between the harmonics. For the symmetric case, shown in Figure 4.3.6a, the minimum in harmonic response occurs at the free corrosion potential. However, as the ratio of the reverse cathodic to forward anodic Tafel coefficients is increased, the potential at which the response is a minimum for each harmonic is increased. This phenomenon is shown in Figure 4.3.8 and has been suggested as the basis for a corrosion monitor (Gill, Callow, and Scantlebury [1983]).

Precise measurements of the current responses at each of a number of harmonics can be used, in conjunction with Eq. (16), to evaluate the unknown terms in Eq. (11). A complete description of the faradic current–voltage response is sufficient to define the anodic partial current (the corrosion rate) at any potential; the parameters needed are the forward and reverse Tafel coefficients, the free corrosion current, and the free corrosion potential.

In applying the harmonic method to corrosion rate monitoring, three major sources of interferences must be overcome. A major source of error is due to the presence of an uncompensated series electrolyte resistance. The harmonic current responses due to the corrosion process are transformed by such a resistance into a voltage that appears as an input perturbation at the harmonic frequency, leading to an erroneous harmonic current response. A second source of error appears at large values of anodic or cathodic polarization, where the measured dc (and thus low-frequency impedance) response may be largely dominated by diffusional processes. Since the genesis of the harmonic response is considered to be in the faradic processes, it is necessary to deconvolve the diffusional from the charge-transfer impedance terms. A practical, although approximate, solution to the problem of uncompensated resistance and diffusional impedance is to completely determine the equivalent circuit for the corroding electrode by performing the impedance (fundamental or harmonic) study over a wide range of frequencies and mathematically correcting the data set retrospectively (McKubre and Syrett [1984]).

A more insidious problem is the limitation on precision imposed by the vanishingly small magnitude of the anodic (corrosion) component compared with the cathodic partial current at large values of cathodic bias, due to the exponential form of the current–voltage relationships for the anodic and cathodic half-reactions.

Methods by which these limitations can be minimized and the parameters of interest calculated are described by McKubre and Syrett [1984]. The ratio of the harmonic admittance to the fundamental admittance (both corrected as described above) is used to evaluate the desired corrosion parameters. Equation (15) can be expressed in the more appropriate form by noting that

$$\frac{_nY}{_mY} = \frac{_nI}{_1V} \cdot \frac{_1V}{_mI} = \frac{_nI}{_mI} \cdot \frac{I_{fc}}{_mI} \tag{17}$$

$$\frac{_0Y}{_1Y} = \frac{[\exp(\beta_a\eta)](_0C^+) - [\exp(-\beta_c\eta)](_0C^-)}{[\exp(\beta_a\eta)](_1C^+) - [\exp(-\beta_c\eta)](_1C^-)} \tag{18}$$

$$\frac{_2Y}{_1Y} = \frac{[\exp(\beta_a\eta)](_2C^+) - [\exp(-\beta_c\eta)](_2C^-)}{[\exp(\beta_a\eta)](_1C^+) - [\exp(-\beta_c\eta)](_1C^-)} \tag{19}$$

$$\frac{_3Y}{_1Y} = \frac{[\exp(\beta_a\eta)](_3C^+) - [\exp(-\beta_c\eta)](_3C^-)}{[exp(\beta_a\eta)](_1C^+) - [\exp(-\beta_c\eta)](_1C^-)} \tag{20}$$

The constants $_hC^{\pm}$ can be evaluated using Eqs. (17) and (18); the unknown parameters η, β_a, β_c are calculated from the best fit of the measured admittance ratios to this system of equations.

4.3.5 Kramers–Kronig Transforms

At this point it is fitting to ask the question: "How do I know that my impedance data are correct?" This question is particularly pertinent in view of the rapid expansion in the use of impedance spectroscopy over the past decade and because more complex electrochemical and corroding systems are being probed. These give rise to a variety of impedance spectra in the complex plane, including those that exhibit pseudoinductance and intersecting loops in the Nyquist domain. By merely inspecting the experimental data, it is not possible to ascertain whether or not the data are valid or have been distorted by some experimental artifact. However, this problem can be addressed by using the Kramers–Kronig (KK) transforms (Kramers [1929], Kronig [1926], Tyagai and Kolbasov [1972], Van Meirhaeghe, Dutoit, Cardon, and Gomes [1976], Bode [1945], Macdonald and Brachman [1956]), as recently described by Macdonald and Urquidi-Macdonald [1985] and Urquidi-Macdonald, Real, and Macdonald [1986].

The derivation of the KK transforms (Bode [1945]) is based on the fulfillment of four general conditions of the system:

1. *Causality.* The response of the system is due only to the perturbation applied and does not contain significant components from spurious sources.
2. *Linearity.* The perturbation/response of the system is described by a set of linear differential laws. Practically, this condition requires that the impedance be independent of the magnitude of the perturbation.
3. *Stability.* The system must be stable in the sense that it returns to its original state after the perturbation is removed.
4. The impedance must be finite-valued at $\omega \to 0$ and $\omega \to \infty$ and must be a continuous and finite-valued function at all intermediate frequencies.

If the above conditions are satisfied, the KK transforms are purely a mathematical result and do not reflect any other physical property or condition of the system.

These transforms have been used extensively in the analysis of electrical circuits (Bode [1945]), but only rarely in the case of electrochemical systems (Tyagai and Kolbasov [1972], Van Meirhaeghe et al. [1976]).

The KK transforms may be stated as follows:

$$Z'(\omega) - Z'(\infty) = \left(\frac{2}{\pi}\right) \int_0^\infty \frac{xZ''(x) - \omega Z''(\omega)}{x^2 - \omega^2} \, dx \tag{21}$$

$$Z'(\omega) - Z'(0) = \left(\frac{2\omega}{\pi}\right) \int_0^\infty \left[\left(\frac{\omega}{x}\right) Z''(x) - Z''(\omega)\right] \cdot \frac{1}{x^2 - \omega^2} \, dx \tag{22}$$

$$Z''(\omega) = -\left(\frac{2\omega}{\pi}\right) \int_0^\infty \frac{Z'(x) - Z'(\omega)}{x^2 - \omega^2} \, dx \tag{23}$$

$$\phi(\omega) = \left(\frac{2\omega}{\pi}\right) \int_0^\infty \frac{\ln|Z(x)|}{x^2 - \omega^2} \, dx \tag{24}$$

where $\phi(\omega)$ is the phase angle, Z' and Z'' are the real and imaginary components of the impedance, respectively, and ω and x are frequencies. Therefore, according to Eq. (4), the polarization resistance simply becomes

$$R_p = \left(\frac{2}{\pi}\right) \int_0^\infty \left[\frac{Z''(x)}{x}\right] dx \approx \left(\frac{2}{\pi}\right) \int_{x_{min}}^{x_{max}} \left[\frac{Z''(x)}{x}\right] dx \tag{25}$$

where x_{max} and x_{min} are the maximum and minimum frequencies selected such that the error introduced by evaluating the integral over a finite bandwidth, rather than over an infinite bandwidth, is negligible.

To illustrate the application of the KK transformation method for validating polarization resistance measurements in particular and for verifying impedance data in general, we consider the case of TiO$_2$-coated carbon steel corroding in

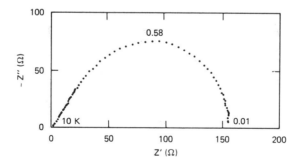

FIGURE 4.3.9. Complex plane impedance plot for TiO$_2$-coated carbon steel in HCl–KCl solution (pH = 2) at 25°C. The parameter is frequency is hertz. (From D. D. Macdonald and M. Urquidi-Macdonald, Application of Kramers–Kronig Transforms to the Analysis of Electrochemical Impedance Data: I. Polarization Resistance, *J. Electrochem. Soc.* **132**, 2316–2319, 1985. Reprinted by permission of the publisher, The Electrochemical Society, Inc.)

HCl–KCl solution (pH = 2) at 25°C (McKubre [1985]). The complex plane diagram for this case is shown in Figure 4.3.9, illustrating that at high frequencies the locus of points is linear, but that at low frequencies the locus curls over to intersect the real axis. Application of Eq. (25) predicts a polarization resistance of 158.2 Ω compared with a value of 157.1 Ω calculated from the high- and low-frequency intercepts on the real axis (Macdonald and Urquidi-Macdonald [1985]).

By using the full set of transforms, as expressed by Eqs. (21)–(24), it is possible to transform the real component into the imaginary component and vice versa (Macdonald and Urquidi-Macdonald [1985], Urquidi-Macdonald, Real, and Macdonald [1986]). These transforms therefore represent powerful criteria for assessing the validity of experimental impedance data. The application of these transforms to the case of TiO_2-coated carbon steel is shown in Figures 4.3.10 and 4.3.11. The accuracy of the transform was assessed by first analyzing synthetic impedance data calculated from an equivalent electrical circuit. An average error between the "experimental" and "transformed" data of less than 1% was obtained. In this case, the residual error may be attributed to the algorithm used for evaluating the integrals in Eqs. (21)–(23). A similar level of precision was observed on transforming McKubre's (1985) extensive data set for TiO_2-coated carbon steel in HCl/KCl (Macdonald and Urquidi-Macdonald [1985], Urquidi-Macdonald, Real, and Macdonald [1985]). Not all impedance data are found to

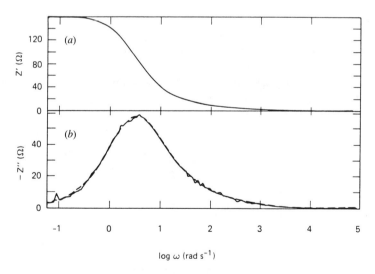

log ω (rad s⁻¹)

FIGURE 4.3.10. Kramers–Kronig transforms of impedance data for TiO_2-coated carbon steel in HCl/KCl solution (pH = 2) at 25°C: (a) Real impedance component vs. log ω. (b) Comparison of the experimental imaginary impedance component (- - -) with $Z''(\omega)$ data (——) obtained by KK transformation of the real component. (From D. D. Macdonald and M. Urquidi-Macdonald, Application of Kramers–Kronig Transforms to the Analysis of Electrochemical Impedance Data: I. Polarization Resistance, *J. Electrochem. Soc.* **132**, 2316–2319, 1985. Reprinted by permission of the publisher, The Electrochemical Society, Inc.)

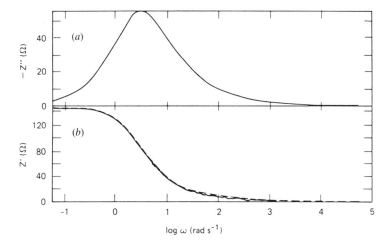

FIGURE 4.3.11. Kramers-Kronig transforms of impedance data for TiO_2-coated carbon steel in HCl/KCl solution (pH = 2) at 25°C: (*a*) Imaginary impedance component vs. log ω. (*b*) Comparison of the experimental real impedance component (- - -) with $Z''(\omega)$ data (———) obtained by KK transformation of the imaginary component. (From D. D. Macdonald and M. Urquidi-Macdonald, Application of Kramers–Kronig Transforms to the Analysis of Electrochemical Impedance Data: I. Polarization Resistance, *J. Electrochem. Soc.* **132,** 2316–2319, 1985. Reprinted by permission of the publisher, The Electrochemical Society, Inc.)

transform as well as those for the equivalent electrical circuit and the TiO_2-coated carbon steel system referred to above. For example, Urquidi-Macdonald, Real, and Macdonald [1986] recently applied the KK transforms (21)–(23) to the case of an aluminum alloy corroding in 4-M KOH at temperatures between 25 and 60°C and found that significant errors occured in the transforms that could be attributed to interfacial instability as reflected in the high corrosion rate.

4.3.6   Corrosion Mechanisms

4.3.6.1   *Active Dissolution*

Impedance spectroscopy has been applied extensively in the analysis of the mechanism of corrosion of iron and other metals in aqueous solutions. Typical work of this kind is that reported by Keddam et al. [1981], who sought to distinguish between various mechanisms that had been proposed for the electrodissolution of iron in acidified sodium sulfate solutions. Since this particular study provides an excellent review of how impedance spectroscopy is used to discern reaction mechanism, the essential features of the analysis are described below.

As the result of analyzing a large number of possible mechanisms for the dissolution of iron, Keddam et al. [1981] concluded that the most viable mechanism for this reaction involves three intermediate species

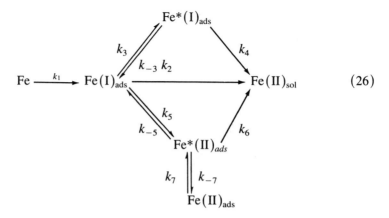

in which steps 4 and 6 are written in complete form as

$$\overset{k_4}{Fe^*(I)_{ads} + Fe \rightarrow Fe^*(I)_{ads} + Fe(II)_{sol} + 2e^-} \tag{27}$$

and

$$\overset{k_6}{Fe^*(II)_{ads} + Fe \rightarrow Fe^*(II)_{ads} + Fe(II)_{sol} + 2e^-} \tag{28}$$

In setting up the reaction model for this case, Keddam et al. assumed that the elementary steps obey Tafel's law, that the transfer coefficient (α) has a value between 0 and 1 and is independent of potential, and that the coverage by adsorbed species obeys the Langmuir isotherm. Designating the fractional coverages by the species $Fe(I)_{ads}$, $Fe^*(I)_{ads}$, and $Fe^*(II)_{ads}$ to be θ_1, θ_2, and θ_3, respectively, and that of the passivating species $Fe(II)_{ads}$ as θ_4, and assuming no overlap, then the current flowing across the interface may be expressed as

$$I = F\left[k_1 \Sigma + (k_2 + k_5)\theta_1 + 2k_4\theta_2 + (2k_6 - k_{-5})\theta_3\right] \tag{29}$$

where

$$\Sigma = 1 - \theta_1 - \theta_2 - \theta_3 - \theta_4 \tag{30}$$

and k_i is the rate constant for the ith step defined by

$$k_i = k_{o,i} \exp\left(\frac{z\alpha F}{RT} \cdot E\right) \tag{31}$$

Mass balance relationships involving the adsorbed species results in the following expressions for the time dependencies of θ_1, θ_2, θ_3, and θ_4:

$$\beta_1 \frac{d\theta_1}{dt} = k_1 \Sigma - (k_2 + k_3 + k_5)\theta_1 + k_{-3}\theta_2 + k_{-5}\theta_3 \tag{32}$$

$$\beta_2 \frac{d\theta_2}{dt} = k_3\theta_1 - k_{-3}\theta_2 \tag{33}$$

$$\beta_3 \frac{d\theta_3}{dt} = k_5\theta_1 - (k_{-5} + k_7)\theta_3 + k_{-7}\theta_4 \tag{34}$$

$$\beta_4 \frac{d\theta_4}{dt} = k_7\theta_3 - k_{-7}\theta_4 \tag{35}$$

where β is a constant that links the surface fractions to surface concentrations (mole/cm^{-2}). The value for β is $\sim 10^{-8}$ mole/cm^2, which corresponds to about one monolayer. The steady state is characterized by $d\theta_i/dt = 0$, in which case

$$\bar{\theta}_1 = \frac{k_1 k_{-3} k_{-5} k_{-7}}{D} \tag{36}$$

$$\bar{\theta}_2 = \frac{k_1 k_3 k_{-5} k_{-7}}{D} \tag{37}$$

$$\bar{\theta}_3 = \frac{k_1 k_{-3} k_5 k_{-7}}{D} \tag{38}$$

$$\bar{\theta}_4 = \frac{k_1 k_{-3} k_5 k_7}{D} \tag{39}$$

where

$$D = k_1 k_{-3} k_5 k_7 + \left\{ k_1 \left[k_3 k_{-5} + k_{-3} (k_5 + k_{-5}) \right] + k_2 k_{-3} k_{-5} \right\} k_{-7} \tag{40}$$

and hence the steady state current becomes

$$\bar{I} = 2 F (k_2 \bar{\theta}_1 + k_4 \bar{\theta}_2 + k_6 \bar{\theta}_3) \tag{41}$$

In order to derive the faradic impedance (Z_F) we note that for sinusoidal variations in the potential and in the surface coverages of reaction intermediates we may write

$$\delta E = |\delta E| e^{j\omega t} \tag{42}$$

$$\delta \theta_i = |\delta \theta_i| e^{j\omega t} \tag{43}$$

$$j = \sqrt{-1} \tag{44}$$

Thus, from Eq. (41) and defining Z_F as

$$Z_F = \frac{\delta E}{\delta I} \tag{45}$$

we obtain the following expression for the faradic impedance:

$$\frac{1}{Z_F} = \frac{1}{R_T} - F\left[(k_1 - k_2 - k_5)\frac{d\theta_1}{dE} + (k_1 - 2k_4)\frac{d\theta_2}{dE} \right.$$

$$\left. + (k_1 + k_{-5} - 2k_6)\frac{d\theta_3}{dE} + k_1\frac{d\theta_4}{dE}\right] \tag{46}$$

where

$$\frac{1}{R_t} = F\left[(b_1 + b_2)k_2\bar{\theta}_1 + 2b_4k_4\bar{\theta}_2 + [(b_5 + b_{-5})k_{-5} + 2b_6k_6]\bar{\theta}_3\right] \tag{47}$$

$$b_i = \alpha_i F/RT \tag{48}$$

The faradic impedance is readily obtained by first deriving expressions for $d\theta_i/dt$. This is done by taking the total differentials of Eqs. (32)–(35). For example, in the case of Eq. (33) we write

$$\delta\left(\beta_2\frac{d\theta_2}{dt}\right) = k_3\delta\theta_1 + \theta_1\delta k_3 - k_{-3}\delta\theta_2 - \theta_2\delta k_{-3} \tag{49}$$

Since

$$\delta\left(\beta_2\frac{d\theta_2}{dt}\right) = \beta_2\frac{d(\delta\theta_2)}{dt} = \beta_2 j\omega\delta\theta_2 \tag{50}$$

and

$$\delta k_i = b_i k_i \delta E \tag{51}$$

we obtain

$$(k_{-3} + j\omega\beta_2)\frac{\delta\theta_{-2}}{\delta E} - k_3\frac{\delta\theta_1}{\delta E} - \theta_1 b_3 k_3 + \theta_2 b_{-3}k_{-3} = 0 \tag{52}$$

Additional linear simultaneous equations may be generated from Eqs. (32), (34), and (35), and the set may be solved for $\delta\theta_1/\delta E$, $\delta\theta_2/\delta E$, $\delta\theta_3/\delta E$, and $\delta\theta_4/\delta E$. These values are then substituted into Eq. (46) to calculate the faradaic impedance, which in turn yields the interfacial impedance as

$$Z_T = Z_F/(1 + j\omega C_{dl} Z_F) \tag{53}$$

where C_{dl} is the double layer capacitance.

Experimental and simulated [Eq. (29)] steady state current–voltage curves for iron in Na_2SO_4–H_2SO_4 solutions as a function of pH are shown in Figure 4.3.12,

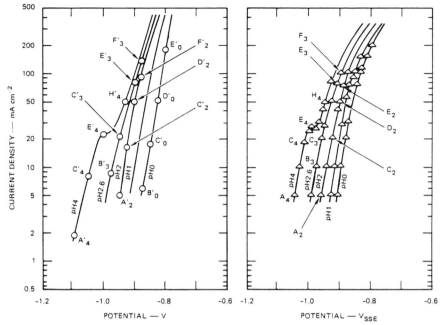

FIGURE 4.3.12. Steady state polarization curves for iron in $NaSO_4$–H_2SO_4 solutions according to Keddam et al. [1981]: (*a*) Simulated curves. (*b*) Experimental data. Rotating disk electrode (rotating speed = 1600 rpm, diameter = 3 mm), $T = 25 \pm 0.2°C$. (From M. Keddam, O. R. Mattos, and H. J. Takenouti, Reaction Model for Iron Dissolution Studied by Electrode Impedance: Determination of the Reaction Model, *J. Electrochem. Soc.* **128**, 257–274, 1981. Reprinted by permission of the publisher, The Electrochemical Society, Inc.)

and simulated and experimental complex plane impedance diagrams at various points on these curves are shown in Figures 4.3.13 and 4.3.14, respectively (Keddam et al. [1981]). The close agreement between the steady state polarization curves is immediately apparent, even to the extent that the inflection in the curve for pH 4 is accurately reproduced. Examination of the impedance diagrams in Figures 4.3.13 and 4.3.14 show that the mechanism selected by Keddam et al. [1981] is capable of reproducing the essential features of the diagrams, including the number and type of relaxations, but not the details. However, the latter depend strongly upon the values selected for the rate constants, and are also probably affected by the isotherm selected for describing the adsorption of intermediate species onto the surface.

4.3.6.2 Active–Passive Transition

The sudden transition of a metal–solution interface from a state of active dissolution to the passive state is a phenomenon of great scientific and technological interest. This transition has been attributed to the formation of either a monolayer (or less) of adsorbed oxygen on the surface or to the coverage of the surface by a three-dimensional corrosion product film. In either case, the reactive metal is

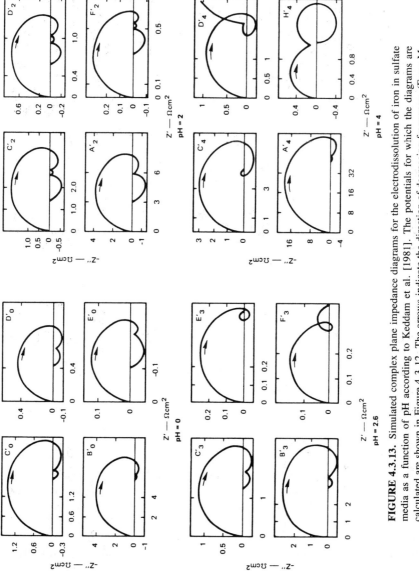

FIGURE 4.3.13. Simulated complex plane impedance diagrams for the electrodissolution of iron in sulfate media as a function of pH according to Keddam et al. [1981]. The potentials for which the diagrams are calculated are shown in Figure 4.3.12. The arrows indicate the direction of decreasing frequency. (From M. Keddam, O. R. Mattos, and H. J. Takenouti, Reaction Model for Iron Dissolution Studied by Electrode Impedance: Determination of the Reaction Model, *J. Electrochem. Soc.* **128**, 257–274, 1981. Reprinted by permission of the publisher, The Electrochemical Society, Inc.)

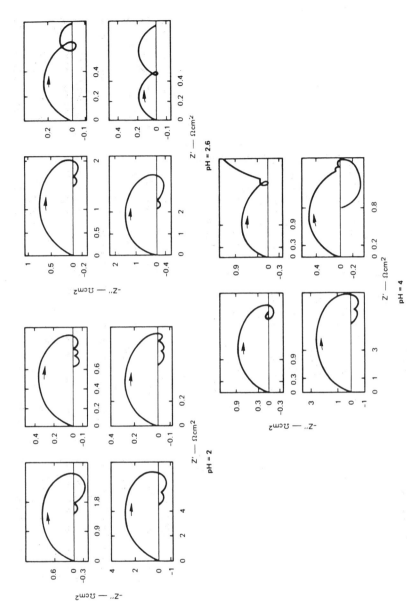

FIGURE 4.3.14. Experimental complex plane impedance diagrams for iron in sulfate media as a function of pH according to Keddam et al. [1981]. The potentials at which the diagrams were measured are shown in Figure 4.3.12. The arrows indicate the direction of decreasing frequency. (From M. Keddam, O. R. Mattos, and H. J. Takenouti, Reaction Model for Iron Dissolution Studied by Electrode Impedance: Determination of the Reaction Model, *J. Electrochem. Soc.* **128**, 257–274, 1981. Reprinted by permission of the publisher, The Electrochemical Society, Inc.)

shielded from the aqueous environment, and the current drops sharply to a low value that is determined by the movement of ions or vacancies across the film.

The changes that typically occur in the complex-plane impedance diagram on increasing the potential through an active-to-passive transition are shown in Figure 4.3.15 (Keddam et al. [1984]). At point A, the high-frequency arm of the impedance is typical of a resistive–capacitive system, but the impedance locus terminates in a negative resistance as $\omega \rightarrow 0$. This, of course, is consistent with the negative slope of the steady state polarization curve. At higher potentials, the high-frequency locus is again dominated by an apparent resistive–capacitance response (see Section 4.3.6.3), but the low-frequency arm is not observed to terminate at the real axis in this case because of the very high value for the polarization resistance (horizontal I vs. E curve). The origin of the negative resistance can be ac-

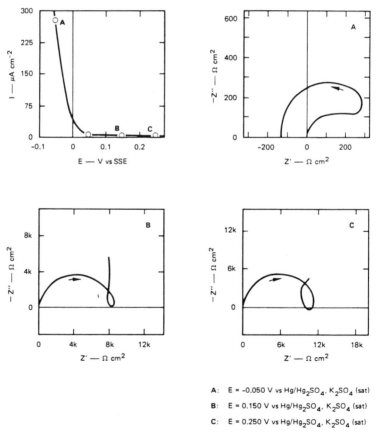

A: E = -0.050 V vs $Hg/Hg_2SO_4 \cdot K_2SO_4$ (sat)

B: E = 0.150 V vs $Hg/Hg_2SO_4 \cdot K_2SO_4$ (sat)

C: E = 0.250 V vs $Hg/Hg_2SO_4 \cdot K_2SO_4$ (sat)

FIGURE 4.3.15. Steady-state polarization curve and complex plane impedance diagrams at selected potentials through the active-to-passive transition for iron in 1M H_2SO_4 as reported by Keddam, Lizee, Pallotta, and Takenouti [1984]. The arrows indicate the direction of decreasing frequency. (From M. Keddam, O. R. Mattos, and H. J. Takenouti, Reaction Model for Iron Dissolution Studied by Electrode Impedance: Determination of the Reaction Model, *J. Electrochem. Soc.* **128**, 257–274, 1981. Reprinted by permission of the publisher, The Electrochemical Society, Inc.)

counted for theoretically (Keddam et al. [1984]) in terms of an increasing coverage of the surface by an adsorbed intermediate as the potential is increased. Thus, the low-frequency loop exhibited in Figure 4.3.15a is due to relaxations involving these surface species.

Epelboin and coworkers [1972] and Diard and LeGorrec [1979] have recognized a serious shortcoming of classical potentiostatic methods for investigating the active-to-passive transition. The problem arises because a potentiostat has a load line of negative slope in the I vs. E plane and hence is incapable of effectively defining the current–voltage characteristics of a metal–solution interface in the active-to-passive transition region. To overcome this limitation of potentiostatic control, Epelboin et al. [1972, 1975] and Diard and LeGorrec [1979] devised potential control instruments having negative output impedances, which are characterized by load lines having positive (and controllable) slopes. These negative impedance converters (NICs) have allowed "Z-shaped" active-to-passive transitions to be studied and the impedance characteristics to be determined, as shown by the data plotted in Figure 4.3.16. In contrast to the case shown in Figure 4.3.15a, the active-to-passive transition shown in Figure 4.3.16, as determined using a NIC, exhibits a change in the sign of dI/dE from negative to positive to negative as the current decreases from the active to the passive state. This change in sign, as reflected in the shape of the Z-shaped polarization curves, has been explained by Epelboin et al. [1975] in terms of coupling between mass transfer and surface reactions, although other explanations have also been advanced (Law and Newman [1979]).

4.3.6.3 *The Passive State*

The phenomenon of passivity is enormously important in corrosion science and engineering, since it is responsible for the relatively low corrosion rates that are observed for most engineering metals and alloys. It is not surprising, therefore, that passivity has been studied extensively using a wide variety of techniques, including IS. A brief account of these impedance studies is given below.

In discussing this subject it is convenient to delineate the processes that occur at the film–solution interface and those that take place within a passive film (Figure 4.3.17). In the first case, the processes are essentially ion exchange phenomena with the possibility of solution phase transport, whereas the second processes involve only transport. The movement of charged species within the film (anion vacancies $V_0^{\cdot\cdot}$ and cation vacancies $V_M^{x'}$) occurs, however, under the influence of both concentration and electrical potential gradients, with the electrical effects probably dominating, at least in the case of thin films. Accordingly, any analysis of the impedance characteristic of passive films must consider electromigration as well as diffusional transport.

The total impedance of the system of interphases shown in Figure 4.3.17 may be written as

$$Z_T = Z_{m/f} + Z_f + Z_{f/s} \tag{54}$$

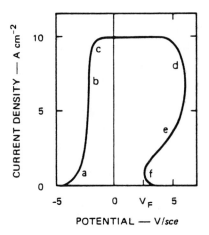

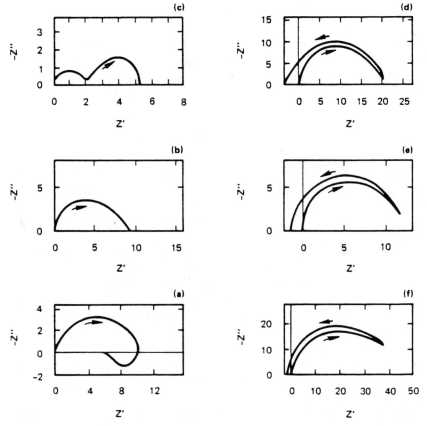

FIGURE 4.3.16. Impedance spectra for iron in 1M H_2SO_4 at various potentials within the active dissolution and active-to-passive transition regions as determined using a negative impedance converter (NIC). Impedance values are given in ohms (electrode diameter = 0.5 cm), and the arrows indicate the direction of decreasing frequency. (After Epelboin et al. [1975]).

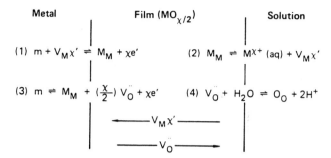

FIGURE 4.3.17. Schematic of physicochemical processes that occur within a passive film according to the point defect model. Here m = metal atom, M_M = metal cation in cation site, O_O = oxygen ion in anion site, $V_M^{x'}$ = cation vacancy, $V_O^{\cdot\cdot}$ = anion vacancy. During film growth, cation vacancies are produced at the film–solution interface but are consumed at the metal–film interface. Likewise, anion vacancies are formed at the metal–film interface but are consumed at the film–solution interface. Consequently, the fluxes of cation vacancies and anion vacancies are in the directions indicated.

where $Z_{m/f}$, Z_f, and $Z_{f/s}$ are the impedances associated with the metal–film interface, the film, and the film–solution interface, respectively. Because the elements are connected in series, the largest impedance will dominate the total impedance of the system. However, the impedance elements are frequency-dependent, so that each of the elements may dominate over different frequency ranges.

Metal–Film Interface. According to Armstrong and Edmondson (1973), the impedance of the metal–film interface can be described in terms of a capacitance (C_∞') in parallel with two charge transfer resistances, one for the transfer of electrons (R_e) and the other for the transfer of cations (R_c) from the metal to the film. Accordingly,

$$Z_{m/f} = \frac{R_e R_c (R_e + R_c)}{(R_e + R_c)^2 + \omega^2 C^2 R_e^2 R_c^2} - j \frac{\omega\, CR_e^2 R_c^2}{(R_e + R_c)^2 + \omega^2 C^2 R_e^2 R_c^2} \quad (55)$$

For most systems, particularly for diffuse metal–oxide junctions, we assume that the resistance to the movement of electrons across the interface is small compared with the resistance to the movement of cations, so that

$$R_e \ll R_c \quad (56)$$

In this case, Eq. (55) reduces to

$$Z_{m/f} = \frac{R_e}{1 + \omega^2 C^2 R_c^2} - j \frac{\omega\, CR_e^2}{1 + \omega^2 C^2 R_e^2} \quad (57)$$

Furthermore, the capacitance associated with this interface is probably that due to the space charge layer within the oxide. Therefore, over the frequency range of most interest to corrosion scientists (10^{-4}–10^4 Hz), $1/CR_e \gg \omega$, so that

$$Z_{m/f} \sim R_e \tag{58}$$

Accordingly, under these conditions, the impedance of the metal–film interface is likely to appear as a (small) frequency-independent resistance due to the transfer of electrons between the two phases.

The Film. A quantitative analysis of the impedance of a passive film has been reported by Chao, Lin, and Macdonald [1982], and the essential features of this treatment are reproduced here. The treatment is based upon their previously proposed (Chao et al. [1981], Lin et al. [1981]) point defect model for the growth and breakdown of passive films; the essential features of which are depicted in Figure 4.3.17.

In this model, it is assumed that the total current that is detected in an external circuit upon application of a voltage is the sum of four components: (1) electronic current due to the transport of electrons (e'); (2) electronic current due to the flow of electron holes (h); (3) ionic current due to the transport of anion vacancies ($V_O^{..}$); and (4) ionic current due to the movement of cation vacancies ($V_H^{x'}$)

$$I = I_{e'} + I_{h'} + I_{V_0} + I_{V_M} \tag{59}$$

Therefore,

$$1/Z_f = 1/Z_e + 1/Z_h + 1/Z_0 + 1/Z_M \tag{60}$$

The total impedance of the film is therefore described in terms of the transport of vacancies in parallel with the electron and hole resistances, provided that electron or hole exchange processes do not occur at the film–solution interface. This situation exists in the absence of any redox couples in the solution.

The movement of anion and cation vacancies within the film under the influence of concentration (C) and electrical potential (ϕ) gradients is determined by Fick's second law:

$$\frac{\partial C}{\partial t} = D\frac{\partial^2 C}{\partial x^2} - DqK\frac{\partial C}{\partial x} \tag{61}$$

where

$$K = \epsilon F/RT \tag{62}$$

$$\epsilon = -d\phi/dx \tag{63}$$

Here q is the charge on the moving species ($-\chi$ for cation vacancies and $+2$ for oxygen vacancies for an oxide film of stoichiometry $MO_{\chi/2}$), D is the diffusivity, and F, R, and T have their usual meanings. The current observed in an external conductor due to the movement of the vacancies is given by Fick's first law, as applied to the metal–film interface:

$$I = qFJ = qF\left(-D\frac{\partial C}{\partial x} + DqKC\right)_{m/f} \tag{64}$$

According to the point defect model, and under conditions where the various equations can be linearized with respect to the applied ac voltage (V_{ac}), the concentration of vacancies at the metal–film and film–solution interfaces may be expressed as (Chao, Lin, and Macdonald [1982])

$$C_{V_0}(m/f) = \left[C_{V_0}(m/f)\right]_{dc} \cdot \frac{2F(1-\alpha)}{RT} \cdot V_{ac} \tag{65}$$

$$C_{V_M}(m/f) = \left[C_{V_M}(m/f)\right]_{dc} \cdot \frac{\chi F(\alpha-1)}{RT} \cdot V_{ac} \tag{66}$$

$$C_{V_0}(f/s) = \left[C_{V_0}(f/s)\right]_{dc} \cdot \frac{2F\alpha}{RT} \cdot V_{ac} \tag{67}$$

$$C_{V_M}(f/s) = \left[C_{V_M}(f/s)\right]_{dc} \cdot \frac{\chi F\alpha}{RT} \cdot V_{ac} \tag{68}$$

where the quantities in square brackets are constants, which are related to the thermodynamic parameters for the interfacial reactions shown in Figure 4.3.17.

The set of Eqs. (61)–(68) are readily solved by Laplace transformation to yield the impedance

$$Z_f(j\omega) = \overline{V}_{ac}/\overline{I}_{ac}, \quad s = j\omega, \quad j = \sqrt{-1} \tag{69}$$

where s is the Laplace frequency. The impedance so calculated is found to have the form

$$Z_T = \left(\frac{\sigma_M \sigma_0}{\sigma_M + \sigma_0}\right)\omega^{-1/2}(1-j) \tag{70}$$

where σ_0 and σ_M are given by

$$\sigma_0 = RT/F^2(32D)^{1/2}\left\{\left[C_{V_0}(m/f)\right]_{dc}(1-\alpha)\right\} \tag{71}$$

$$\sigma_M = RT/F^2(2\chi^4 D^{1/2}\left\{(\alpha-1)\left[C_{V_M}(m/f)\right]_{dc}\right\} \tag{72}$$

The reader will recognize Eq. (70) as being of the form of a Warburg impedance for two parallel moving species. Two limiting cases may be defined: (1) movement of cations vacancies alone ($\sigma_0 \gg \sigma_M$, $Z_T \to Z_0$), and (2) movement of anion vacancies alone ($\sigma_M \gg \sigma_0$, $Z_T \to Z_M$). Accordingly,

$$Z_T = \sigma_M \omega^{-1/2}(1 - j) \qquad \text{(cation vacancies)} \qquad (73)$$

$$Z_T = \sigma_0 \omega^{-1/2}(1 - j) \qquad \text{(anion vacancies)} \qquad (74)$$

Substitution of the appropriate expression for $[C_{V_0}(m/f)]_{dc}$ into Eq. (71) yields the Warburg coefficients for the movement of oxygen ion vacancies as

$$\sigma_0 = \frac{1}{I_{dc}} \left(\frac{D_0}{2}\right)^{1/2} \cdot \frac{\epsilon}{1 - \alpha} \qquad (75)$$

The form of the equation for σ_0 is particularly interesting, because it suggests that if the electric field strength (ϵ) and α are constants (as assumed in the point defect model), then the product $\sigma_0 I_{dc}$ should be independent of the applied voltage across the system and the thickness of the film.

A plot of $-Z''$ vs. Z' for selected values of the various parameters contained in Eq. (74) is shown in Figure 4.3.18. As expected, the impedance locus is a straight line when

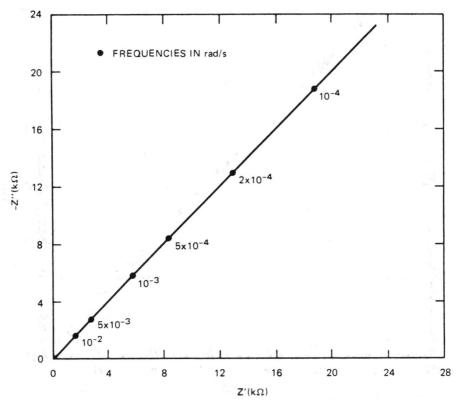

FIGURE 4.3.18. Complex impedance plane predicted by Eq. (74). $D_0 = 10^{-21}$ cm^2/s, $I_{dc} = 1$ μA/cm^2, $\epsilon = 10^6$ V/cm, $\alpha = 0.88$, area = 1 cm^2.

$$\omega > D(F^2\epsilon^2/R^2T^2) \tag{76}$$

However, for sufficiently low frequencies (Chao, Lin, and Macdonald [1982]), the impedance is predicted to intercept the real axis at a value of

$$R_{\omega=0} = \frac{RT}{4F^2DK\left\{2\left[C_{V_0}(m/f)\right]_{dc}(1-\alpha)-A'\right\}} \tag{77}$$

where

$$A' = A(RT/F\overline{V}_{ac}) \tag{78}$$

with

$$A = -\left(\frac{F\overline{V}_{ac}}{RT}\right)$$

$$\cdot \frac{\left[C_{V_0}(f/s)\right]_{dc}\alpha + \left[C_{V_0}(m/f)\right]_{dc}(1-\alpha)\exp\left[KL-(K^2+s/DL)^{1/2}\right]}{\exp(KL)\sinh\left[(K^2+s/DL)^{1/2}\right]}$$

$$\tag{79}$$

These equations show that the most critical parameters in determining the value of $R_{\omega=0}$ is the diffusivity of oxygen vacancies ($D \equiv D_0$) and the film thickness L; $R_{\omega=0}$ increases roughly exponentially with L and with $1/D$. Similar arguments can be made in the case of the transport of cation vacancies across a passive film.

The diagnostic features of this analysis have been used by Chao, Lin, and Macdonald [1982] in their investigation of the growth of passive films on nickel and Type 304 stainless steel in borate and phosphate buffer solutions. Typical complex plane and Randles's plane plots for nickel in 0.1N Na_2HPO_4 (pH = 9.1) and in 0.15 N H_3BO_3/0.15N $Na_2B_4O_7$ (pH = 8.7) solutions are shown in Figures 4.3.19 and 4.3.20. The data shown in Figure 4.3.19 reveal a linear impedance locus at low frequencies, with a partially resolved semicircle at high frequencies. This latter characteristic is attributed to relaxations occurring at the film–solution interface, as discussed later. The Randles plots shown in Figure 4.3.20 provide further evidence for the low-frequency Warburg response predicted by Eqs. (73) and (74). These plots also show that the experimentally measured Warburg coefficient $\sigma = dZ'/d\omega^{-1/2}$ is independent of the film thickness (as measured ellipsometrically) and of the applied voltage. Furthermore, the values of σ obtained for the phosphate (23.1 $k\Omega/s^{1/2}$) and borate (8.53 $k\Omega/s^{1/2}$) environments differ by nearly the factor of 3, as do the passive currents. This is predicted by Eqs. (75) and (76), assuming that negligible differences exist between the passive films formed in these two solutions.

A more extensive impedance analysis of passive films formed on nickel in borate buffer solution has been reported by Liang, Pound, and Macdonald [1984]. In

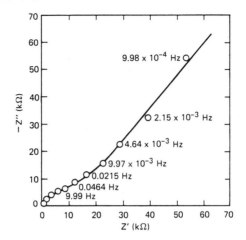

FIGURE 4.3.19. Complex plane impedance plot for Ni(111) in phosphate buffer ([PO$_4$] = 0.1 M) at 25°C. E = 0.1 V (SCE), pH = 10. The frequency at which each point was measured is indicated. Electrode area = 0.998 cm^2.

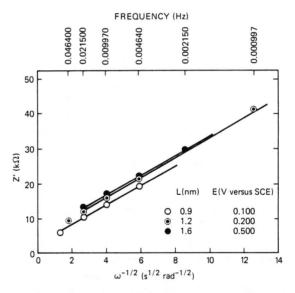

FIGURE 4.3.20. Randles plot of Z' vs. $\omega^{-1/2}$ for Ni(111) passivated in 0.1M phosphate solution (pH = 9). L = film thickness measured ellipsometrically. Electrode area = 0.998 cm^2.

this study, impedance data were obtained for passive nickel over a wide range of applied potential and pH (Figure 4.3.21), and these data serve as a good test of the constancy of σI_{dc}, as predicted by Eqs. (75) and (76). The data (Fig. 4.3.21) show that this product is indeed constant, within experimental error, thereby supporting the original hypothesis of the point defect model that the electric field strength is independent of film thickness and applied voltage.

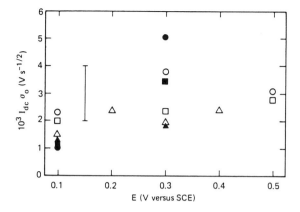

FIGURE 4.3.21. Plot of $I_{dc}\sigma_0$ vs. potential for Ni(111) in phosphate buffers ([PO$_4$] = 0.1 M) as a function of pH. Open circles = pH 7, closed circles = pH 8, open squares = pH 9, closed squares = pH 10, open triangles = pH 11, closed triangles = pH 12. T = 25°C, I = estimated error for each point, electrode area = 0.998 cm^2.

An important application of the data shown in Figure 4.3.21 and of the equation for the Warburg coefficient [Eq. (75)] is in the calculation of the diffusivity for anion vacancies within the film. In the case of passive polycrystalline nickel in borate and phosphate buffer solutions, Chao, Lin, and Macdonald (1982) computed a value of 1.3×10^{-21} cm^2/s for the diffusivity of oxygen ion vacancies. In a later study by Liang, Pound, and Macdonald [1984], a somewhat higher (and possibly more reliable) value of 1.5×10^{-19} cm^2/s for this same quantity was calculated from the data shown in Figure 4.3.21 for passive films formed on single-crystal nickel (100) in borate buffer solution. The principal problem with these calculations lies in the accurate measurement of the passive current I_{dc}. Experience shows that this quantity can vary over several orders of magnitude depending upon how the surface is prepared, the method by which it is measured (potentiostatic vs. potentio-dynamic techniques), and possibly the means by which the passive film is formed. Nevertheless, the values for D_0 given above are consistent with data extrapolated from high temperatures for a variety of oxides, and they appear to be eminently reasonable from a physicochemical viewpoint (Chao, Lin, and Macdonald [1982]).

The data shown in Figure 4.3.22 for Type 304 stainless steel appear to contradict the findings reported above for nickel. However, as noted by Chao et al. [1982], the potential-dependent and film-thickness-dependent Warburg coefficient can be accounted for by the fact that the composition of the passive film also changes with potential. A study of passive films on Fe-25 Ni-XCr alloys by Silverman, Cragnolino, and Macdonald [1982] indicated that the quantity $\epsilon/(1 - \alpha)$ changes very little as the Fe/Cr ratio is varied over a wide range, but diffusivity data for oxygen ion vacancies in iron and chromium oxides at elevated temperatures vary by many orders of magnitude, and in a manner that provides a qualitative explanation of the Randles plots shown in Figure 4.3.22.

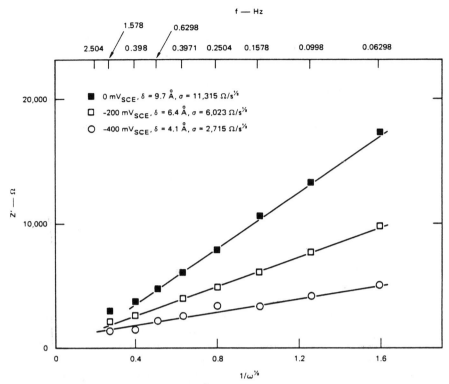

FIGURE 4.3.22. Dependence of Z' on $\omega^{-1/2}$ for Type 304SS passivated in 0.1N Na_2HPO_4 solution (pH 9.1). (From C.-Y. Chao, L. F. Lin, and D. D. Macdonald, A Point Defect Model for Anodic Passive Films; III. Impedance Response, *J. Electrochem Soc.* **129**, 1874–1879, 1982. Reprinted by permission of the publisher, The Electrochemical Society, Inc.)

Film–Solution Interface. The most comprehensive treatment of the impedance characteristics of the film–solution interface of a passive film is that reported by Armstrong and Edmondson [1973]. Their treatment essentially considers the ion exchange properties of an interface (Fig. 4.3.23) by addressing the movement of anions and cations between the film surface and the solution as the applied potential is modulated over a wide frequency range.

Armstrong and Edmondson [1973] begin their analysis by noting that the time dependence of the excess of cations over anions in the surface layer (Fig. 4.3.23) is given as

$$\frac{d\Gamma}{dt} = V_1 + V_2 - V_3 - V_4 \tag{80}$$

in which rates V_1, V_2, V_3, and V_4 can be expanded linearly as Taylor series in applied potential and excess cation concentration. Therefore,

FILM   SOLUTION

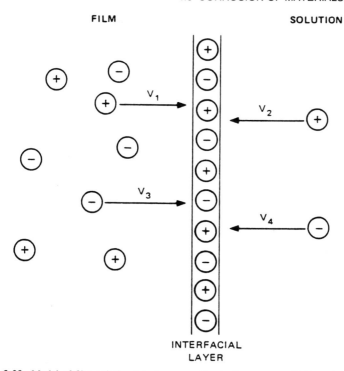

INTERFACIAL
LAYER

FIGURE 4.3.23. Model of film–solution interface according to Armstrong and Edmondson [1973]. (Reprinted with permission from R. D. Armstrong and K. Edmondson, The Impedance of Metals in the Passive and Transpassive Regions, *Electrochim. Acta* **18**, 937–943, 1973. Copyright 1973 Pergamon Journals Ltd.)

$$V_i = V_{io} + \left(\frac{\partial V_i}{\partial E}\right)_\Gamma \Delta E \, e^{j\omega t} + \left(\frac{\partial V_i}{\partial \Gamma}\right)_E \Delta\Gamma e^{j\omega t}, \quad i = 1\text{-}4 \qquad (81)$$

Equations (80) and (81) may be combined to yield

$$\Delta\Gamma = \frac{\left[\left(\frac{\partial V_i}{\partial E}\right)_\Gamma + \left(\frac{\partial V_2}{\partial E}\right)_\Gamma - \left(\frac{\partial V_3}{\partial E}\right)_\Gamma - \left(\frac{\partial V_3}{\partial E}\right)_\Gamma\right]\Delta E}{j\omega - \left(\frac{\partial V_1}{\partial \Gamma}\right)_E - \left(\frac{\partial V_2}{\partial \Gamma}\right)_E + \left(\frac{\partial V_3}{\partial \Gamma}\right)_E + \left(\frac{\partial V_4}{\partial \Gamma}\right)_E} \qquad (82)$$

If the number of electrons that flow through an external circuit upon the transfer of one species in reaction i (Fig. 4.3.23) is n_i, then the faradic admittance of the film–solution interface is

$$Y_{f/s} = \frac{n_1 F V_1 + n_2 F V_2 + n_3 F V_3 + n_4 F V_4}{\Delta E \exp(j\omega t)} \qquad (83)$$

which upon substitution of Eq. (81) becomes

$$Y_{f/s} = \sum_{i=1}^{4} n_i F \left(\frac{\partial V_i}{\partial E} \right)_\Gamma + \Delta\Gamma \sum_{i=1}^{4} n_i F \left(\frac{\partial V_i}{\partial \Gamma} \right)_E \tag{84}$$

where $\Delta\Gamma$ is given by Eq. (82). According to Armstrong and Edmondson [1973], it is convenient to define infinite-frequency charge-transfer resistances as

$$1/R_{\infty1} = n_1 F \left(\frac{\partial V_1}{\partial E} \right)_\Gamma + n_2 F \left(\frac{\partial V_2}{\partial E} \right)_\Gamma \tag{85}$$

$$1/R_{\infty2} = n_3 F \left(\frac{\partial V_3}{\partial E} \right)_\Gamma + n_4 F \left(\frac{\partial V_3}{\partial E} \right)_\Gamma \tag{86}$$

and resistances at zero frequency as

$$1/R_{01} = \left[\left(\frac{\partial V_1}{\partial E} \right)_\Gamma + \left(\frac{\partial V_2}{\partial E} \right)_\Gamma - \left(\frac{\partial V_3}{\partial E} \right)_\Gamma - \left(\frac{\partial V_4}{\partial E} \right)_\Gamma \right]$$
$$\cdot \left[n_1 F \left(\frac{\partial V_1}{\partial \Gamma} \right)_E + n_2 F \left(\frac{\partial V_2}{\partial \Gamma} \right)_E \right] \Big/ k \tag{87}$$

$$1/R_{02} = \left[\left(\frac{\partial V_1}{\partial E} \right)_\Gamma + \left(\frac{\partial V_2}{\partial E} \right)_\Gamma - \left(\frac{\partial V_3}{\partial E} \right)_\Gamma - \left(\frac{\partial V_4}{\partial E} \right)_\Gamma \right]$$
$$\cdot \left[n_3 F \left(\frac{\partial V_3}{\partial \Gamma} \right)_E + n_4 F \left(\frac{\partial V_4}{\partial \Gamma} \right)_E \right] \Big/ k \tag{88}$$

where k defines the relaxation time τ as

$$\tau = 1/k \tag{89}$$

with

$$k = \left(\frac{\partial V_3}{\partial \Gamma} \right)_E + \left(\frac{\partial V_4}{\partial \Gamma} \right)_E - \left(\frac{\partial V_1}{\partial \Gamma} \right)_E - \left(\frac{\partial V_2}{\partial \Gamma} \right)_E \tag{90}$$

Accordingly, the faradic admittance becomes

$$Y_{f/s} = \frac{1}{R_{\infty1}} + \frac{1}{R_{\infty2}} + \frac{k}{k + j\omega} \cdot \left(\frac{1}{R_{01}} + \frac{1}{R_{02}} \right) \tag{91}$$

However, Armstrong and Edmondson [1973] claim that, in most cases, oxygen will be in equilibrium between the interface and the solution, a condition that can only be satisfied if $R_{\infty 2} = -R_{02}$. Thus, Eq. (91) becomes

$$Y_{f/s} = \frac{1}{R_{\infty 1}} + \frac{k}{k + j\omega} \cdot \frac{1}{R_{01}} + \frac{1}{R_{\infty 2}} \cdot \frac{j\omega}{k + j\omega} \tag{92}$$

and the total interfacial impedance is written as

$$Z_{f/s} = \frac{1}{Y_{f/s} + j\omega C_{\infty}} \tag{93}$$

where C_{∞} is the double-layer capacitance. The complex-plane impedance loci that can be generated by Eq. (93) according to the relative values of k and ω have been explored by Armstrong and Edmondson [1973], and their results are summarized below.

(1) Large k ($k \gg \omega$). In this case, Eq. (92) reduces to

$$Y_{f/s} = \frac{1}{R_{\infty 1}} + \frac{1}{R_{01}} + \frac{j\omega}{kR_{\infty 2}} \tag{94}$$

and the impedance locus takes the form of a single semicircle in the complex plane (Fig. 4.3.24) resulting from a resistance $R_{\infty 1} R_{01} / (R_{\infty 1} + R_{01})$ in parallel with the capacitance $C_{\infty} + kR_{\infty 2}$.

(2) Small k ($k \ll \omega$), in which case Eq. (92) becomes

$$Y_{f/s} = \frac{1}{R_{\infty 1}} + \frac{1}{R_{\infty 2}} - \frac{jk}{\omega R_{01}} \tag{95}$$

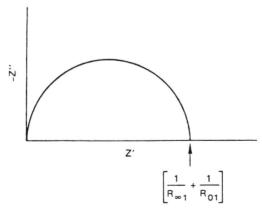

FIGURE 4.3.24. Film–solution interface according to Eq. (89) for large k. (Reprinted with permission from R. D. Armstrong and K. Edmondson, The Impedance of Metals in the Passive and Transpassive Regions, *Electrochim. Acta* **18**, 937–943, 1973. Copyright 1973 Pergamon Journals Ltd.)

indicating that two semicircles will appear in the complex plane according to the relative values of A and B:

$$A = \frac{R_{\infty 1} R_{01}}{R_{\infty 1} + R_{01}} \tag{96}$$

$$B = \frac{R_{\infty 1} R_{\infty 2}}{R_{\infty 1} + R_{\infty 2}} \tag{97}$$

Thus, for $A > B > 0$ and $B > A > 0$, the complex plane impedance loci shown in Figures 4.3.25a and b are obtained, whereas for $B > 0 > A$ that shown in Figure 4.3.25c results. The latter case occurs because R_{01} may be positive or negative, depending upon the relative values of the differentials contained in Eq. (87). Provided that R_{01} is negative, but that $|R_{\infty 1}| > |R_{01}|$, second-quadrant behavior at low frequencies is predicted, terminating in a negative resistance at $\omega \to 0$. The low-frequency inductive response predicted in the second case ($B > A > 0$) is also of considerable practical and theoretical interest, because fourth-quadrant behavior is frequently observed experimentally.

At this point it is worthwhile pausing to consider the properties of the total impedance of the interphase system consisting of the metal–film interface, the film,

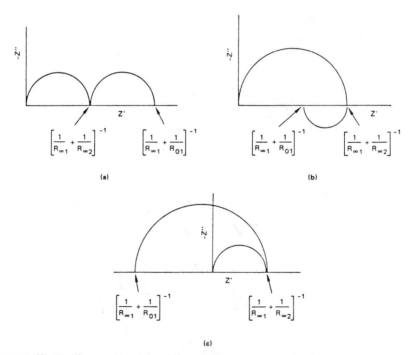

FIGURE 4.3.25. Complex plane impedance diagrams for film–solution interface according to Eq. (93) for small k. (After Armstrong and Edmondson [1973])

and the film–solution interface. According to Eq. (54) and subsequent expressions for $Z_{m/f}$, Z_f, and $Z_{f/s}$, the total impedance becomes

$$Z_T = R_e + W + \frac{X}{X^2 + Y^2} - j\left(W + \frac{Y}{X^2 + Y^2}\right) \tag{98}$$

$$W = \frac{\sigma_M \sigma_0 \omega^{-1/2}}{\sigma_M + \sigma_0} \tag{99}$$

$$X = \frac{1}{R_{\infty 1}} + \frac{1}{k^2 + \omega^2}\left(\frac{\omega^2}{R_{\infty 2}} + \frac{k^2}{R_{01}}\right) \tag{100}$$

$$Y = \frac{\omega k}{k^2 + \omega^2}\left(\frac{1}{R_{\infty 2}} - \frac{1}{R_{01}}\right) + \omega C_\infty \tag{101}$$

These rather complicated expressions predict that a variety of impedance loci in the complex plane might be observed, depending upon the frequency range employed and the relative values of the parameters contained in Eqs. (98)–(101).

By way of illustration, we calculate complex impedance diagrams for the case of a passive film in which only anion vacancies are mobile and for which k [Eq. (90)] is large. Thus, for $k \gg \omega$, $\sigma_M \gg \sigma_0$, and assuming that no redox reactions occur at the film–solution interface, then the total impedance becomes

$$Z_t = \left(\frac{a}{a^2 + \omega^2 b^2} + \sigma_0 \omega^{-1/2}\right) - j\left(\frac{\omega b}{a^2 + \omega^2 b^2} + \sigma_0 \omega^{-1/2}\right) \tag{102}$$

where

$$a = \frac{1}{R_{\infty 1}} + \frac{1}{R_{01}} \tag{103}$$

$$b = C_\infty + \frac{1}{kR_{\infty 2}} \tag{104}$$

Impedance diagrams for a passive film, computed using Eq. (102), are displayed in Figure 4.3.26. These diagrams show a partially resolved semicircle at high frequencies and a low-frequency Warburg response. These same general features are exhibited by the impedance data for passive Ni(III) in phosphate buffer solution, as shown in Figure 4.3.19. The calculated impedance spectra (Fig. 4.3.26) show that as the kinetics of the interfacial ion exchange processes become slower (increasing R_{01}), the impedance becomes increasingly dominated by the nondiffusional component. In the limit of sufficiently slow interfacial reactions but fast transport of vacancies across the film, the impedance locus takes the form of a semicircle, which is similar to that expected for a purely capacitive (dielectric) response.

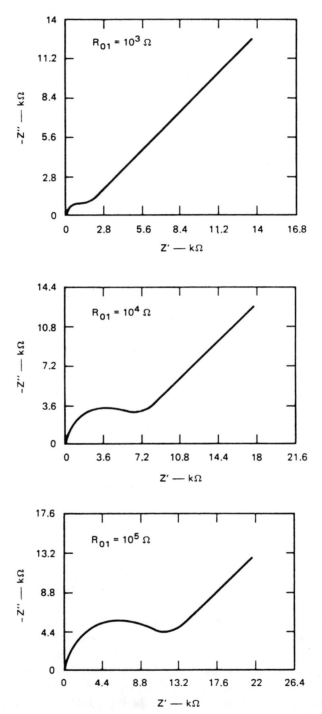

FIGURE 4.3.26. Theoretical complex plane impedance diagrams for a passive film according to Eq. (102). $C_\infty = 5 \times 10^{-5}$ F, $R_{\infty 1} = 10^4$ Ω, $R_{\infty 2} = 1$ Ω, $\sigma_0 = 10^3$ Ω/s$^{1/2}$.

A major shortcoming of the theory developed above is that the various processes (e.g., ion exchange and vacancy transport) are assumed to be uncoupled. More realistically, the kinetic expression given by Eq. (80), for example, should be employed as a boundary condition for the solution of Eq. (61). In this way, coupling, which is expected to become most apparent at intermediate frequencies where the vacancy transport and ion exchange processes are of comparable importance, may be a significant factor in determining the impedance spectrum. Such coupling is included in the approach described in Section 2.2.3.3.

4.3.7 Equivalent Circuit Analysis

Because electrochemical impedance techniques have their genesis in electrical engineering, great emphasis has been placed in the past on identifying an equivalent "electrical circuit" for the interface. Although this exercise is useful, in that the equivalent circuit is capable of mimicking the behavior of the system, the circuits that often are adopted are too simplistic to be of any interpretive value. Thus, the impedance data plotted in Figure 4.3.9, for example, cannot be interpreted in terms of a simple RC circuit of the type shown in Figure 4.3.27, if for no other reason that the circuit does not delineate the partial anodic and cathodic reactions. Also, this circuit strictly yields a semicircle in the complex plane that is centered on the real axis, whereas the locus of the experimental data is clearly neither semicircular nor centered on the abscissa. Nevertheless, an equivalent circuit of the type shown in Figure 4.3.27 is frequently assumed, and values are often calculated for the various components. It is important to realize, however, that the choice of an "incorrect" equivalent circuit has no consequence for the measured polarization resistance or corrosion rate, since the resistance $|Z|_{\omega \to 0} - |Z|_{\omega \to \infty}$ is independent of the form of the reactive component.

The shape of the impedance locus plotted in Figure 4.3.9 suggests that a more appropriate equivalent circuit is an electrical transmission line of the type shown in Figure 4.3.28. Although extensive use has been made of transmission lines in the past (de Levie [1964, 1965a,b, 1967], McKubre [1976], Atlung and Jacobsen [1976], Park and Macdonald [1983], Lenhart et al. [1984]) to model rough surfaces, porous electrodes, corroding interfaces, polarization in soils and clays, and coated surfaces, they have not been universally embraced by corrosion scientists and electrochemists. The particular transmission line shown in Figure 4.3.28 was

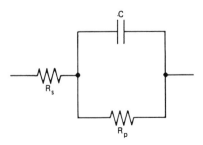

FIGURE 4.3.27 Simple electrical equivalent circuit for a corroding interface. Note that this circuit does not delineate the partial anodic or cathodic reactions.

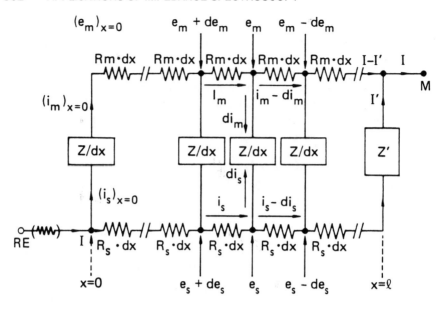

FIGURE 4.3.28. Discretized form of the transmission line model. e_m and e_s are the potentials in the magnetite and solution phases, respectively. Here i_m and i_s are the currents in the magnetite and solution phases, respectively; I and I' are the total current and the current flowing across the metal–solution interface and base of the pore, respectively; RE and M designate the reference electrode and metal (working electrode) locations, respectively. (Reprinted with permission from J. R. Park and D. D. Macdonald, Impedance Studies of the Growth of Porous Magnetite Films on Carbon Steel in High Temperature Aqueous Systems, *Corros. Sci.* **23**, 295, 1983. Copyright 1983 Pergamon Journals Ltd.)

developed by Park and Macdonald [1983] and by Lenhart, Macdonald, and Pound [1984] to model porous magnetite films on carbon steel in high-temperature chloride solutions and to describe the degradation of porous $Ni(OH)_2$–NiOOH battery electrodes in alkaline solution upon cyclic charging and discharging. We will discuss the first case in some detail, because it is a good example of how a transmission line can provide a physical picture of the processes that occur with a complex corrosion reaction.

The work of Park and Macdonald [1983] was performed in an attempt to understand the phenomenon of "denting" corrosion in pressurized water reactor (PWR) steam generators (Garnsey [1979]). Briefly, this phenomenon, which is represented schematically in Figure 4.3.29, occurs because of the rapid growth of magnetite in the crevice between an Inconel 600 steam generator tube and carbon steel the support plate. The growing crystalline magnetite can deform or "dent" the tube and may result in leakage from the primary (radioactive) circuit to the secondary (nonradioactive) side via stress corrosion cracks that nucleate and grow in the highly stressed regions of the tubes. Previous work by Potter and Mann [1965], among others (see Park [1983]), had indicated that the growth of magnetite on carbon steel in simulated tube–support plate crevice environments is kinetically linear; that is, the growing film offers no protection to the underlying metal. How-

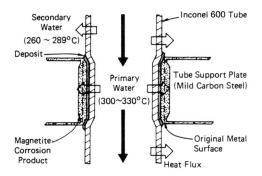

FIGURE 4.3.29. Schematic representation of denting corrosion in PWR steam generators.

ever, this conclusion was based on ex situ analyses of weight loss and film growth, neither of which yield real-time estimates of the rate of corrosion in the environment of interest.

In the work conducted by Park and Macdonald [1983], impedance spectroscopy was used to measure the polarization resistance of corroding carbon steel in a variety of chloride (1M)-containing solutions at temperatures from 200 to 270°C. Typical complex plane impedance spectra for one such system as a function of time are shown in Figure 4.3.30. In all cases (different exposure times), the impedance spectra are characteristic of a system that can be represented by an electrical transmission line. Also of interest is the observation (Park [1983]) that the inverse of the polarization resistance, which is proportional to the instantaneous corrosion rate, increases with increasing exposure time. Accordingly, the corrosion process is kinetically autocatalytic, rather than being linear, as was previously reported. This autocatalytic behavior has been found for a wide range of solutions that simulate crevice environments, particularly those containing reducible cations, such as Cu^{2+}, Ni^{2+}, and Fe^{3+}. The large amount of experimental polarization resistance data generated in the study showed that these reducible cations greatly accelerate the corrosion rate and yield autocatalytic behavior, with the effect lying in the order $Cu^{2+} > Ni^{2+} > Fe^{3+}$. The enhanced rates were found to be far greater than could be accounted for by hydrolysis (to form H^+) alone, and it was concluded that reduction of the foreign cations was largely responsible for the rapid corrosion of the steels. However, the autocatalytic nature of the reaction was also partly attributed to the buildup of chloride in the porous film. This buildup occurs for the reason of electrical neutrality, in response to the spatial separation of the anodic (iron dissolution at the bottom of the pores) and cathodic (H^+ or cation reduction at the magnetite–solution interface) processes.

The model that was adopted to account for the fast growth of porous magnetite on carbon steel is shown schematically in Figure 4.3.31, and an ideal single pore is depicted in Figure 4.3.32. This single pore can be represented electrically by the continuous transmission line, as in Figure 4.3.33, and by the discretized transmission line shown in Figure 4.3.28, where Z is the impedance per unit length down the pore wall and Z' is the impedance of the metal–solution interface at the

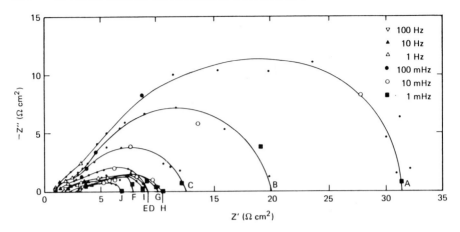

FIGURE 4.3.30. Impedance diagrams at the corrosion potential as a function of time for carbon steel exposed to a 0.997m NaCl + 0.001m FeCl$_3$ + 3500 ppb O$_2$ solution at 250°C. A = 11 h, B = 22 h, C = 35 h, D = 46 h, E = 58 h, F = 70 h, G = 86.5 h, H = 112 hr, I = 136 h, J = 216 h.

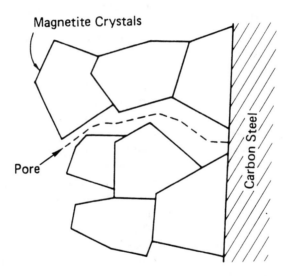

FIGURE 4.3.31. Schematic of fast growth of magnetite in acidic chloride solutions at elevated temperatures. (Reprinted with permission from J. R. Park and D. D. Macdonald, Impedance Studies of the Growth of Porous Magnetite Films on Carbon Steel in High Temperature Aqueous Systems, *Corros. Sci.* **23**, 295, 1983. Copyright 1983 Pergamon Journals Ltd.)

bottom of the pore. The impedance of a single pore is readily derived using Kirchhoff's equations to yield

$$Z_p = \frac{R_m R_s l}{R_m + R_s} + \frac{2\gamma^{1/2} R_m R_s + \gamma^{1/2}(R_m^2 + R_s^2)C + \delta R_s^2 S}{\gamma^{1/2}(R_m + R_s)(\gamma^{1/2}S + \delta C)} \qquad (105)$$

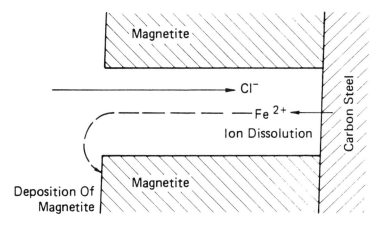

FIGURE 4.3.32. Ideal single-pore model for the growth of porous magnetite film in acidic chloride solutions at high temperature. (Reprinted with permission from J. R. Park and D. D. Macdonald, Impedance Studies of the Growth of Porous Magnetite Films on Carbon Steel in High Temperature Aqueous Systems, *Corros. Sci.* **23**, 295, 1983. Copyright 1983 Pergamon Journals Ltd.)

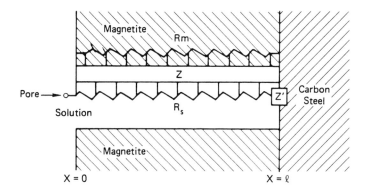

FIGURE 4.3.33. Transmission line model for a single one-dimensional pore: R_m = magnetite resistance per unit length of a pore, R_s = solution resistance unit length of a pore (Ω/cm), Z = pore wall–solution interface impedance for unit length of a pore (Ω/cm), Z' = metal–solution interface impedance at the base of a pore (Ω), X = distance from the mouth of the pore along the pore axis (cm). (Reprinted with permission from J. R. Park and D. D. Macdonald, Impedance Studies of the Growth of Porous Magnetite Films on Carbon Steel in High Temperature Aqueous Systems, *Corros. Sci.* **23**, 295, 1983. Copyright 1983 Pergamon Journals Ltd.)

where

$$\gamma = \frac{R_m + R_s}{Z} \tag{106}$$

$$\delta = \frac{R_m + R_s}{Z'} \tag{107}$$

$$C = \cosh \left(\gamma^{1/2} l \right) \tag{108}$$

$$S = \sinh \left(\gamma^{1/2} l \right) \tag{109}$$

Here R_m is the resistance of the magnetite per unit length of the pore, R_s is the corresponding quantity for the solution, and l is the length of the pore. For an oxide consisting of n independent parallel pores, the total impedance of the film then becomes

$$Z_T = Z_p / n \tag{110}$$

To calculate the total impedance, it is necessary to assume models for the pore wall–solution and pore base–solution interfacial impedances (Z and Z', respectively). In the simplest case, both interfaces are assumed to exist under charge transfer control, in which case Z and Z' can be represented by the equivalent circuit shown in Figure 4.3.27. Typical impedance spectra calculated for various values of the resistance at the pore base are shown in Figure 4.3.34. These theoretical curves generally exhibit the features displayed by the experimental data (Fig. 4.3.30). Similar calculations have been carried out for porous films involving mass transfer control within the pores, and impedance spectra that compare well with experiment have again been generated (Park and Macdonald [1983], Park [1983]).

Coatings. An effective means of reducing the rate of corrosion of a metal is to protect the surface with a coating. However, coatings generally are not impervious

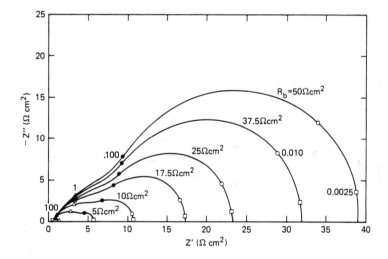

FIGURE 4.3.34. Calculated impedance spectra for a porous magnetite film on carbon steel in high-temperature aqueous sodium chloride solution as a function of the resistance at the base of the pore. N (no. of pores/cm^2) = 10^4, ρ (solution) = 18.5 Ω-cm, ρ (Fe$_3$O$_4$) = 116 Ω-cm, θ (film porosity) = 0.01, L (film thickness) = 0.025 cm, R_W (pore wall–solution resistance) = 10^3 Ω-cm^2, C_w (capacitance at pore wall–solution interface) = 2×10^{-2}/cm^2, c_b (capacitance at pore base) = 7×10^{-3} F/cm^2.

to water or even ions, so that corrosion reactions still proceed at the metal–coating interface, albeit at a low rate. Because the rate of corrosion depends upon the transport of corrosive species through the film, in addition to the reactions occurring at the metal surface, it is evident that the overall process of attack is a very complicated one. It is therefore not surprising that coated metals have been studied using impedance spectroscopy.

Perhaps one of the more extensive studies of this type is that reported by Mansfeld, Kendig, and Tsai [1982]. These workers evaluated polybutadiene coatings on carbon steel and aluminum alloys that had been subjected to different surface treatments, including phosphating for the steel and exposure to a conversion coating in the case of the aluminum alloys. Typical Bode plots ($\log |Z|$ and θ vs. $\log \omega$) for 1010 carbon steel that had been degreased, phosphated, and then coated with polybutadiene (8 ± 2 μm thick), are shown in Figure 4.3.35 for two times after exposure to 0.5-N NaCl solution at ambient temperature. The low-frequency behavior of the phase angle ($\sim \pi/8$) suggests the existence of a transport process through the coating that is best described in terms of a transmission line (see earlier discussion in this section). However, the authors chose to analyze their data in terms of a classical Randles-type equivalent electrical circuit containing a normal semiinfinite Warburg diffusional impedance element. Randles plots were then used to evaluate Warburg coefficients (σ), and polarization resistances (R_p) were calculated directly from the impedance magnitude at $\omega \to 0$ and $\omega \to \infty$ [Eq. 4)]. The R_p and σ parameters are plotted in Figure 4.3.36 as a function of time for a variety of surface pretreatments (including phosphating), as listed in Table 4.3.1. Both parameters are observed to decrease with time after an initial increase, at

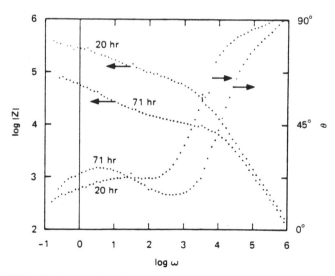

FIGURE 4.3.35. Bode plots for phosphated and coated 1010 carbon steel in 0.5N NaCl solution. (From F. Mansfield, M. W. Kendig, and S. Tsai, Evaluation of Organic Coating Metal Systems by AC Impedance Techniques. Reprinted with permission from *Corrosion*, Vol. 38, p. 482, 1982, NACE, Houston, TX.)

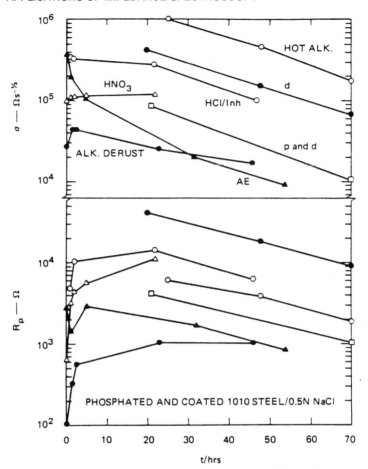

FIGURE 4.3.36. Time dependence of resistance R_p and Warburg coefficient σ for different surface pretreatment procedures applied to 1010 carbon steel. (From F. Mansfield, M. W. Kendig, and S. Tsai, Evaluation of Organic Coating Metal Systems by AC Impedance Techniques. Reprinted with permission from *Corrosion*, Vol. 38, p. 482, 1982, NACE, Houston, TX.)

least in the case of R_p. The simultaneous decrease in R_p and σ was rationalized in terms of a model involving the development of ionically conducting paths through the polybutadiene coating as the corrosion reaction proceeds at the metal–coating interface.

4.3.8 Other Impedance Techniques

In keeping with the theme established at the beginning of this section that impedance techniques can be used to analyze many cause-and-effect phenomena, a number of other transfer functions have been defined. Three such functions, the electrochemical hydrodynamic impedance (EHI), fracture transfer function (FTF), and the electrochemical mechanical impedance (EMI) are discussed briefly below.

TABLE 4.3.1. Surface Pretreatment Procedures Used by Mansfeld et al. [1982] Prior to Applying Polybutadiene Coatings to 1010 Carbon Steel

Pretreatment	Procedure
Degrease (d)	Trichloroethylene, 15 min, TT-C-490, method II
Polish (p)	Wet polish, 600-grit SiC
p + d	
d + hot alkaline cleaning	5 min, 100°C, TT-C-490 method III
d + alkaline derusting	10 min, 50°C, TT-C-490, method V, type III
p + d + HCl/Inh[a]	5 min, RT, 42 v/o HCl + 5.8 g/liter 2-butyne-1,4-diol
d + anodic etch (AE)	2 min, RT, 30 v/o H_2SO_4, 0.1 A/cm^2
d + conc. HNO_3	30 min, RT
Zinc phosphate coating	6.4 g/liter ZnO, 14.9 g/l H_3PO_4, 4.1 g/liter HNO_3, 95°C, 30 min

[a]Inhibitor. See Mansfeld *et al.* [1982].

4.3.8.1 Electrochemical Hydrodynamic Impedance (EHI)

In the EHI technique (Deslouis et al. [1980, 1982], Bonnel et al. [1983]), the rate of mass transport of reactants to, or products from, an electrochemical interface is modulated (sine wave), and the response current or potential is monitored at constant potential or constant current, respectively, depending upon whether the transfer function is determined potentiostatically or galvanostatically. Experimentally, the transfer function is most easily evaluated using a rotating-disk electrode, because of the ease with which the rotational velocity can be modulated. For a fast redox reaction occurring at the surface of a disk, the potentiostatic and galvanostatic electrochemical hydrodynamic admittances are given as follows:

$$Y_{HD}^I = \left(\frac{\delta I}{\delta \Omega}\right)_V = -\frac{3}{2\Omega_0} \exp\left(-0.26\,pj\right) \cdot \frac{C_\infty - C_0}{1.288}$$

$$\cdot \frac{3^{1/3}}{\delta} \cdot \frac{K_1(s)}{Ai\,(s)} \cdot D \cdot \frac{Z_D}{Z} \tag{111}$$

$$Y_{HD}^V = \left(\frac{\delta V}{\delta \Omega}\right)_I = -\frac{3}{2\Omega_0} \exp\left(-0.26pj\right) \cdot \frac{C_\infty - C_0}{1.288}$$

$$\cdot \frac{K_1(s)}{Ai'\,(s)} \cdot R_T(k_f - k_b) \tag{112}$$

where Ω_0 is the mean angular velocity of the disk, $p = \omega/\Omega_0$, ω is the modulation frequency, $Ai(s)$ and $Ai'(s)$ are the Airy function of the first kind and its first derivative, $s = 1.56\,jpSc^{1/3}$, Sc (Schmidt number) $= \nu/D$, ν is the kinematic viscosity, D is the diffusivity, $j = \sqrt{-1}$, Z_D is the diffusion impedance, $R_1 =$ the charge transfer resistance, k_f and k_b are the forward and backward rate constants for the redox reaction, Z is the overall impedance, and $K_1(s)$ is given by

$$K_1(s) = -\sum_0^\infty \frac{s^n}{n!} \int_0^\infty \frac{\delta^n Ai}{\delta\xi^n} \left[\xi_2 \exp\left(-\frac{\xi^3}{6}\right) + \xi^4 \rho_0 (\xi) \exp\left(\frac{\xi^3}{6}\right) \right] d\xi \quad (113)$$

The application of EHI analysis to a corroding system has been described by Bonnel et al. [1983] in their recent study of the corrosion of carbon steel in neutral chloride solutions. An example of their data is shown in Figure 4.3.37 for carbon steel soon after immersion into air-saturated 3% NaCl solution. The steel surface is believed to be free of solid corrosion products under these conditions. The best fit of Eq. (112) to the experimental data shown in Figure 4.3.37 (broken line) is found for $Sc = 615$ and $D(O_2) = 1.63 \times 10^{-5}$ cm^2/s. The value for the diffusivity of oxygen was found to be consistent with the steady state current–voltage curve

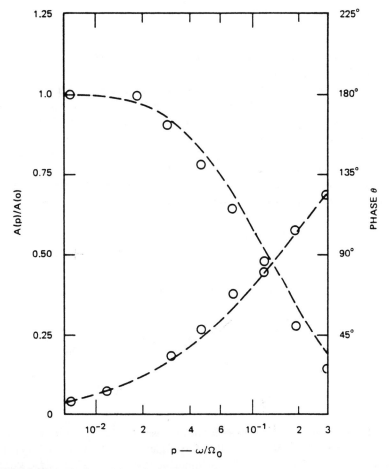

FIGURE 4.3.37. Bode plots of the reduced EHI modulus and phase angle as a function of reduced frequency (ω/Ω_0) for carbon steel in 3% NaCl solution. The EHI was measured in the galvanostatic mode [Eq. (112)] at a mean rotational velocity (Ω_0) of 1000 rpm and at $I = 0$ (open-circuit conditions). (After Bonnel et al. [1983]).

obtained for this system. A significant advantage claimed for the EHI technique is that it may be used to detect the presence of a porous corrosion product film on a corroding surface (Bonnel et al. [1983]), but in general the analysis is complicated, and the method has yet to demonstrate any clear advantages over more conventional techniques for examining coupled mass transfer–charge-transfer phenomena.

4.3.8.2 Fracture Transfer Function (FTF)

Recently, Chung and Macdonald [1981] defined a transfer function for the propagation of a crack through an elastic–plastic material under transient loading conditions. Designating the stress intensity as $K_I(t)$ and the instantaneous crack velocity as $da(t)/dt$ (both time-dependent quantities), the fracture transfer function may be defined as

$$\overline{H}_F(s) = \frac{\mathcal{L}[K_I - K_{ISCC}]}{\mathcal{L}[da(t)/dt]} \tag{114}$$

where K_{ISCC} is the critical stress intensity for slow crack growth, and $\mathcal{L}$ designates the Laplace transform. The transfer function $\overline{H}_F(s)$ is a function of the Laplace frequency s.

Application of this analysis to transient crack growth data for AISI 4340 steel in 3% NaCl solution (Chung [1983]) is shown in Figure 4.3.38. It is seen that the transfer function $\overline{H}_F(s)$ increases linearly with the Laplace frequency s. Accordingly, the electrical analog for crack propagation is simply a resistance and inductance in series (Fig. 4.3.39). However, the data shown in Figure 4.3.38 also reveal that the inductance, but not the resistance, is a function of the applied stress intensity; that is,

$$\overline{H}_F(s) = R + sL(K_I) \tag{115}$$

It is clear, therefore, that the system is nonlinear, so that $\overline{H}_F(s)$ is not expected to obey the Kramers–Kronig transforms.

The parameter R provides a measure of the resistance to crack propagation, such that $1/R$ is proportional to the steady state crack velocity. The R values obtained from the intercept at $s = 0$ indicate that the steady state crack velocity is independent of stress intensity, a finding that is consistent with the fact that the applied stress intensity correlation for this system.

The inductance contained in the equivalent circuit shown in Figure 4.3.39 corresponds to an energy adsorption process, presumably due to the plastic deformation of the steel matrix in front of the crack tip. The size of the plastically deformed region, and hence the amount of energy deposited in the matrix in front of the crack tip, is known to increase with increasing K_I (Chung [1983]), and this accounts for the observed increase in $L(K_I)$.

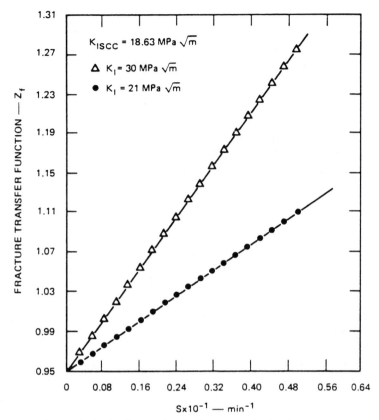

FIGURE 4.3.38. Fracture transfer function vs. Laplace frequency.

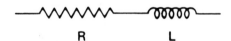

FIGURE 4.3.39. Series RL equivalent circuit used to represent the growth of a stress corrosion crack through AISI 4340 steel in 3% NaCl solution.

4.3.8.3 Electrochemical Mechanical Impedance

Instead of the purely mechanical case discussed above, it is also possible to define an impedance in terms of the electrochemical response of a fractured specimen to a mechanical input that causes crack propagation.

One case in which this analysis can be usefully applied is that of the electrochemically assisted (stress corrosion) cracking of metals submerged in an electrolyte and subjected to a cyclic mechanical load or stress, corresponding to vibrational effects (Eiselstein, McKubre, and Caligiuri [1983, 1985]). With some simplification, crack growth due to cyclical film rupture at the growing crack tip, followed by metal dissolution, can be considered as described below.

Figure 4.3.40 shows schematically the case of a crack growing into a metal

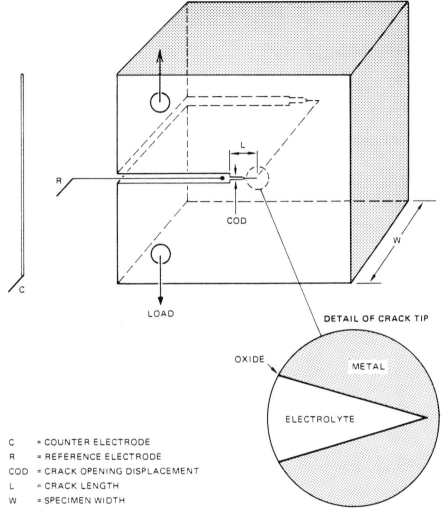

FIGURE 4.3.40. CT specimen geometry for determining the electrochemical/mechanical impedance for the propagation of a crack.

under constant current but cyclic load conditions. The loading conditions can be controlled to achieve a sinusoidal variation in the crack opening displacement D:

$$D = \overline{D} + \tilde{d} \sin (\omega t) \tag{116}$$

where $\overline{D}$ is the mean crack opening and $\tilde{d}$ the amplitude of the superimposed sinusoidal perturbation. Both $\overline{D}$ and $\tilde{d}$ will increase with crack length. To a first approximation, the area exposed at the crack tip can be considered to be proportional to the crack opening angle,

$$A = \gamma_W \tan (D/L) \tag{117}$$

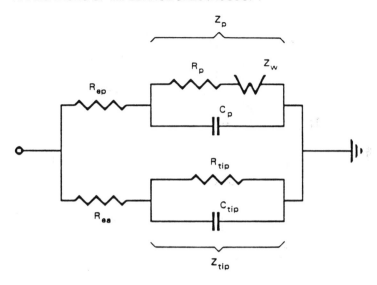

R_{ep} = ELECTROLYTE RESISTANCE TO CRACK WALLS

R_p = CHARGE TRANSFER RESISTANCE AT PASSIVE WALLS

C_p = DOUBLE LAYER CAPACITANCE AT PASSIVE OXIDE

Z_W = DIFFUSIONAL IMPEDANCE AT PASSIVE WALLS

Z_p = INTERFACIAL IMPEDANCE AT PASSIVE WALLS

R_{ea} = ELECTROLYTE RESISTANCE TO CRACK TIP

R_{tip} = CHARGE TRANSFER RESISTANCE AT EXPOSED METAL

C_{tip} = DOUBLE LAYER CAPACITANCE OF EXPOSED METAL

Z_{tip} = INTERFACIAL IMPEDANCE AT CRACK TIP

FIGURE 4.3.41. Simplified equivalent circuit for a stress corrosion crack.

$$A \approx \gamma_W (D/L) = \frac{\gamma W}{L} \left[\overline{D} + \tilde{d} \sin (\omega t) \right] \tag{118}$$

where γ is a proportionality constant determined by the mechanical properties of the metal and the oxide film (Eiselstein et al. [1985]) and W is the specimen width.

Figure 4.3.41 shows the approximate* equivalent circuit for the inside of the crack, with the impedance of the oxide-passivated walls appearing in parallel with the impedance element due to dissolution of exposed metal at the crack tip. The variation in crack opening displacement, described by Eq. (116), results in a sinusoidal perturbation at the exposed tip area such that

$$C_{tip} = C_{tip}^0 A \approx \frac{C_{tip}^0 \gamma W}{L} \left[\overline{D} + \tilde{d} \sin (\omega t) \right] \tag{119}$$

*Because of the potential distribution along the crack, this equivalent circuit should more properly be represented as a nonuniform, finite transmission line for the oxide wall impedance, with the crack tip as a terminating impedance. Such a case is shown in Figure 4.3.33.

$$R_{\text{tip}} = \frac{R_{\text{tip}}^0}{A} \approx \frac{R_{\text{tip}}^0 L}{\gamma W \left[\overline{D} + \tilde{d} \sin (\omega t)\right]} \qquad (120)$$

Without going through the algebra, for the equivalent circuit shown in Figure 4.3.41 we can obtain an expression for the electrochemical–mechanical impedance (Z_{em}) for this system, defined as the ratio of the ac voltage that appears at the reference electrode to the crack opening displacement due to the sinusoidal load variation under dc galvanostatic conditions:

$$Z_{\text{em}} = \frac{\tilde{V}_{\text{ref}}}{\tilde{d}} = \overline{V}_{\text{tip}} \left[R_{e,p} + Z_p\right] \left[\frac{1}{R_{\text{tip}}^0} + j\omega C_{\text{tip}}^0\right] \frac{\gamma W}{L} \qquad (121)$$

Under optimal conditions, Z_{em}, measured over a range of frequences, can be used to deconvolve the equivalent circuit parameters in a manner analogous to that of electrochemical impedance analysis. The advantages of this method, in the example given, are that measurements can be made under dynamic load conditions and that the perturbation is imposed at the point of principal interest, the crack tip.

Figure 4.3.42 presents the real vs. imaginary components of the electrochemical–mechanical impedance response measured for an HY80 steel specimen of geometry shown in Figure 4.3.40 immersed in 3.5 wt% NaCl. Due to the form of Eq. (121), these plots are somewhat more complex than a conventional Nyquist

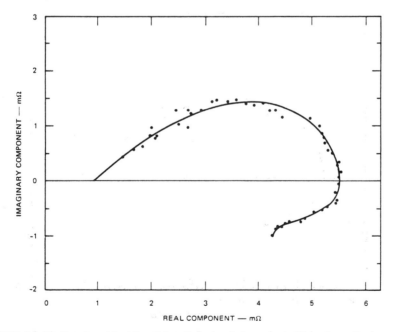

FIGURE 4.3.42. Complex plane plot of the electrochemical–mechanical impedance for the propagation of a crack through HY80 steel in 3.5% NaCl solution at 25°C under sinusoidal loading conditions.

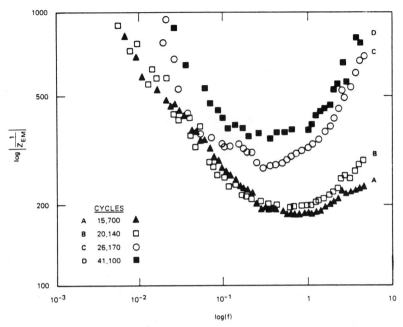

FIGURE 4.3.43. Log ($|1/Z_{em}|$) vs. log(frequency) for an HY80 specimen of the geometry shown in Figure 4.3.38, exposed to 3.5 wt % NaCl at 25°C.

plot. Nevertheless, these data are amenable to standard methods of electrical analysis. We observe that the processes at the growing crack tip dominate at low frequencies, and the properties of greatest interest, R_{tip} and C_{tip}, can be deconvolved at limitingly low frequencies.

Figure 4.3.43 shows the logarithm of the electrochemical–mechanical admittance ($Y_{em} = 1/Z_{em}$) plotted vs. log frequency, for a typical data set, as a function of the number of mechanical load cycles experienced by the specimen during crack growth. These data show some evolution in the crack tip parameters (the low-frequency, descending portion of the curves), but considerably more changes appear in the crack wall impedance, reflecting the increased area with increasing crack length.

This section was written by Digby D. Macdonald and Michael C. H. McKubre

REFERENCES

H. Abe, M. Esashi, and T. Matsuo
  [1979]   ISFET's Using Inorganic Gate Thin Films, *IEEE Trans. Electron Devices* **ED-26,** 1939–1944.

P. Abelard and J. F. Baumard
  [1982]   Study of the d.c. and a.c. Electrical Properties of an Yttria-Stabilized Zirconia Single Crystal [$(ZrO_2)_{0.88}$–$(Y_2O_3)_{0.12}$], *Phys. Rev.* **B26,** 1005–1017.
  [1984]   Dielectric Relaxation in Alkali Silicate Glasses: A New Interpretation, Solid State Ionics **14,** 61–65.

M. Abramowitz and I. A. Stegun
  [1965]   *Handbook of Mathematical Functions*, NBS Publication 55, p. 376.

K. El Adham and A. Hammou
  [1983]   "Grain Boundary Effect" on Ceria-Based Solid Solutions, *Solid State Ionics* **9-10,** 905–912.

S. Alexander, J. Bernasconi, W. R. Schneider, and R. Orbach
  [1981]   Excitation Dynamics in Random One-Dimensional Systems, *Rev. Mod. Phys.* **53,** 175–198.

D. P. Almond, G. K. Duncan, and A. R. West
  [1983]   The Determination of Hopping Rates and Carrier Concentrations in Ionic conductors by a New Analysis of ac Conductivity, Solid State Ionics **8,** 159–164.

D. P. Almond, C. C. Hunter, and A. R. West
  [1984]   The Extraction of Ionic Conductivities and Hopping Rates from A. C. Conductivity Data, *J. Mat. Sci.* **19,** 3236–3248.

D. P. Almond and A. R. West
  [1981]   Measurement of Mechanical and Electrical Relaxations in Beta-Alumina, *Solid State Ionics* **3/4,** 73–77.
  [1983a]   Mobile Ion Concentrations in Solid Electrolytes from an Analysis of AC Conductivity, *Solid State Ionics* **9-10,** 277–282.
  [1983b]   Impedance and Modulus Spectroscopy of "Real" Dispersive Conductors, *Solid State Ionics* **11,** 57–64.

D. P. Almond, A. R. West, and R. J. Grant
  [1982]   Temperature Dependence of the A. C. Conductivity of Na Beta-Aluminum, *Solid State Comm.* **44,** 1277–1280.

W. I. Archer and R. D. Armstrong
  [1980]   The Application of A. C. Impedance Methods to Solid Electrolytes, in *Electrochemistry*, Chemical Society Specialist Periodical Reports **7,** 157–202.

R. D. Armstrong
  [1974]   The Metal Solid Electrolyte Interphase, *J. Electroanal. Chem.* **52,** 413–419.

317

R. D. Armstrong, M. F. Bell, and A. A. Metcalfe
[1978] The AC Impedance of Complex Electrochemical Reactions, in *Electrochemistry, Chemical Society Specialist Periodical Reports* **6,** 98–127.

R. D. Armstrong, T. Dickinson, and J. Turner
[1973] Impedance of the Sodium β-alumina Interphase, *J. Electroanal. Chem.* **44,** 157–167.

R. D. Armstrong, T. Dickinson, and P. M. Willis
[1974] The A. C. Impedance of Powdered and Sintered Solid Ionic Conductors, *J. Electroanal. Chem.* **53,** 389–405.

R. D. Armstrong and K. Edmondson
[1973] The Impedance of Metals in the Passive and Transpassive Regions, *Electrochim. Acta* **18,** 937–943.

R. D. Armstrong and M. Henderson
[1972] Impedance Plane Display of a Reaction with an Adsorbed Intermediate, *J. Electroanal. Chem.* **39,** 81–90.

R. D. Armstrong and R. Mason
[1973] Double Layer Capacity Measurement Involving Solid Electrolytes, *J. Electroanal. Chem.* **41,** 231–241.

R. D. Armstrong, W. P. Race, and H. R. Thirsk
[1968] Determination of Electrode Impedance over an Extended Frequency Range by Alternating-Current Bridge Methods, *Electrochim. Acta* **13,** 215.

R. D. Armstrong et al.
[1977] R. D. Armstrong, M. F. Bell and A. A. Metcalfe, The Anodic Dissolution of Molybdenum in Alkaline Solutions. Electrochemical Measurements, *J. Electroanal. Chem.* **84,** 61; K. L. Bladen, The Anodic Dissolution of Lead in Oxygenated and Deoxygenated Sulphuric Acid Solutions, *J. Appl. Electrochem.* **7,** 345; A. A. Metcalfe, The Anodic Dissolution of Silver into Silver Rubidium Iodide-Impedance Measurements, *J. Electroanal. Chem.* **88,** 187–192.

S. Atlung and T. Jacobsen
[1976] On the AC-Impedance of Electroactive Powders: γ-Manganese Dioxide, *Electrochim. Acta* **21,** 575–584.

S. P. S. Badwal
[1983] New Electrode Materials for Low Temperature Oxygen Sensors, *J. Electroanal. Chem.* **146,** 425–429.
[1984] Kinetics of the Oxygen Transfer Reaction at the $(Cu_{0.5}Sc_{0.5})O_{2\pm x}$/YSZ Interface by Impedance Spectroscopy, *J. Electroanal. Chem.* **161,** 75–91.

S. P. S. Badwal and H. J. de Bruin
[1978] Polarization at the $(U_{0.7}Y_{0.3})O_{2+x}$/YSZ Interface by Impedance Dispersion Analysis, *Aust. J. Chem.* **31,** 2337–2347.

S. P. S. Badwal, M. J. Bannister, and M. J. Murray
[1984] Non-Stoichiometric Oxide Electrodes for Solid State Electrochemical Devices, *J. Electroanal. Chem.* **168,** 363–382.

P. R. Barabesh and R. S. C. Cobbald
[1982] Dependence on Interface State Properties of Electrolyte: Silicon Dioxide–Silicon Structures, *IEEE Trans. Elect. Devices* **ED-29,** 102–108.

A. J. Bard and L. R. Faulkner
[1980] *Electrochemical Methods: Fundamentals and Applications*, Wiley, New York.

G. Barker
[1958] Square-Wave Polarography. III. Single Drop Polarography, *Anal. Chim. Acta* **18,** 118.
[1969] Noise connected with Electrode Processes, *J. Electroanal. Chem.* **21,** 127.

G. Barker, R. Faircloth, and A. Gardner
 [1958]   Use of Faradaic Rectification for the Study of Rapid Electrode Processes, IV. Theoretical Aspects of Square Wave Polarography, *Nature* **181**, 247–248.

H. H. Bauer
 [1964]   Theory of Faradaic Distortion: Equation for the 2nd-Harmonic Current, *Australian J. Chem.* **17**, 715.

J. E. Bauerle
 [1969]   Study of Solid Electrolyte Polarization by a Complex Admittance Method, *J. Phys. Chem. Solids* **30**, 2657–2670.

N. M. Beekmans and L. Heyne
 [1976]   Correlation between Impedance, Microstructure, and Compositions of Calcia-Stabilized Zirconia, *Electrochim. Acta* **21**, 303–310.

J. G. Berberian and R. H. Cole
 [1969]   Null Instrument for Three- and Four-Terminal Admittance Measurements at Ultra-low Frequencies, *Rev. Sci. Instrum.* **40**, 811.

P. Bergveld
 [1970]   Development of an Ion-Sensitive Solid State Device for Neurophysiological Measurements, *IEEE Trans. Biomed. Eng.* **BME-17**, 70–71.

P. Bergveld and N. F. De Rooij
 [1981]   History of Chemically Sensitive Semiconductor Devices, *Sensors and Actuators,* **1**, 5–15.

H. Bernard
 [1981]   *Microstructure et Conductivité de la Zircone Stabilisée Frittée*, Report CEA-R-5090, Commissariat a l'Energie Atomique, CEN-Saclay, France

J. Bernasconi, H. U. Beyeler, S. Strassler, and S. Alexander
 [1979]   Anomalous Frequency-Dependent Conductivity in Disordered One-Dimensional System, *Phys. Rev. Lett.* **42**, 819–822.

U. Bertocci
 [1979]   AC Induced Corrosion: The Effect of an Alternating Voltage on Electrodes under Charge Transfer Control, *Corrosion* **35**, 211–215.

U. Bertocci and J. L. Mullen
 [1981]   Effect of Large Voltage Modulations on Electrodes under Charge Transfer Control, in *Electrochemical Corrosion Testing*, ed. F. Mansfeld and U. Bertocci, American Society of Testing and Materials, San Francisco, pp. 365–380.

P. Bindra, M. Fleischmann, J. W. Oldham, and D. Singleton
 [1973]   Nucleation, *Faraday Discuss. Chem. Soc.* **56**, 180.

R. L. Birke
 [1971]   Operational Admittance of an Electrode Process where an *a priori* Dependence of Rate Constant on Potential Is Not Assumed, *J. Electroanal. Chem.* **33**, 201–207.

G. Blanc, C. Gabrielli, and N. Keddam
 [1975]   Measurement of the Electrochemical Noise by a Cross Correlation Method, *Electrochim. Acta* **20**, 687.

R. Boddy
 [1965]   The Structure of the Semiconductor–Electrolyte Interface, *J. Electroanal. Chem.* **10**, 199–244.

H. W. Bode
 [1945]   *Network Analysis and Feedback Amplifier Design*, Van Nostrand, New York, Ch. 14.

E. V. Bohn
 [1963]   *The Transform Analysis of Linear Systems*, Addison-Wesley, Reading, Mass., 1963.

N. Bonanos and E. P. Butler
    [1985]   Ionic Conductivity of Monoclinic and Tetragonal Yttria–Zirconia Single Crystals, *J. Mat. Sci. Lett.* **4,** 561–564.

N. Bonanos and E. Lilley
    [1980]   Ionic Conductivity of Suzuki Phases, *Solid State Ionics* **1,** 223–230.
    [1981]   Conductivity Relaxations in Single Crystals of Sodium Chloride Containing Suzuki Phase Precipitates, *J. Phys. Chem. Solids* **42,** 943–952.

N. Bonanos, R. K. Slotwinski, B. C. H. Steele, and E. P. Butler
    [1984a]   Electrical Conductivity/Microstructural Relationships in Aged CaO and CaO + MgO Partially Stabilized Zirconia, *J. Mat. Sci.* **19,** 785–793.
    [1984b]   High ionic conductivity in polycrystalline tetragonal Y_2O_3–ZrO_2, *J. Mat. Sci. Lett.* **3,** 245–248.

A. Bonnel, F. Dabosi, C. Deslouis, M. Duprat, M. Keddam, and B. Tribollet
    [1983]   Corrosion Study of a Carbon Steel in Neutral Chloride Solution by Impedance Techniques, *J. Electrochem. Soc.* **130,** 753–761.

P. Bordewijk
    [1975]   Defect-Diffusion Models of Dielectric Relaxation, *Chem. Phys. Lett.* **32,** 592–596.

C. J. F. Böttcher and P. Bordewijk
    [1978]   *Theory of Electric Polarization*, Vol. II, *Dielectrics in Time-Dependent Fields*, Elsevier, Amsterdam.

P. H. Bottelberghs and G. H. J. Broers
    [1976]   Interfacial Impedance Behavior of Polished and Paint Platinum Electrodes at Na_2WO_4–Na_2MoO_4 Solid Electrolytes, *J. Electroanal. Chem.* **67,** 155–167.

B. A. Boukamp
    [1984]   A Microcomputer Based System for Frequency Dependent Impedance/Admittance Measurements, *Solid State Ionics* **11,** 339–346.

L. Bousse
    [1982]   Ph.D. thesis, Twente University of Technology, the Netherlands.

L. Bousse and P. Bergveld
    [1983]   On the Impedance of the Silicon Dioxide/Electrolyte Interface, *J. Electroanal. Chem.* **152,** 25–39.

A. D. Brailsford and D. K. Hohnke
    [1983]   The Electrical Characterization of Ceramic Oxides, *Solid State Ionics* **11,** 133–142.

B. Breyer and H. H. Bauer
    [1964]   *Alternating Current Polarography and Tensammetry*, Interscience, New York.

R. Brown, D. E. Smith, and G. Booman
    [1968]   Operational Amplifier Circuits Employing Positive Feedback for IR Compensation. I. Theoretical Analysis of Stability and Bandpass Characteristics, *Anal. Chem.* **40,** 1411.

P. G. Bruce and A. R. West
    [1983]   The A. C. Conductivity of Polycrystalline LISICON $Li_{2+2x}Zn_{1-x}GeO_4$ and a Model for Intergranular Constriction Resistances, *J. Electrochem. Soc.* **130,** 662–669.

P. G. Bruce, A. R. West, and D. P. Almond
    [1982]   A New Analysis of ac Conductivity Data in Single Crystal Beta-Alumina, *Solid State Ionics* **7,** 57–60.

R. P. Buck
    [1980]   Impedance Diagrams Applied to the Investigation of Ion-Selective Electrodes, *Hungarian Sci. Instr.* **49,** 7–23.
    [1982]   The Impedance Method Applied to the Investigation of Ion-Selective Electrodes, *Ion-Selective Electrode Rev.* **4,** 3–74.

A. J. Burggraaf, M. van Hemert, D. Scholten, and A. J. A. Winnubst
  [1985] Chemical Composition of Oxidic Interfaces in relation with Electric and Electrochemical Properties, *Mater. Sci. Monogr.* **28B** (Reactivity of Solids), pp. 797–802.

E. P. Butler and N. Bonanos
  [1985] The Characterization of Zirconia Engineering Ceramics by a.c. Impedance Spectroscopy, *Mat. Sci. Eng.* **71**, 49–56.

B. D. Cahan and C.-T. Chen
  [1982] The Nature of the Passive Film on Iron. II. AC Impedance Studies, *J. Electrochem. Soc.* **129**, 474–480.

L. M. Callow, J. A. Richardson, and J. L. Dawson
  [1976] Corrosion Monitoring Using Polarization Resistance Measurements, Parts I and II, *Brit. Corrosion J.* **11**, 123, 132.

R. Calvert
  [1948] A new technique in bridge measurements, *Electron. Eng.* **20**, 28–29.

H. S. Carslaw and J. C. Jaeger
  [1959] *Conduction of Heat in Solids*, 2nd Ed., Clarendon Press, Oxford.

C. W. Carter
  [1925] Impedance of Networks Containing Resistances and Two Reactances, *Bell Sys. Tech. J.* **4**, 387–401.

M. Casciola and D. Fabiani
  [1983] Ionic Conduction and Dielectric Properties of Anydrous Alkali Metal Salt Forms of Zirconium Phosphase, *Solid State Ionics* **11**, 31–38.

H. Chang and G. Jaffé
  [1952] Polarization in Electrolytic Solutions, Part I, Theory, *J. Chem. Phys.* **20**, 1071–1077.

C.-Y. Chao, L. F. Lin, and D. D. Macdonald
  [1981] A Point Defect Model for Anodic Passive Films. I. Film Growth Kinetics, *J. Electrochem. Soc.* **128**, 1187–1194.

  [1982] A Point Defect Model for Anodic Passive Films. III. Impedance Response, *J. Electrochem. Soc.* **129**, 1874–1879.

D.-T. Chin and S. Venkatesh
  [1979] A Study of Alternating Voltage Modulation on the Polarization of Mild Steel, *J. Electrochem. Soc.* **126**, 1908–1913.

H. Chung
  [1983] Transient Crack Growth in AISI 4340 Steel in Sodium Chloride Solution, Ph.D. thesis, Ohio State University.

H. Chung and D. D. Macdonald
  [1981] Corrosion and Corrosion Cracking of Materials for Water Cooled Reactors, Report to the Electric Power Research Institute, Palo Alto, CA, Project RP1166-1.

K. S. Cole
  [1972] *Membranes, Ions, and Impulses*, University of California Press, Berkeley.

K. S. Cole and R. H. Cole
  [1941] Dispersion and Absorption in Dielectrics. I. Alternating Current Characteristics, *J. Chem. Phys.* **9**, 341–351.

M. Colloms
  [1983] *Computer Controlled Testing and Instrumentation*, Pentech Press, London.

P. Colonomos and R. G. Gordon
  [1979] Bounded Error Analyses of Experimental Distributions of Relaxation Times, *J. Chem. Phys.* **71**, 1159–1166.

B. E. Conway, K. Tellefsen, and S. Marshall
    [1984]   Lattice Models, Interactions, and Solvent Effects in Electrochemical Monolayer Formation, in *The Chemistry and Physics of Electrocatalysis*, ed. J. D. E. McIntyre, M. J. Weaver, and E. B. Yaeger, The Electrochemical Society Inc., Pennington, N.J. **84-12**, pp. 15–33.

J. W. Cooley and J. W. Tukey
    [1965]   An Algorithm for the Machine Computation of Complex Fourier Series, *Math. Comp.* **19**, 297.

E. J. Crain
    [1970]   *Laplace and Fourier Transforms for Electrical Engineers*, Holt, Rinehart and Winston, New York.

S. C. Creason and D. E. Smith
    [1972]   Fourier Transform Faradaic Admittance Measurements: Demonstration of the Applicability of Random and Pseudo-Random Noise as Applied Potential Signals, *J. Electroanal. Chem.* **36**, 1.
    [1973]   Fourier Transform Faradaic Admittance Measurements: On the Use of High-Precision Data for Characterization of Very Rapid Electrode Process Kinetic Parameters, *J. Electroanal. Chem.* **47**, 9.

L. S. Darken
    [1948]   Diffusion, Mobility, and their Interrelation through Free Energy in Binary Metallic Systems, *Trans. AIME* **175**, 184–201.

D. W. Davidson
    [1961]   Dielectric Relaxation in Liquids. I. The Representation of Relaxation Behavior, *Can. J. Chem.* **39**, 571–594.

D. W. Davidson and R. H. Cole
    [1951]   Dielectric Relaxation in Glycerol, Propylene Glycol, and *n*-Propanol, *J. Chem. Phys.* **19**, 1484–1490.

C. de Boor
    [1978]   *A Practical Guide to Splines*, Springer-Verlag, New York; see pp. 249 ff.

P. Debye
    [1929]   *Polar Molecules*, Chemical Catalogue Company, New York, p. 94.

P. Debye and H. Falkenhagen
    [1928]   Dispersion of the Conductivity and Dielectric Constants of Strong Electrolytes, *Phys. Z.* **29**, 121–132, 401–426.

L. C. De Jonghe
    [1979]   Grain Boundaries and Ionic Conduction in Sodium Beta-Alumina, *J. Mat. Sci.* **14**, 33–48.

P. Delahay
    [1954]   *New Instrumental Methods in Electrochemistry*, Interscience, New York.

P. Delahay, M. Senda, and C. H. Weis
    [1960]   Faradaic Rectification and Electrode Processes, *J. Phys. Chem.* **64**, 960.

R. de Levie
    [1963]   Porous Electrodes in Electrolyte Solutions, I, *Electrochim. Acta* **8**, 751.
    [1964]   On Porous Electrodes in Electrolyte Solutions, II, *Electrochim. Acta* **9**, 1231–1245.
    [1965]   The Influence of Surface Roughness of Solid Electrodes on Electrochemical Measurements, *Electrochim. Acta* **10**, 113–130.
    [1967]   Electrochemical Response of Porous and Rough Electrodes, in *Advances in Electrochemistry and Electrochemical Engineering*, Vol. 6, ed. P. Delahay and C. W. Tobias, Interscience, New York, pp. 329–397.

R. de Levie and D. Vukadin

[1975]   Dipicrylamine Transport across Ultrathin Phosphatidylethanolamine Membrane, *J. Electroanal. Chem.* **62**, 95–109.

C. Deslouis, C. Gabrielli, Ph. Sainte-Rose Franchine, and B. Tribollet

[1980]   Relationship between the Electrochemical Impedance and the Electrohydrodynamical Impedance measured using a Rotating Disk Electrode, *J. Electroanal. Chem.* **107**, 193–195.

[1982]   Electrohydrodynamical Impedance on a Rotating Disk Electrode, *J. Electrochem. Soc.* **129**, 107–118.

J. Devay and L. Meszaros

[1980]   Study of the Rate of Corrosion of Metals by a Faradaic Distortion Method. II. Effect of the Ohmic Resistance and Capacity of the Double Layer on the Measurement of the Harmonic Components of the Faradaic Currents, *Acta Chim. Acad. Sci. Hung.* **104**, 311.

J. P. Diard and B. LeGorrec

[1979]   Use of a New Generator with Negative Impedance for Studying the Anodic Behaviour of Zinc in an Alkaline Medium, *Electroanal. Chem.* **103**, 363–374.

J. L. Diot, J. Joseph, J. R. Matin, and P. Clechet

[1985]   pH Dependence of the Si/SiO$_2$ Interface State Density for EOS Systems, *Electroanal. Chem.* **193**, 75–88.

K. Doblhofer and A. A. Pilla

[1971]   LaPlace Plane Analysis of the Faradaic and Non-Faradaic Impedance of the Mercury Electrode, *J. Electroanal. Chem.* **39**, 91.

J. Drennan and E. P. Butler

[1982]   Does Alumina Act as a Grain Boundary Scavenger in Zirconia? *J. Am. Ceram. Soc.* **65(11)**, Comm. C-194-195.

S. S. Dukhin and V. N. Shilov

[1974]   *Dielectric Phenomena and the Double Layer in Disperse Systems and Polyelectrolytes* (English Translation), Keter, Jerusalem.

E. C. Dutoit, R. L. Van Meirhaeghe, F. Cardon, and W. P. Gomes

[1975]   Investigation on the Frequency-Dependence of the Impedance of the Nearly Ideally Polarizable Semiconductor Electrodes CdSe, CdS, and TiO$_2$, *Ber. Bunsen Gesell.* **79**, 1206–1213.

L. E. Eiselstein, M. C. H. McKubre, and R. D. Caligiuri

[1983]   *Prediction of Crack Growth in Aqueous Environments*, First Annual Report by SRI to the Office of Naval Research, N00014-82-K-0343.

[1985]   *Electrical and Mechanical Impedance Studies of Crack Growth*, Final Report by SRI to the Office of Naval Research.

H. Engstrom and J. C. Wang

[1980]   Automatic Multifrequency Measurements of the Complex Impedance of Fast Ion Conductors, *Solid State Ionics* **1**, 441–459.

I. Epelboin, C. Gabrielli, M. Keddam, and H. Tanenouti

[1975]   A Model of the Anodic Behaviour of Iron in Sulphoric Acid Medium, *Electrochim. Acta* **20**, 913–916.

I. Epelboin, C. Gabrielli, M. Keddam, J.-C. Lestrade, and H. Takenouti

[1972]   Passivation of Iron in Sulfuric Acid Medium, *J. Electrochem. Soc.* **119**, 1632–1637.

I. Epelboin and M. Keddam

[1970]   Faradaic Impedances: Diffusion Impedance and Reaction Impedance, *J. Electrochem. Soc.* **117**, 1052–1056.

D. R. Franceschetti
   [1981]   Small-Signal A. C. Response Theory for Systems Exhibiting Impurity Oxygen Transport, *Solid State Ionics* **5**, 613–616.
   [1982]   *Physics of Electrochemical Systems*, Lecture Notes, State University, Utrecht, Netherlands, January, pp. 1–165.
   [1984]   Small-Signal AC Response Theory for Systems Exhibiting Diffusion and Trapping of an Electroactive Species, *J. Electroanal. Chem.* **178**, 1–9.

D. R. Franceschetti and J. R. Macdonald
   [1977]   Electrode Kinetics, Equivalent Circuits, and System Characterization: Small-Signal Conditions, *J. Electroanal. Chem.* **82**, 271–301.
   [1979a]   Numerical Analysis of Electrical Response: Statics and Dynamics of Space-Charge Regions at Blocking Electrodes, *J. Appl. Phys.* **50**, 291–302.
   [1979b]   Numerical Analysis of Electrical Response: Biased Small-Signal A. C. Response for Systems with One or Two Blocking Electrodes, *J. Electroanal. Chem.* **100**, 583–605.
   [1979c]   Diffusion of Neutral and Charged Species under Small-Signal A. C. Conditions, *J. Electroanal. Chem.* **101**, 307–316.
   [1980]   Coupling of Electrode Kinetics and Space Charge Dynamics in Unsupported Systems, in *Proc. Third Symposium on Electrode Processes*, ed. S. Bruckenstein, J. D. McIntyre, B. Miller, and E. Yeager, the Electrochemical Society 1980 Proceedings **80-3**, 94–114.
   [1982]   Small-Signal A-C Response Theory for Electrochromic Thin Films, *J. Electrochem. Soc.* **129**, 1754–1756.

D. R. Franceschetti and P. C. Shipe
   [1984]   Bulk Conductivity and Polarization of Ionic Crystals Exhibiting Defect Exchange between Inequivalent Sites, *Solid State Ionics* **11**, 285–291.

A. D. Franklin
   [1975]   Electrode Effects in the Measurement of Ionic Conductivity, *J. Am. Ceram. Soc.* **58**, 465–473.

A. D. Franklin and H. J. de Bruin
   [1983]   The Fourier Analysis of Impedance Spectra of Electroded Solid Electrolytes, *Phys. Stat. Sol. (a)* **75**, 647–656.

R. J. Friauf
   [1954]   Polarization Effects in the Ionic Conductivity of Silver Bromide, *J. Chem. Phys.* **22**, 1329–1338.

H. Fricke
   [1932]   The Theory of Electrolytic Polarization, *Philos. Mag.* **14**, 310–318.
   [1953]   The Maxwell–Wagner Dispersion in a Suspension of Ellipsoids, *J. Phys. Chem.* **57**, 934–937.

K. Funke
   [1986]   Debye–Hückel-Type Relaxation Processes in Solid Ionic Conductors: The Model, *Solid State Ionics* **18–19**, 183–190.

R. M. Fuoss and J. G. Kirkwood
   [1941]   Electrical Properties of Solids. VIII. Dipole Moments in Polyvinyl Chloride–Diphenyl Systems, *J. Amer. Chem. Soc.* **63**, 385–394.

C. Gabrielli
   [1981]   *Identification of Electrochemical Processes by Frequency Response Analysis*, Monograph Reference 004/83, Solartron Instrumentation Group, Farnsborough, England.

C. Gabrielli and M. Keddam
   [1974]   Recent Progress in the Measurement of Electrochemical Impedances in a Sinusoidal System, *Electrochim. Acta* **19**, 355.

R. Garnsey
    [1979]   Boiler Corrosion and the Requirement for Feed- and Boiler-Water Chemistry Control in Nuclear Steam Generators, *Nucl. Energy* **18**, 117.

R. C. Garvie, R. H. Hannink, and R. T. Pascoe
    [1975]   Ceramic Steel? *Nature* **258**, 703–704.

Y. Gefen, A. Aharony and S. Alexander
    [1983]   Anomalous Diffusion on Percolating Clusters, *Phys. Rev. Lett.* **50**, 77–80.

H. Gerisicher and W. Mehl
    [1955]   Zum Mechanismus der Kathodischen Wasserstoffabscheidung an Quecksilber, Silber, und Kupfer, *Z. Elektochem.* **59**, 1049–1060.

J. S. Gill, L. M. Callow, and J. D. Scantlebury
    [1983]   Corrosion Measurements Derived from Small Perturbation Nonlinearity. I. Harmonic Analysis, *Corrosion*, **39**, 61–66.

S. H. Glarum
    [1960]   Dielectric Relaxation of Isoamyl Bromide, *J. Chem. Phys.* **33**, 639–643.

S. H. Glarum and J. H. Marshall
    [1980]   The AC Response of Iridium Oxide Films, *J. Electrochem. Soc.* **127**, 1467–1474.

S. Goldman
    [1950]   *Transformation Calculus and Electrical Transients*, Prentice-Hall, Englewood Cliffs, N.J.

D. C. Grahame
    [1952]   Mathematical Theory of the Faradaic Admittance, *J. Electrochem. Soc.* **99**, 370c–385c.

A. Guerrero, E. P. Butler, and P. L. Pratt
    [1979]   The Suzuki Phase in $NaCl:Cd^{2+}$ and $NaCl:Mg^{2+}$, *J. Physique Colloque* **7**, C6/363–366.

T. K. Gupta, J. H. Bechtold, R. C. Kuznicki, L. H. Kadoff, and B. R. Rossing
    [1977]   Stabilization of Tetragonal Phase in Polycrystalline Zirconia, *J. Mat. Sci.* **12**, 2421–2426.

R. Haak and D. Tench
    [1984]   Electrochemical Photocapacitance Spectroscopy Method for Characterization of Deep Levels and Interface States in Semiconductor Materials, *J. Electrochem. Soc.* **131**, 275–283.

R. Haak, C. Ogden, and D. Tench
    [1982]   Electrochemical Photocapacitance Spectroscopy: A New Method for Characterization of Deep Levels in Semiconductors, *J. Electrochem. Soc.* **129**, 891–893.

B. Hague
    [1957]   *AC Bridge Methods*, Pitman, London.

J. W. Hartwell
    [1971]   A Procedure for Implementing the Fast Fourier Transform on Small Computers, *IBM J. Res. Develop.* **15**, 355–363.

L. Heyne
    [1983]   Interfacial Effects in Mass Transport in Ionic Solids, in *Mass Transport in Solids*, ed. F. Beniere and C. R. A. Catlow, Plenum Press, New York, pp. 425–456.

K. Hladky, L. M. Callow, and J. L. Dawson
    [1980]   Corrosion Rates from Impedance Measurements: An Introduction, *Brit. Corrosion J.*, **13**, 1, 20.

C. Ho

[1980] Application of AC Techniques to Kinetic Studies of Electrochemical Systems, Ph.D. thesis, Stanford University.

C. Ho, I. D. Raistrick, and R. A. Huggins

[1980] Application of AC Techniques to the Study of Lithium Diffusion in Tungsten Trioxide Thin Films, *J. Electrochem. Soc.* **127**, 343–350.

I. M. Hodge, M. D. Ingram, and A. R. West

[1975] A New Method for Analyzing the a.c. Behavior of Polycrystalline Solid Electrolytes, *J. Electroanal. Chem.* **58**, 429–432.

[1976] Impedance and Modulus Spectroscopy of Polycrystalline Solid Electrolytes, *J. Electroanal. Chem.* **74**, 125–143.

K. Holub, G. Tessari, and P. Delahay

[1967] Electrode Impedance without *a priori* Separation of Double-Layer Charging and Faradaic Process, *J. Phys. Chem.* **71**, 2612–2618.

A. Hooper

[1977] A Study of the Electrical Properties of Single Crystal and Polycrystalline β-Alumina Using Complex Plane Analysis, *J. Phys. D. Appl. Phys.* **10**, 1487–1497.

B. D. Hughes, M. F. Shlesinger, and E. W. Montroll

[1981] Random Walks with Self Similar Clusters, *Proc. Nat. Acad. Sci.* **78**, 3287–3291.

I. Iami and P. Delahay

[1962] Faradaic Rectification and Electrode Processes. III. Experimental Methods for High Frequencies and Application to the Discharge of Mercurous Ion, *J. Phys. Chem.* **66**, 1108–1113.

N. Ibl

[1983a] Fundamentals of Transport Phenomena, in *Comprehensive Treatise of Electrochemistry*, Vol. 6, ed. E. Yaeger, J. O'M. Bockris, B. E. Conway, and S. Sarangapani, Plenum Press, New York, pp. 1–63.

[1983b] Current Distribution, in *Comprehensive Treatise of Electrochemistry*, Vol. 6, *Electrodics: Transport*, ed. E. Yaeger, J. O'M. Bockris, B. E. Conway, and S. Sarangapani, Plenum Press, New York, p. 239.

G. Jaffé

[1952] Theory of Conductivity of Semiconductors, *Phys. Rev.* **85**, 354–363.

J. Janata and R. J. Huber, eds.

[1985] *Solid State Chemical Sensors*, Academic Press, New York.

H. B. Johnson and W. L. Worrell

[1982] Lithium and Sodium Intercalated Dichalcogenides: Properties and Electrode Applications, *Synthetic Metals* **4**, 225–248.

W. B. Johnson, N. J. Tolar, G. R. Miller, and I. B. Cutler

[1969] Electrical and Mechanical Relaxation in CaF_2 Doped with NaF, *J. Phys. Chem. Solids* **30**, 31–42.

A. K. Jonscher

[1974] Hopping Losses in Polarisable Dielectric Media, *Nature* **250**, 191–193.

[1975a] Physical Basis of Dielectric Loss, *Nature* **253**, 717–719.

[1975b] The Interpretation of Non-Ideal Dielectric Admittance and Impedance Diagrams, *Phys. Stat. Sol.* (a) **32**, 665–676.

[1975c] New Interpretation of Dielectric Loss Peaks, *Nature* **256**, 566–568.

[1977] The "Universal" Dielectric Response, *Nature* **267**, 673–679.

[1980] The Universal Dielectric Response: A Review of Data and their New Interpretation, *Physics of Thin Films* **11**, 205–317.

[1983] *Dielectric Relaxation in Solids*, Chelsea Dielectrics Press, London.

A. K. Jonscher and J.-M. Reau
[1978]   Analysis of the complex impedance data for beta-PbF2, *J. Mat. Sci.* **13**, 563–570.

M. Keddam, O. R. Mattos, and H. J. Takenouti
[1981]   Reaction Model for Iron Dissolution Studied by Electrode Impedance. Determination of the Reaction Model, *J. Electrochem. Soc.* **128**, 257–266; **128**, 266–274.

M. Keddam, J.-F. Lizee, C. Pallotta, and H. J. Takenouti
[1984]   Electrochemical Behaviour of Passive Iron in Acid Medium, *J. Electrochem. Soc.* **131**, 2016–2024.

H. Keiser, K. D. Beccu, and M. A. Gutjahr
[1976]   Abschatzung der Porenstrukter Poroser Elektroden aus Impedanzmessung, *Electrochim. Acta* **21**, 539–543.

M. Kimura, T. Osaki, and S. Toshima
[1975]   Interfacial Capacity between a Silver Bromide Solid Electrolyte and a Blocking Electrode, *Bull. Chem. Soc. Japan* **48**, 830–834.

J. G. Kirkwood and R. M. Fuoss
[1941]   Anomalous Dispersion and Dielectric Loss in Polar Polymers, *J. Chem. Phys.* **9**, 329–340.

M. Kleitz, H. Bernard, E. Fernandez, and E. Schouler
[1981]   Impedance Spectroscopy and Electrical Resistance Measurements on Stabilized Zirconia, in *Advances in Ceramics*, Vol. 3, ed. A. H. Heuer and L. W. Hobbs, American Ceramic Society, Columbus, Ohio, pp. 310–336.

M. Kleitz and J. H. Kennedy
[1979]   Resolution of Multicomponent Impedance Diagrams, in *Fast Ion Transport in Solids*, ed. P. Vashishta, J. N. Mundy, and G. K. Shenoy eds., North-Holland, Amsterdam, pp. 185–188.

K. Kobayashi, H. Kayajima, and T. Masaki
[1981]   Phase Change and Mechanical Properties of Zirconia-Yttria Solid Electrolyte after Aging, *Solid State Ionics* **3/4**, 489–493.

R. Kohlrausch
[1854]   Theorie des Elektrischen Rückstandes in der Leidener Flasche, *Pogg. Ann. der Phys. und Chemie* **91**, 179–214.

A. Kornyshev, W. Schmickler, and M. Vorotyntsev
[1982]   Nonlocal Electrostatic Approach to the Problem of a Double Layer at a Metal–Electrolyte Interface, *Phys. Rev.* **B25**, 5244–5256.

H. A. Kramers
[1929]   Theory of Dispersion in the X-Ray Region, *Physik. Z.* **30**, 52.

R. de L. Kronig
[1926]   The Theory of Dispersion of X-Rays, *J. Opt. Soc. Am.* **12**, 547.

F. Kruger
[1903]   *Z. Physik. Chem.*, **45**, 1.

R. B. Laibowitz and Y. Gefen
[1984]   Dynamic Scaling near the Percolation Threshold in Thin Au Films, *Phys. Rev. Lett.* **53**, 380–383.

V. Lanteri, A. H. Heuer, T. E. Mitchell
[1983]   Tetragonal Precipitation in the System Y_2O_3–ZrO_2, in *Proceedings of the 41st Annual Meeting of the Electron Microscopy Society of America*, ed. G. W. Bailey, San Francisco Press, San Francisco, pp. 58–59.

328 REFERENCES

S. Lànyi
    [1975]   Polarization of Ionic Crystals with Incompletely Blocking Electrodes, *J. Phys. Chem. Solids* **36,** 775-781.

I. R. Lauks and J. N. Zemel
    [1979]   The Silicon Nitride/Silicon Ion Sensitive Semiconductor Electrode, *IEEE Trans. Electron Devices*, **ED-26,** 1959-1964.

C. G. Law and J. J. Newman
    [1979]   A Model for the Anodic Dissolution of Iron in Sulfuric Acid, *J. Electrochem. Soc.* **126,** 2150-2155.

A. Le Mehaute
    [1984]   Transfer Processes in Fractal Media, *J. Statistical Physics* **36,** 665-676.

A. Le Mehaute and G. Crepy
    [1983]   Introduction to Transfer and Motion in Fractal Media: The Geometry of Kinetics, *Solid State Ionics* **9-10,** 17-30.

S. J. Lenhart, D. D. Macdonald, B. G. Pound
    [1984]   Restructuring of Porous Nickel Electrodes, *Proc. 19th IECEC Conference*, San Francisco, p. 875.

F. Leroy, P. Gareil, and R. Rosset
    [1982]   Gas and Ion Selective Field Effect Transistor Electrodes, *Analysis* **10,** 351-366.

L. M. Levinson
    [1981]   Grain Boundary Phenomena in Electronic Ceramics, in *Advances in Ceramics*, Vol. 1, ed. L. M. Levinson, American Ceramic Society, Columbus, Ohio.

R. Liang, B. Pound, and D. D. Macdonald
    [1984]   Impedance Analysis of Passive Films on Single Crystal Nickel, *Ext. Abst. Electrochem. Soc.* **84-2,** 356.

E. Lilley and J. E. Strutt
    [1979]   Bulk and Grain Boundary Ionic Conductivity in Polycrystalline β''-Alumina, *Phys. Stat. Sol.* (a) **54,** 639-650.

L. F. Lin, C.-Y. Chao, and D. D. Macdonald
    [1981]   A Point Defect Model for Anodic Passive Films, II, Chemical Breakdown, *J. Electrochem. Soc.* **128,** 1194-1198.

C. P. Lindsey and G. D. Patterson
    [1980]   Detailed Comparison of the Williams–Watts and Cole–Davidson Functions, *J. Chem. Phys.* **73,** 3348-3357.

S. H. Liu
    [1985]   Fractal Model for the ac Response of a Rough Interface, *Phys. Rev. Lett.* **55,** 529-532.

J. Llopis and F. Colom
    [1958]   Study of the Impedance of a Platinum Electrode Acting as Anode, *Proc. 8th meeting of the CITCE*, Madrid, 1956, Butterworths, London, pp. 414-427.

D. D. Macdonald
    [1977]   *Transient Techniques in Electrochemistry*, Plenum Press, New York.
    [1978a]   An Impedance Interpretation of Small Amplitude Cyclic Voltammetry. I. Theoretical Analysis for a Resistive–Capacitive System, *J. Electrochem. Soc.* **125,** 1443-1449.
    [1978b]   An Impedance Interpretation of Small Amplitude Cyclic Voltammetry. II. Theoretical Analysis of Systems that Exhibit Pseudoinductive Behavior. *J. Electrochem. Soc.* **125,** 1977-1981.
    [1978c]   A Method for Estimating Impedance Parameters for Electrochemical Systems That Exhibit Pseudoinductance, *J. Electrochem. Soc.* **125,** 2062-2064.

D. D. Macdonald and M. C. H. McKubre

[1981] Electrochemical Impedance Techniques in Corrosion Science, in *Electrochemical Corrosion Testing*, ed. Florian Mansfield and Ugo Bertocci, STP 727, ASTM, Philadelphia.

[1982] Impedance Measurements in Electrochemical Systems, in *Modern Aspects of Electrochemistry*, Vol. 14, ed. J. O'M. Bockris, B. E. Conway, and R. E. White, Plenum Press, New York.

D. D. Macdonald and M. Urquidi-Macdonald

[1986] Distribution Functions for the Breakdown of Passive Films, *Electrochim. Acta* **31**, 1079–1086.

[1985] Application of Kramers–Kronig Transforms to the Analysis of Electrochemical Impedance Data. I. Polarization Resistance, *J. Electrochem. Soc.* **132**, 2316–2319.

J. R. Macdonald

[1953] Theory of a.c. Space-Charge Polarization Effects in Photoconductors, Semiconductors, and Electrolytes, *Phys. Rev.* **92**, 4–17.

[1954] Static Space-Charge Effects in the Diffuse Double Layer, *J. Chem. Phys.* **22**, 1317–1322.

[1955] Capacitance and Conductance Effects in Photoconducting Alkali Halide Crystals, *J. Chem. Phys.* **23**, 275–295.

[1962] Restriction on the Form of Relaxation-Time Distribution Functions for a Thermally Activated Process, *J. Chem. Phys.* **36**, 345–349.

[1963] Transient and Temperature Response of a Distributed, Thermally Activated Process, *J. Appl. Phys.* **34**, 538–552.

[1971a] Electrical Response of Materials Containing Space Charge with Discharge at the Electrodes, *J. Chem. Phys.* **54**, 2026–2050.

[1971b] The Impedance of a Galvanic Cell with Two Plane-Parallel Electrodes at a Short Distance, *J. Electroanal. Chem.* **32**, 317–328.

[1974a] Binary Electrolyte Small-Signal Frequency Response, *J. Electroanal. Chem.* **53**, 1–55.

[1974b] Simplified Impedance/Frequency-Response Results for Intrinsically Conducting Solids and Liquids, *J. Chem. Phys.* **61**, 3977–3996.

[1975] Binary Electrolyte Small-Signal Frequency Response, *J. Electroanal. Chem.* **53**, 1–55.

[1976a] Complex Rate Constant for an Electrochemical System Involving an Adsorbed Intermediate, *J. Electroanal. Chem.* **70**, 17–26.

[1976b] Interpretation of AC Impedance Measurements in Solids, in *Superionic Conductors*, ed. G. D. Mahan and W. L. Roth, Plenum Press, New York, pp. 81–97.

[1984] Note on the Parameterization of the Constant-Phase Admittance Element, *Solid State Ionics* **13**, 147–149.

[1985a] Ambiguity in Models of Small-Signal Response of Dielectric and Conducting Systems and a New Umbrella Model, *Bull. Am. Phys. Soc.* **30**, 587.

[1985b] New Aspects of Some Small-Signal ac Frequency Response Functions, *Solid State Ionics* **15**, 159–161.

[1985c] Frequency Response of Unified Dielectric and Conductive Systems Involving an Exponential Distribution of Activation Energies, *J. Appl. Phys.* **58**, 1955–1970.

[1985d] Generalizations of ''Universal Dielectric Response'' and a General Distribution-of-Activation-Energies Model for Dielectric and Conducting Systems, *J. Appl. Phys.* **58**, 1971–1978.

[1987] Relaxation is Systems with Exponential or Gaussian Distributions of Activation Energies, J. Appl. Phys. **61**, 700–713.

J. R. Macdonald and M. K. Brachman

[1956] Linear System Integral Transform Relations, *Rev. Mod. Phys.* **28**, 393–422.

J. R. Macdonald and G. B. Cook

[1984]   Analysis of Impedance Data for Single Crystal Na Beta-Alumina at Low Temperatures, *J. Electroanal. Chem.* **168**, 335–354. Also presented at a meeting of the South-
eastern Section of the American Physical Society, November 1983, and at the Sixth
Australian Electrochemistry Conference, Geelong, February, 1984.

[1985]   Reply to Comments by Almond and West on Na β-Alumina Immittance Data Analysis, *J. Electroanal. Chem.* **193**, 57–74.

J. R. Macdonald and D. R. Franceschetti

[1978]   Theory of Small-Signal A-C Response of Solids and Liquids with Recombining Mobile Charge, *J. Chem. Phys.* **68**, 1614–1637.

[1979a]   Compact and Diffuse Double Layer Interaction in Unsupported System Small-Signal
Response, *J. Electroanal. Chem.* **99**, 283–298.

[1979b]   A Method for Estimating Impedance Parameters for Electrochemical Systems That
Exhibit Pseudoinductance, *J. Electrochem. Soc.* **126**, 1082–1083.

J. R. Macdonald, D. R. Franceschetti, and A. P. Lehnen

[1980]   Interfacial Space Charge and Capacitance in Ionic Crystals: Intrinsic Conductors, *J.
Chem. Phys.* **73**, 5272–5293.

J. R. Macdonald, D. R. Franceschetti, and R. Meaudre

[1977]   Electrical Response of Materials with Recombining Space Charge, *J. Phys. C: Solid
State Phys.* **10**, 1459–1471.

J. R. Macdonald and J. A. Garber

[1977]   Analysis of Impedance and Admittance Data for Solids and Liquids, *J. Electrochem.
Soc.* **124**, 1022–1030.

J. R. Macdonald and R. L. Hurt

[1986]   Analysis of Dielectric or Conductive System Frequency Response Data Using the Wil-
liams-Watts Function, J. Chem. Phys. **84**, 496–502.

J. R. Macdonald, A. Hooper, and A. P. Lehnen

[1982]   Analysis of Hydrogen-Doped Lithium Nitride Admittance Data, *Solid State Ionics* **6**,
65–77.

J. R. Macdonald and C. A. Hull

[1984]   Pseudo Reaction Rate in the AC Response of an Electrolytic Cell, *J. Electroanal.
Chem.* **165**, 9–20.

J. R. Macdonald and S. W. Kenkel

[1985]   Comparison of Two Recent Approaches towards a Unified Theory of the Electrical
Double Layer, *Electrochim. Acta* **30**, 823–826.

J. R. Macdonald, J. Schoonman, and A. P. Lehnen

[1981]   Three Dimensional Perspective Plotting and Fitting of Immittance Data, *Solid State
Ionics* **5**, 137–140.

[1982]   The Applicability and Power of Complex Nonlinear Least Squares for the Analyses
of Impedance and Admittance Data, *J. Electroanal. Chem.* **131**, 77–95.

P. B. Macedo, C. T. Moynihan, and R. Bose

[1972]   The Role of Ionic Diffusion in Polarization in Vitreous Ionic Conductors, *Physics and
Chem. of Glasses* **13**, 171–179.

P. B. Macedo, R. Bose, V. Provenzano, and T. A. Litowitz

[1972]   In *Amorphous Materials*, ed, R. W. Douglas and B. Ellis, Interscience, London, see
p. 251.

B. Mandelbrot

[1983]   *The Fractal Geometry of Nature*, Freeman, New York.

F. Mansfeld

[1976]   Electrochemical Monitoring of Atmospheric Corrosion Phenomena, *Adv. Corr. Sci. Tech.* **6,** 163.

F. Mansfeld, M. W. Kendig, and S. Tsai

[1982]   Evaluation of Organic Coating Metal Systems by AC Impedance Techniques, *Corrosion* **38,** 478–485.

S. W. Martin and C. A. Angell

[1986]   D. C. and A. C. Conductivity in Wide Composition Range $Li_2O-P_2O_5$ Glasses, *J. Non-Cryst. Solids* **83,** 185–207.

N. Matsui

[1981]   Complex-Impedance Analysis for the Development of Zirconia Oxygen Sensors, *Solid State Ionics* **3/4,** 525–529.

A. Matsumoto and K. Higasi

[1962]   Dielectric Relaxation of Nonrigid Molecules at Lower Temperature, *J. Chem. Phys.* **36,** 1776–1780.

J. C. Maxwell

[1881]   A Treatise on Electricity and Magnetism, 2nd Ed., Clarendon Press, Oxford.

J. F. McCann and S. P. S. Badwal

[1982]   Equivalent Circuit Analysis of the Impedance Response of Semiconductor/Electrolyte/Counterelectride Cells, *J. Electrochem. Soc.* **129,** 551–559.

N. G. McCrum, B. E. Read, and G. Williams

[1967]   *Anelastic and Dielectric Effects in Polymeric Solids*, Wiley, London, see p. 111.

M. C. H. McKubre

[1976]   An Impedance Study of the Membrane Polarization Effect in Simulated Rock Systems, Ph.D. thesis, Victoria University, Wellington, N.Z.

[1981]   The Use of Harmonic Analysis to Determine Corrosion Rates, Paper number 135 in the 160th meeting of the Electrochemical Society, Denver, Colorado.

[1983]   *The Electrochemical Measurement of Corrosion Rates in Cathodically Protected Systems*, EPRI Report CS-2858 on Project 1689-7, SRI International, Menlo Park, Calif.

[1985]   Unpublished data.

M. C. H. McKubre and G. J. Hills

[1979]   A Digitally Demodulated Synchronous Detector for Fast, High Precision Impedance Measurement, unpublished work; L. E. A. Berlouis, An Impedance Study of Flow-Through Porous Electrodes, Ph.D. thesis, The University, Southampton, England.

M. C. H. McKubre and D. D. Macdonald

[1984]   Electronic Instrumentation for Electrochemical Studies, in *A Comprehensive Treatise of Electrochemistry*, ed. J. O'M. Bockris, B. E. Conway, and E. Yeager, Plenum Press, New York, pp. 1–98.

M. C. H. McKubre and H. B. Sierra-Alcazar

[1985]   *Sulfur Side Supporting Studies for the Sodium Sulfur Battery*, Interim Report on EPRI Contract 128-2, SRI International, Menlo Park, Calif.

M. C. H. McKubre and B. C. Syrett

[1986]   Harmonic Impedance Spectroscopy for the Determination of Corrosion Rates in Cathodically Protected Systems. Corrosion Monitoring in Industrial Plants Using Nondestructive Testing and Electrochemical Methods, ASTM STP 908, ed. G. C. Moran and P. Labine, American Society for Testing and Materials, Philadelphia, pp. 433–458.

R. E. Meredith and C. W. Tobias

[1962] Conduction in Heterogeneous Systems, in *Advances in Electrochemistry and Electrochemical Engineering*, ed. C. W. Tobias, Vol. 2, Interscience, New York, p. 15.

L. Meszaros and J. Devay

[1982] Study of the Rate of Corrosion by a Faradaic Distortion Method, IV, Application of the Method for the Case of Reversible Charge Transfer Reaction. *Acta Chim. Acad. Sciences*, Hungary, **109**, 241–244.

S. P. Mitoff

[1968] Properties Calculations for Heterogeneous Systems, in *Advances in Materials Research*, Vol. 3, ed. H. Herman, Wiley-Interscience, New York, pp. 305–329.

J. Mizusaki, K. Amano, S. Yamuchi, and K. Fueki

[1983] Response and Electrode Reaction of Zirconia Oxygen Gas Sensor, in *Analytical Chemistry Symposium Series*, Vol. 17, Elsevier, New York, pp. 279–284.

E. W. Montroll and J. T. Bendler

[1984] On the Levy (or Stable) Distribution and the Williams–Watts Model of Dielectric Relaxation, *J. Statistical Physics* **34**, 129–162.

E. W. Montroll and M. F. Shlesinger

[1984] On the Wonderful World of Random Walks, in *Non-Equilibrium Phenomena. II. From Statistics to Hydrodynamics*, ed. J. L. Lebowitz and E. W. Montroll, Elseview, New York, pp. 1–121.

E. W. Montroll and G. H. Weiss

[1965] Random Walks on Lattices, II, *J. Math. Phys.* **6**, 167–181.

C. I. Mooring and H. L. Kies

[1977] AC Voltammetry at Large Amplitudes: A Simplified Theoretical Approach, *J. Electroanal. Chem.* **78**, 219.

C. T. Morse

[1974] A Computer Controlled Apparatus for Measuring AC Properties of Materials over the Frequency Range 10^{-5} to 10^{+5} Hz, *J. Phys. E* **7**, 657–662.

N. F. Mott

[1939] Copper-Cuprous Oxide Photocells, *Proc. Royal Soc.* (London), **A171**, 281–285.

C. T. Moynihan, L. P. Boesch, and N. L. Laberge

[1973] Decay Function for the Electric Field Relaxation in Vitreous Ionic Conductors, *Physics and Chem. Glasses* **14**, 122–125.

H. Näfe

[1984] Ionic Conductivity of ThO_2- and ZrO_2-based Electrolytes between 300 and 2000 K, *Solid State Ionics* **13**, 255–263.

J. S. Newman

[1973] *Electrochemical Systems*, Prentice-Hall, Englewood Cliffs, N.J.

K. L. Ngai

[1979] Universality of Low-frequency Fluctuation, Dissipation, and Relaxation Properties of Condensed Matter. I. Comments, *Solid State Phys.* **9**, 127–140.

[1980] Universality of Low-frequency Fluctuation, Dissipation, and Relaxation Properties of Condensed Matter. II. Comments, *Solid State Phys.* **9**, 141–155.

K. L. Ngai, A. K. Jonscher, and C. T. White

[1979] On the Origin of the Universal Dielectric Response in Condensed Matter, *Nature* **277**, 185–189.

K. L. Ngai and C. T. White

[1979] Frequency dependence of dielectric loss in condensed matter, *Phys. Rev.* **B20**, 2475–2486.

E. H. Nicollian and J. R. Brew
   [1982]   *MOS (Metal Oxide Semiconductor Technology) Physics and Technology*, Wiley, New York.

E. H. Nicollian and A. Goetzberger
   [1967]   The Si–SiO$_2$ Interface: Electrical Properties as Determined by the Metal–Insulator–Silicon Conductance Technique, *Bell Syst. Tech. J.* **46**, 1055–1133.

I. M. Novoseleskii, N. N. Gudina, and Yu. I. Fetistov
   [1972]   Identical Equivalent Impedance Circuits, *Sov. Electrochem* **8**, 546–548.

K. B. Oldham
   [1957]   Faradaic Rectification: Theory and Application to the Hg^{++}/Hg Electrode, *Trans. Faraday Soc.* **53**, 80–90.

L. Onsager
   [1926]   Theory of Electrolytes, I, *Phys. Z.* **27**, 388–392.
   [1927]   Theory of Electrolytes, II, *Phys. Z.* **28**, 277–298.

J. R. Park
   [1983]   The Mechanism of the Fast Growth of Magnetite on Carbon Steel under PWR Crevice Conditions, Ph.D. thesis, Ohio State University.

J. R. Park and D. D. Macdonald
   [1983]   Impedance Studies of the Growth of Porous Magnetite Films on Carbon Steel in High Temperature Aqueous Systems, *Corros. Sci.* **23**, 295.

G. Pfister and H. Scher
   [1978]   Dispersive (Non-Gaussian) Transient Transport in Disordered Solids, *Adv. Phys.* **27**, 747–798.

A. A. Pilla
   [1970]   Transient Impedance Technique for the Study of Electrode Kinetics: Application to Potentiostatic Methods, *J. Electrochem. Soc.* **117**, 467.
   [1972]   Electrochemistry, in *Computers in Chemistry and Instrumentation*, Vol. 2, ed. J. S. Mattson, H. B. Mark, Jr., and H. C. Macdonald, Jr., Marcel Dekker, New York, p. 139.
   [1975]   Introduction to Laplace Plane Analysis, *Information Chemistry*, Eds. S. Fujiwara and H. B. Mark, Japan Science Society, pp. 181–193.

E. C. Potter and G. M. W. Mann
   [1965]   The Fast Linear Growth of Magnetite on Mild Steel in High Temperature Aqueous Conditions, *Corros. J.* **1**, 26.

B. G. Pound and D. D. Macdonald
   [1985]   *Development of Noise Impedance Techniques for the Measurement of Corrosion Rates*, Final Report by SRI International on Sohio Contract NT-MEM-2219, August.

R. W. Powers
   [1984]   The Separability of Inter- and Intragranular Resistivities in Sodium Beta-Alumina Type Ceramics, *J. Mat. Sci.* **19**, 753–760.

R. W. Powers and S. P. Mitoff
   [1975]   Analysis of the Impedance of Polycrystalline Beta-alumina, *J. Electrochem. Soc.* **122**, 226–231.

I. D. Raistrick
   [1983]   Lithium Insertion Reactions in Tungsten and Vanadium Oxide Bronzes, *Solid State Ionics* **9/10**, 425–430.

I. D. Raistrick, C. Ho, and R. A. Huggins
   [1976]   Ionic Conductivity of Some Lithium Silicates and Aluminosilicates, *J. Electrochem. Soc.* **123**, 1469–1476.

334 REFERENCES

I. D. Raistrick, C. Ho, Y.-W. Hu, and R. A. Huggins
[1977]   Ionic Conductivity and Electrode Effects on β-PbF$_2$, *J. Electroanal. Chem.* **77**, 319–337.

D. O. Raleigh
[1976]   The Electrochemical Double Layer in Solid Electrolytes, in *Electrode Processes in Solid State Ionics*, ed. M. Kleitz and J. Dupuy, Reidel, Dordrecht/Boston, pp. 119–147.

A. C. Ramamurthy and S. K. Rangarajan
[1977]   Fournier Polarography Revisited, *J. Electroanal. Chem.* **77**, 267–286.

J. E. B. Randles
[1947]   Kinetics of Rapid Electrode Reactions, *Disc. Farad. Soc.* **1**, 11–19.

J. E. B. Randles and K. W. Somerton
[1952]   Kinetics of Rapid Electrode Reactions, Parts 3 and 4, *Trans. Farad. Soc.* **48**, 937–950, 951–955.

S. K. Rangarajan
[1974]   A Simplified Approach to Linear Electrochemical Systems. I. The Formalism; II. Phenomenological Coupling; III. The Hierarchy of Special Cases; IV. Electron Transfer through Adsorbed Modes, *J. Electroanal. Chem.* **55**, 297–327, 329–335, 337–361, 363–374.

[1975]   Non Linear Relaxation Methods. III. Current Controlled Perturbations, *J. Electroanal. Chem.* **62**, 31–41.

G. P. Rao and A. K. Mishra
[1977]   Some AC Techniques to Evaluate the Kinetics of Corrosion Kinetics, *J. Electroanal. Chem.* **77**, 121.

D. Ravaine and J.-L. Souquet
[1973]   Application du Trace des Diagrammes D'Impedance Complexe de la Determination de la Conductivité Electrique des Verres Silice-Oxyde Alcalin. *Compt. Rend. Acad. Sci.* (Paris) **277C**, 489–492.

L. E. Reichl
[1980]   *A Modern Course in Statistical Physics*, University of Texas Press, Austin, Tex.

Reticon Corporation
[1977]   *Fast Fourier Transform Module: Product Guide*, Reticon Model R5601, Sunnyvale, Calif.

A. Roos, D. R. Franceschetti, and J. Schoonman
[1984]   The Small Signal AC Response of La$_{1-x}$Ba$_x$F$_{3-x}$ Solid Solutions, *Solid State Ionics* **12**, 485–491.

P. R. Roth
[1970]   Digital Fourier Analysis. Some of the Theoretical and Practical Aspects of Measurements Involving Fourier Analysis by Digital Instrumentation, *Hewlett Packard J.* **21**, 2.

J. Rouxel
[1979]   Recent Progress in Intercalation Chemistry: Alkali Metal in Chalcogenide Hosts, *Rev. Inorg. Chem.* **1**, 245–279.

R. P. Sallen and E. L. Key
[1955]   *IEEE Trans.*, CT-2, 74; D. Hilberman, *IEEE Trans.* **CT-15**, 431 (1968).

J. R. Sandifer and R. P. Buck
[1974]   Impedance Characteristics of Ion Selective Glass Electrodes, *J. Electroanal. Chem.* **56**, 385–398.

D. F. Schanne and E. Ruiz-Ceretti
[1978]   *Impedance Measurements in Biological Cells*, Wiley, New York.

W. Scheider
[1975]   Theory of the Frequency Dispersion of Electrode Polarization: Topology of Networks with Fractional Power Frequency Dependence, *J. Phys. Chem.* **79**, 127–136.

H. Scher and M. Lax
[1973a] Stochastic Transport in Disordered Solids. I. Theory, *Phys. Rev.* **B7**, 4491–4502.
[1973b] Stochastic Transport in Disordered Solids. II. Impurity Conduction, *Phys. Rev.* **B7**, 4502–4519.

H. Scher and E. W. Montroll
[1975]   Anomalous Transit-Time Dispersion in Amorphous Solids, *Phys. Rev.* **B12**, 2455–2477.

M. F. Schlesinger
[1984]   Williams–Watts Dielectric Relaxation: A Fractal Time Stochastic Process, *J. Statist. Phys.* **36**, 639–648.

M. F. Schlesinger and E. W. Montroll
[1984]   On the Williams–Watts Function of Dielectric Relaxation, *Proc. Nat. Acad. Sci.* **81**, 1280–1283.

W. Schottky
[1939]   "Zur Halbeitertheorie der Sperrschiet- und Spitzengleichrichter, *Z. Phys.* **113**, 367–414.
[1942]   Vereinfachte und erweterte Theorie der Randschichte gleichrichter, *Z. Phys.*, **118**, 539–592.

E. J. L. Schouler
[1979]   Etude des Cellules a Oxyde Electrolyte Solide par la Methode des Impédances Complexes: Applications à la Mesure Precise de la Conductivité et a l'Etude de la Réaction d'Electrode à Oxygène, Ph.D. thesis, Institut National Polytechnique de Grenoble.

E. J. L. Schouler, G. Giroud, and M. Kleitz
[1973]   Application selon Bauerle du Tracé des Diagrammes d'Admittance Complexe en Électrochimie des Solides, *J. Chim. Phys.* **70**, 1309–1316.

E. J. L. Schouler, N. Mesbahi, and G. Vitter
[1983]   In Situ Study of the Sintering Process of Yttria Stabilized Zirconia by Impedance Spectroscopy, *Solid State Ionics* **9, 10**, 989–996.

E. J. L. Schouler, M. Kleitz, E. Forest, E. Fernandez, and P. Fabry
[1981]   Overpotential of H_2–H_2O/YSZ Electrodes in Steam Electrolyzers, *Solid State Ionics* **5**, 559–562.

J. Schrama
[1957]   On the Phenomenological Theory of Linear Relaxation Processes, Ph.D. thesis, University of Leiden, Netherlands.

H. G. Scott
[1975]   Phase Relationships in the Zirconia–Yttria System, *J. Mat. Sci.* **10**, 1527–1535.

B. Scrosati
[1981]   Non Aqueous Lithium Cells, *Electrochim. Acta* **26**, 1559–1567.

M. Seralathan and R. de Levie
[1987]   On the Electrode Admittance of Complex Electrode Reactions, to appear in *Electroanalytical Chemistry*, ed. A. Bard, Marcel Dekker, New York.

W.-M. Shen, W. Siripala, and M. Tomkiewicz
> [1986] Electrolyte Electroreflectance Study of Surface Optimization of n-CuInSe$_2$ in Photoelectrochemical Solar Cells, *J. Electrochem. Soc.* **133**, 107–111.

W.-M. Shen, M. Tomkiewicz, and D. Cahen
> [1986] Impedance Study of Surface Optimization of n-CuInSe$_2$ in Photoelectrochemical Solar Cells, *J. Electrochem. Soc.* **133**, 112–116.

H. B. Sierra-Alcazar
> [1976] *The Electrodeposition of Zinc*, Ph.D. thesis, University of Newcastle-upon-Tyne.

H. B. Sierra-Alcazar, A. N. Fleming, and J. A. Harrison
> [1977] A Rapid On-Line Method for Electrochemical Investigations, *Surface Technol.* **6**, 61.

S. Silverman, G. Cragnolino and D. D. Macdonald
> [1982] An Ellipsometric Investigation of Passive Films Formed on Fe-25 Ni-xCr Alloys in Boric Buffer Solution, *J. Electrochem. Soc.* **129**, 2419–2424.

W. M. Siu and R. S. C. Cobbald
> [1979] Basic Properties of the Electrolyte–Silicon Dioxide–Silicon System: Physical and Theoretical Aspects, *IEEE Trans. Electron Devices*, **ED-26**, 1805–1815.

R. K. Slotwinski, N. Bonanos, and E. P. Butler
> [1985] Electrical Properties of MgO + Y$_2$O$_3$ and CaO + Y$_2$O$_3$ Partially Stabilized Zirconias, *J. Mat. Sci. Lett.* **4**, 641–644.

J. H. Sluyters
> [1960] On the Impedance of Galvanic Cells, I, *Rec. Trav. Chim.* **79**, 1092–1100.
> [1963] On the Impedance of Galvanic Cells. V. The Impedance of a Galvanic Cell with Two Plane Parallel Electrodes at a Short Distance, *Rec. Trav. Chim.* **82**, 100–109.

J. H. Sluyters and J. J. C. Oomen
> [1960] On the Impedance of Galvanic Cells. II. Experimental Verification, *Rec. Trav. Chim.* **79**, 1101–1110.

M. Sluyters-Rehbach and J. H. Sluyters
> [1970] Sine Wave Methods in the Study of Electrode Processes, in *Electroanalytical Chemistry*, Vol. 4, ed. A. J. Bard, Marcel Dekker, New York, pp. 1–127.
> [1984] AC Techniques, in *Comprehensive Treatise of Electrochemistry*, Vol. 9, ed. E. Yaeger, J. O'M. Bockris, B. E. Conway, and S. Sarangapani, Plenum Press, New York.

D. E. Smith
> [1966] AC Polarography and Related Techniques: Theory and Practice, in *Electroanalytical Chemistry*, Vol. 1, ed. A. J. Bard, Marcel Dekker, New York, pp. 1–155.
> [1971] Computers in Chemical Instrumentation: Application of On-Line Digital Computers to Chemical Instrumentation, *CRC Crit. Rev. Anal. Chem.* **2**, 248.
> [1976] The Acquisition of Electrochemical Response Spectra by On-Line Fast Fourier Transform: Data Processing in Electrochemistry, *Anal. Chem.* **48**, 221A.

P. H. Smith
> [1939] Transmission Line Calculator, *Electronics* **12**, 29–31.
> [1944] An Improved Transmission Line Calculator, *Electronics* **17**, 130–133, 318–325.

W. Smyrl
> [1985a] Digital Impedance for Faradaic Analysis. I. Introduction to Digital Signal Analysis and Impedance Measurements for Electrochemical and Corrosion Systems, *J. Electrochem. Soc.* **132**, 1551–1555.
> [1985b] Digital Impedance for Faradaic Analysis. II. Electrodissolution of Cu in HCl, *J. Electrochem. Soc.* **132**, 1555–1562.

W. Smyrl and L. L. Stephenson
> [1985] Digital Impedance for Faradaic Analysis. III. Copper Corrosion in Oxygenated 0.1 N HCl, *J. Electrochem. Soc.* **132**, 1563–1567.

Solartron Instrumentation Group

[1977]   *1170 Series Frequency Response Analyzers: Operation Manual*, Farnborough, Hampshire, England.

P. R. Sorensen and T. Jacobsen

[1982]   Conductivity Charge Transfer and Transport Number: An Investigation of the Polymer Electrolyte LiSCN–Poly(Ethylene Oxide), *Electrochim. Acta* **27**, 1671–1675.

U. Staudt and G. Schön

[1981]   Automatic Spectroscopy of Solids in the Low Frequency Range, *Solid State Ionics* **2**, 175–183.

B. C. H. Steele

[1976]   High Temperature Fuel Cells and Electrolysers, in *Electrode Processes in Solid State Ionics*, ed. M. Kleitz and J. Dupuy, Reidel Dordrecht/Boston, pp. 368–384.

B. C. H. Steele and E. P. Butler

[1985]   Electrical Behaviour of Grain Boundaries in Zirconia and Related Ceramic Electrolytes, *Brit. Ceram. Proc.* **36**, 45–55.

B. C. H. Steele, J. Drennan, R. K. Slotwinski, N. Bonanos, and E. P. Butler

[1981]   Factors Affecting the Performance of Oxygen Monitors, in *Advances in Ceramics*, Vol. 3, ed. A. H. Heuer and L. W. Hobbs, American Ceramic Society, Columbus, Ohio, pp. 286–309.

M. Stern and A. L. Geary

[1957]   Electrochemical Polarization. I. Shape of Polarization Curves, *J. Electrochem. Soc.* **104**, 56.

J. E. Strutt, M. W. Weightman, and E. Lilley

[1976]   A Versatile Cell for the Measurement of Ionic Thermo-Currents (ITC), *J. Phys.* **E9**, 683–685.

E. C. Subbarao

[1981]   Zirconia: An Overview, in *Advances in Ceramics*, Vol. 3, ed. A. H. Heuer and L. W. Hobbs, American Ceramic Society, Columbus, Ohio, pp. 1–24.

E. C. Subbarao and H. S. Maiti

[1984]   Solid Electrolytes with Oxygen Ion Conductivity, *Solid State Ionics* **11**, 317–338.

B. C. Syrett and D. D. Macdonald

[1979]   The Validity of Electrochemical Methods for Measuring Corrosion Rates of Copper–Nickel Alloys in Seawater, *Corrosion* **35**, 505–508.

S. M. Sze

[1985]   *Semiconductor Devices, Physics, and Technology*, Wiley, New York.

M. Tomkiewicz

[1979]   Relaxation Spectrum Analysis of Semiconductor–Electrolyte Interface–TiO_2, *J. Electrochem. Soc.* **126**, 2220–2225.

Y.-T. Tsai and D. W. Whitmore

[1982]   Nonlinear Least-Squares Analysis of Complex Impedance and Admittance Data, *Solid State Ionics* **7**, 129–139.

V. A. Tyagai and G. Y. Kolbasov

[1972]   Use of Kramers–Kronig Relations for the Analysis of Low Frequency Impedance, *Elektrokhmiya* **8**, 59.

M. Urquidi-Macdonald, S. Real, and D. D. Macdonald

[1986]   Application of Kramers–Kronig Transforms in the Analysis of Electrochemical Impedance Data, II, Transformations in the Complex Plane, *J. Electrochem. Soc.* **133**, 2018–2024.

338 REFERENCES

L. K. H. van Beek
[1965]   Dielectric Behaviour of Heterogeneous System, in *Progress in Dielectrics*, Vol. 7, ed. J. B. Birks, Heywood Press, London, pp. 69–114.

T. van Dijk and A. J. Burggraaf
[1981]   Grain Boundary Effects on Ionic Conductivity in Ceramic $Gd_xZr_{1-x}O_{2-(x/2)}$ Solid Solutions, *Phys. Stat. Sol.* (a) **63**, 229–240.

R. L. Van Meirhaeghe, E. C. Dutoit, F. Cardon, and W. P. Gomes
[1976]   On the Application of the Kramers–Kronig Relations to Problems Concerning the Frequency Dependence of Electrode Impedance, *Electrochim. Acta* **21**, 39–43.

W. van Weperen, B. P. M. Lenting, E. J. Bijrank and H. W. der Hartog
[1977]   Effect of the Ce^{3+} Concentration on the Reorientation of Dipoles in $SrF_2:Ce^{3+}$, *Phys. Rev.* **B16**, 1953–2958.

M. J. Verkerk and A. J. Burggraaf
[1983]   Oxygen Transfer on Substituted ZrO_2, Bi_2O_3, and CeO_2 Electrolytes with Platinum Electrodes, *J. Electrochem. Soc.* **130**, 78–84.

M. J. Verkerk, B. J. Middlehuis, and A. J. Burggraaf
[1982]   Effect of Grain Boundaries on the Conductivity of High-Purity Zirconium Oxide–Yttrium Oxide Ceramics, *Solid State Ionics* **6**, 159–170.

K. J. Vetter
[1967]   *Electrochemical Kinetics*, Academic Press, New York.

J. Volger
[1960]   Dielectric Properties of Solids in Relation to Imperfections, in *Progress in Semiconductors*, Vol. 4, ed. A. F. Gibson, F. A. Kröger, and R. E. Burgess, Heywood, London, pp. 205–236.

C. Wagner
[1933]   Uber die Natur des Elektrischen Leitvermogens von a Silbersulfide, *Z. Phys. Chemie* **B21**, 42–52.
[1953]   Investigation on Silver Sulfide, *J. Chem. Phys.* **21**, 1819–1827.

K. W. Wagner
[1914]   Explanation of the Dielectric Fatigue Phenomenon on the Basis of Maxwell's Concept, in *Arkiv für Electrotechnik*, ed. H. Schering, Springer-Verlag, Berlin.

K. E. D. Wapenaar, H. G. Koekkoek, and J. van Turnhout
[1982]   Low-Temperature Ionic Conductivity and Dielectric Relaxation Phenomena in Fluorite-Type Solid Solutions, *Solid State Ionics* **7**, 225–242.

K. E. D. Wapenaar and J. Schoonman
[1981]   Small Signal AC Response of $Ba_{1-x}La_xF_{2+x}$ Crystals, *Solid State Ionics* **2**, 253–263.

E. Warburg
[1899]   Uber das Verhalten sogenannter unpolarisierbarer Electroden gegen Wechselstrom, *Ann. Phys. Chem.* **67**, 493–499.

G. H. Weiss and R. J. Rubin
[1983]   Random Walks: Theory and Selected Applications, *Adv. Chem. Phys.* **52**, 363–505.

W. Weppner and R. A. Huggins
[1977]   Determination of the Kinetic Parameters of Mixed Conducting Electrodes and Application to the System Li_3Sb, *J. Electrochem. Soc.* **124**, 1569–1578.

A. R. West
[1983]   Private communication to J. R. Macdonald.

S. Whittingham
[1978]   Chemistry of Intercalation Compounds: Metal Guests in Chalcogenide Hosts, *Prog. Solid State Chem.* **12**, 41–49.

M. S. Whittingham and R. A. Huggins
  [1971]   Measurement of Sodium Ion Transport in β-Alumina Using Reversible Solid Electrodes, *J. Chem. Phys.* **54,** 414–416.

G. Williams and D. C. Watts
  [1970]   Non-Symmetrical Dielectric Relaxation Behavior Arising from a Simple Empirical Decay Function, *Trans. Faraday Soc.* **66,** 80–85.

G. Williams, D. C. Watts, S. B. Dev, and A. M. North
  [1970]   Further Considerations of Non-Symmetrical Dielectric Relaxation Behavior Arising from a Simple Empirical Decay Functions, *Trans. Farad. Soc.* **67,** 1323–1335.

J. M. Wimmer, H. C. Graham, and N. M. Tallan
  [1974]   Microstructural and Polyphase Effects, in *Electrical Conduction in Ceramics and Glasses*, Part B, ed. N. M. Tallan, Marcel Dekker, New York, pp. 619–652.

T. Wong and M. Brodwin
  [1980]   Dielectric Relaxation Phenomena Associated with Hopping Ionic Conduction, *Solid State Comm.* **36,** 503–508.

T. Zeuthen
  [1978]   Potentials and Small-Signal Impedance of Platinum Micro-Electrodes in Vivo and Vitro, *Med. Biol. Eng. Comput.* **16,** 483–488.

INDEX